THE INTERNATION
MONOGRAPHS ON

C000058663

THE INTERNATIONAL SERIES OF MONOGRAPHS ON CHEMISTRY

1. J. D. Lambert: *Vibrational and rotational relaxation in gases*
2. N. G. Parsonage and L. A. K. Staveley: *Disorder in crystals*
3. G. C. Maitland, M. Rigby, E. B. Smith, and W. A. Wakeham: *Intermolecular forces: their origin and determination*
4. W. G. Richards, H. P. Trivedi, and D. L. Cooper: *Spin-orbit coupling in molecules*
5. C. F. Cullis and M. M. Hirschler: *The combustion of organic polymers*
6. R. T. Bailey, A. M. North, and R. A. Pethrick: *Molecular motion in high polymers*
7. Atta-ur-Rahman and A. Basha: *Biosynthesis of indole alkaloids*
8. J. S. Rowlinson and B. Widom: *Molecular theory of capillarity*
9. C. G. Gray and K. E. Gubbins: *Theory of molecular fluids. Volume 1: Fundamentals*
10. C. G. Gray and K. E. Gubbins: *Theory of molecular fluids. Volume 2: Applications* (in preparation)
11. S. Wilson: *Electron correlation in molecules*
12. E. Haslam: *Metabolites and metabolism: a commentary on secondary metabolism*
13. G. R. Fleming: *Chemical applications of ultrafast spectroscopy*
14. R. R. Ernst, G. Bodenhausen, and A. Wokaun: *Principles of nuclear magnetic resonance in one and two dimensions*
15. M. Goldman: *Quantum description of high-resolution NMR in liquids*
16. R. G. Parr and W. Yang: *Density-functional theory of atoms and molecules*
17. J. C. Vickerman, A. Brown, and N. M. Reed (editors): *Secondary ion mass spectrometry: principles and applications*
18. F. R. McCourt, J. Beenakker, W. E. Köhler, and I. Kuščer: *Nonequilibrium phenomena in polyatomic gases. Volume 1: Dilute gases*
19. F. R. McCourt, J. Beenakker, W. E. Köhler, and I. Kuščer: *Nonequilibrium phenomena in polyatomic gases. Volume 2: Cross-sections, scattering, and rarefied gases*
20. T. Mukaiyama: *Challenges in synthetic organic chemistry*
21. P. Gray and S. K. Scott: *Chemical oscillations and instabilities: non-linear chemical kinetics*
22. R. F. W. Bader: *Atoms in molecules: a quantum theory*
23. J. H. Jones: *The chemical synthesis of peptides*
24. S. K. Scott: *Chemical chaos*
25. M. S. Child: *Semiclassical mechanics with molecular applications*
26. D. T. Sawyer: *Oxygen chemistry*
27. P. A. Cox: *Transition metal oxides: an introduction to their electronic structure and properties*
28. B. R. Brown: *The organic chemistry of aliphatic nitrogen compounds*

Chemical Chaos

STEPHEN K. SCOTT

School of Chemistry
University of Leeds

CLARENDON PRESS · OXFORD

Oxford University Press, Walton Street, Oxford OX2 6DP
Oxford   New York
Athens   Auckland   Bangkok   Bombay
Calcutta   Cape Town   Dar es Salaam   Delhi
Florence   Hong Kong   Istanbul   Karachi
Kuala Lumpur   Madras   Madrid   Melbourne
Mexico City   Nairobi   Paris   Singapore
Taipei   Tokyo   Toronto
and associated companies in
Berlin   Ibadan

Oxford is a trade mark of Oxford University Press

Published in the United States
by Oxford University Press, Inc., New York

British Library Cataloguing in Publication Data
Scott, Stephen K.
Chemical chaos.
1. Chemistry. Chaotic behaviour
I. Title   II. Series
541.39

Library of Congress Cataloging in Publication Data
Scott, Stephen K.
Chemical chaos / Stephen K. Scott.
p.   cm.—(International series of monographs on chemistry; 24)
Includes index.
1. Chemistry, Physical and theoretical.   2. Chaotic behavior in
systems.   I. Title.   II. Series.
QD455.2.S37   1991   541—dc20   90-23152 CIP
ISBN 0 19 855658 6 (Pbk)

Printed by St. Edmundsbury Press,
Bury St. Edmunds, Suffolk

To Hilary and Lucy,
with love

Preface

Every author in this field seems to have a favourite quote. In that spirit, I feel happiest with 'chaos, stamped with a seal, becomes rightful order' due to Ugo Betti in *The Gambler*. The emphasis here is on the order that lies within so-called chaotic behaviour, now that we have learnt to recognize it. Chaos theory would probably not have attained the high profile of the recent years without its 'everyday' title, but the use of such a familiar adjective has its dangers, particularly the attendant connotations of randomness and disorder that are not features of technical chaos. The technical subject involves unpredictability and irreproducibility, recognizing these as quite natural responses even for normal, well-behaved reactions. They are built upon an extreme sensitivity of chemical (or other) systems under some circumstances to their initial conditions. It is not concerned with chance or randomness. Chaos does not suddenly appear in chemical reactions, but almost always develops through one of a small number of strictly ordered, and in many cases now reasonably well-understood, sequences.

The alternative term 'aperiodicity' is perhaps closer, but still has its drawbacks: random noise is not periodic, but is not chaotic. The strictest definition—that chaotic systems are those with positive Lyapounov exponents—has the excellent attribute for mathematics of being completely correct and almost completely unhelpful in terms of clarifying the matter (although perhaps this will mean more after reading the main text). Chaotic reactions follow quite well-defined and 'normal' kinetic rate laws, which are typically simple ordinary differential equations, and may have well-established rate constants, etc. We are dealing with 'deterministic' not 'stochastic' systems.

Returning to 'aperiodicity', this at least gives a hint that a more useful approach is to regard chaotic behaviour simply as a type of oscillatory process, taken to the limit of an infinitely long repeating unit. Certainly chemical oscillations are based on the same mechanistic building blocks as chemical chaos (and other reasonably well-known responses, such as ignition, extinction, and bistability). The question 'What causes chaos?' is both profound and naive. Nothing causes chaos, in the sense that it is simply one of the natural kinetic responses available to chemical reactions. I hope to stress that none of these apparently exotic modes of reaction are particularly rare in chemistry and certainly that they do not require very complex or unusual mechanisms. Indeed, there are only two, simple, basic requirements: non-linearity and feedback (to introduce two more pieces of jargon).

Non-linearity is by no means unusual in chemistry—in fact, it is the linear reactions, those with strictly first-order kinetics, that are the rare ones. Typically the rate of an elementary reaction depends on the product of two concentrations—and so is naturally non-linear (not linear): doubling the concentrations does not double the reaction rate. When more than one elementary process combines to produce a 'real' chemical mechanism, with products from early steps becoming reactants or intermediates in later reactions, all sorts of 'empirical' rate law can emerge with varying degrees of concentration dependence. Feedback is perhaps not quite so widespread, but self-acceleration or self-inhibition can arise through, say, chain branching, autocatalysis, or self-heating in ways that will be revealed in the following chapters. With these, chemistry provides perhaps the cleanest examples of all manner of non-linear behaviour that has so excited mathematicians, physical and biological scientists, engineers, social scientists, and economists in recent years. It often does so with the added bonus of striking visual effects and provides some of the simplest models.

This is not a comprehensive text, even on the restricted field of chaos in chemistry, despite the title. There is another main branch, that of 'quantum chaos', dealing with molecules possessing high degrees of excitation and behaving in curious ways. A weak excuse for not discussing this equally exciting field is that it would have markedly increased the length of the present text. More honestly, such an undertaking requires a greater under-standing of this related but significantly different subject than I have confidence or (at present) energy to develop. Even within the kinetic field I am aware of omissions (and no doubt unaware of some that exist). The Soviet literature is rich, but sadly still less than immediately accessible without a great investment of time and effort. Gregory Yablonski has helped with some suggestions and information here, but I hope that the wonderful events of late 1989 and the brave new administration in the USSR will see wider interactions, face to face, between western and eastern scientists over the next few years, particularly with more relaxed opportunities for full discussions at conferences. I am less optimistic about a similar softening in China, although some contacts there have also helped me. The coverage of biochemical oscillators and chaos has also been restricted here to just a brief taste. Nevertheless there has still been plenty to write about.

I have attempted to provide both a theoretical introduction, as relevant to chemistry, and a presentation of experimental observations. The former occupies Chapters 2 to 6, whilst the latter forms Chapters 7 to 12. Although the discussion of the experimental results is aided by reference to the insights gained from the mathematics, there should be little lost by proceeding straight to any particular chapter of interest in the 'second half', particularly for a first visit.

The preparation of this text has required the guidance of many colleagues to whom I wish to express my thanks. I am extremely fortunate to have a talented and patient research group in Leeds: Alison Tomlin is the longest serving member and has provided much of the material for Chapters 2 and 4; the latter also has contributions from John McGarry and Bo Peng (from West Virginia University), whilst John Leach is responsible for many of the coupled oscillator results in Chapter 6. Paul Ibison provided material for Chapters 7 and 8 whilst Barry Johnson's remarkable experimental results appear prominently in Chapter 9. Tony Edge, Steve Collier, Junli Liu, Hong-tu Feng, and Chris Hildyard have not been quoted directly but have also played their role within the group. More 'senior' colleagues within the extended Leeds family include John Merkin, John Griffiths, Peter Gray (now Cambridge), Dave Needham (University of East Anglia), Mike Pilling, John Brindley, Dave Knapp, and Geoff Cook. I could not have approached this work without much tuition from Ken Showalter (West Virginia) and Rutherford Aris (Minnesota) and slightly more specific help from Lanny Schmidt (Minnesota), Jerzy Maselko (West Virginia), Vilmos Gaspar (Debrecen), Jack Hudson (Virginia) who seems to have contributed to so many of the subfields that appear here, Dan Luss (Houston), Fred Schneider (Wurzburg), Dick Field (Montana), Gregoire Nicolis (Brussels), Patrick De Kepper and Jacques Boissonade (Bordeaux), Harry Swinney (Austin), Brian Gray (Sydney) in his own particular way, Kedma Bar-Eli (Tel Aviv), John Ross (Stanford), Irv Epstein (Brandeis), Ferdi Schuth (Mainz), Ioannis Kevrekidis (Princeton), Nils Jaeger and Peter Plath (Bremen), Pier-Giorgio Lignola (Calabria), Jim Murray (Oxford), Chang-gen Feng (Beijing), and Peter Erdi (Budapest). The staff of Oxford University Press have been supportive and understanding during this project.

Again, Hilary has played a vital role: she has not complained too much that the decorating of our new home has proceeded rather slowly; she has kept me going during slow periods, and even found time to retype several sections when the word processor lost a file. I hope she will forgive my preoccupations and the weekends and bank holidays at work and accept the dedication.

SKS

Leeds
May 1990

Contents

1

Introduction

This quantitative interpretation of experimental results from chemical systems is frequently based on the construction and subsequent analysis of *kinetic mechanisms* or *models*. This process can work in two directions. If we can confidently predict the chemistry, in terms of the individual reactions which are likely to occur between all the possible chemical species in the system, we can then write down the *reaction rate equations* based on the *law of mass action*. The reaction rate equations specify the rates at which the species concentrations change with time, and how these rates depend upon those concentrations. The idea here is to break the overall reaction up into a number of component or *elementary steps*. An example of an elementary step is the following between a hydrogen atom and an oxygen molecule:

$$H + O_2 \rightarrow OH + O. \tag{1.1}$$

The rate of an elementary step is given simply in terms of the product of the concentrations of the participating species (those on the left-hand side of the reaction step). For the above example

$$\text{rate} = k[H][O_2]. \tag{1.2}$$

Elementary steps are typically, but not exclusively, *bimolecular*—involving two species. The reaction is envisaged as occurring in a single collision between the reactants. Bimolecular reactions then give rise to a quadratic dependence of rate on concentration, i.e. to an overall *second-order* process. Equation (1.2) also involves a coefficient k: this is the *reaction rate coefficient* or *reaction rate constant*. Typically, rate constants are independent of concentration but may be sensitive functions of the temperature of the reacting mixture: this temperature dependence can frequently be expressed in the form

$$k(T) = A \exp(-E/RT) \tag{1.3}$$

where T is the thermodynamic (absolute) temperature and R is the universal gas constant, $R = 8.314 \, \text{J K}^{-1} \, \text{mol}^{-1}$. The parameter A is termed the *pre-exponential factor* and represents the high-temperature limit of k: E is known as the *activation energy* and will have units of J mol^{-1}. Many chemical reactions are exothermic (evolve heat): this Arrhenius temperature dependence can then have important consequences.

The overall rate equations are then derived by combining all the rates of the individual elementary steps. Using the hydrogen–oxygen reaction again as an example, we shall see later that an acceptable mechanism for the behaviour of this system over a limited range of pressure and temperature imvolves the following elementary steps:

(0)	$H_2 + O_2 \rightarrow 2OH$	rate $= k_i[H_2][O_2]$
(1)	$OH + H_2 \rightarrow H_2O + H$	rate $= k_1[OH][H_2]$
(2)	$H + O_2 \rightarrow OH + O$	rate $= k_2[H][O_2]$
(3)	$O + H_2 \rightarrow OH + H$	rate $= k_3[O][H_2]$
(4)	$H \rightarrow \frac{1}{2}H_2$	rate $= k_4[H]$.

In each case the 'rate' of the individual elementary step has been derived from law of mass action rules. Notice that for the last step we can allow the production of a non-integer number of product molecules. The rate here is thus defined as the rate at which the reactant species, H atoms, disappears through this step. As two H atoms are required to form an H_2 molecule, the rate of formation of the product is one-half the rate of removal of H. This factor of $\frac{1}{2}$ is known as a *stoichiometric factor*. If we had wished to express the rate of this final step in terms of the rate of formation of H_2, the appropriate form would still only involve the first power of [H]. This is a first-order step and is linear in concentration (it also probably occurs on the surface of the reaction vessel, but that is not important for the present purposes).

We can now construct the reaction rate equations from this scheme for as many of the species as we wish. As examples let us consider the three radical species H, O, and OH. One hydrogen atom is produced in each of steps (1) and (3), and one removed in each of steps (2) and (4). The rate equation for [H] is thus

$$\frac{d[H]}{dt} = k_1[OH][H_2] - k_2[H][O_2] + k_3[O][H_2] - k_4[H]. \quad (1.4)$$

For OH radicals the procedure is much the same, only we note that two OHs are formed in the initiation step (0) so there is a stoichiometric factor of 2 associated with this contribution to the rate equation. Thus

$$\frac{d[OH]}{dt} = 2k_i[H_2][O_2] - k_1[OH][H_2] + k_2[H][O_2] + k_3[O][H_2] \quad (1.5)$$

and for oxygen atoms

$$\frac{d[O]}{dt} = k_2[H][O_2] - k_3[O][H_2]. \quad (1.6)$$

We may note a number of general features of these equations. Any term which removes H in the equation for $d[H]/dt$, or OH in $d[OH]/dt$, or O in $d[O]/dt$ (i.e. any term preceded by a minus sign in the rate equations) also depends on that concentration. Chemically, this is almost self-evident: a species must be directly involved in any reaction which removes it. An important consequence of this, however, is that should any concentration become zero, then all the removal terms (those preceded by a negative stoichiometric coefficient) in the rate equation must also vanish: the derivative must then be either zero or positive, so that concentration will not fall further. Chemical species concentrations do not become negative.

A second feature, typical of most elementary steps but not some of the empirical forms which we will consider later, is that the production terms (those preceded by a positive sign) in any given species rate equation do not involve the concentration of the species being produced.

Table 1.1 contains the full set of the governing reaction rate equations for this system. There are six different chemical species involved in the mechanism

TABLE 1

The governing reaction rate equations for the $H_2 + O_2$ mechanism (0)–(4)

$$\frac{d[H_2]}{dt} = -k_i[H_2][O_2] - k_1[OH][H_2] - k_3[O][H_2] + \tfrac{1}{2}k_4[H]$$

$$\frac{d[O_2]}{dt} = -k_i[H_2][O_2] - k_2[H][O_2]$$

$$\frac{d[H]}{dt} = k_1[OH][H_2] - k_2[H][O_2] + k_3[O][H_2] - k_4[H]$$

$$\frac{d[OH]}{dt} = 2k_i[H_2][O_2] - k_1[OH][H_2] + k_2[H][O_2] + k_3[O][H_2]$$

$$\frac{d[O]}{dt} = k_2[H][O_2] - k_3[O][H_2]$$

$$\frac{d[H_2O]}{dt} = k_1[OH][H_2]$$

Matrix of stoichiometric coefficients

	H_2	O_2	H	OH	O	H_2O	Σ_1	Σ_2
(0)	-1	-1	0	$+2$	0	0	0	0
(1)	-1	0	$+1$	-1	0	$+1$	0	0
(2)	0	-1	-1	$+1$	$+1$	0	0	0
(3)	-1	0	$+1$	$+1$	-1	0	0	0
(4)	$+\tfrac{1}{2}$	0	-1	0	0	0	0	0

we can write down six rate equations. However, there is some redundancy in this set. In particular two of the rate equations can be written simply as linear combinations of the remaining four. This can be seen by considering the matrix of stoichiometric coefficients, also given in Table 1.1. In this matrix the rows represent the five reaction steps and the columns give the appropriate stoichiometric cofficients for the six species in these steps. In general we can use the notation v_{ij} to represent the stoichiometric coefficient of the jth species in the ith reaction: thus v_{0,H_2} refers to H_2 in reaction (0) and because one hydrogen molecule disappears in this step, $v_{0,H_2} = -1$.

The final two columns in the matrix are given by the sums of the elements across each row in the following linear combinations:

$$\sum_{i,1} = 2v_{i,H_2} + v_{i,H} + v_{i,OH} + 2v_{i,H_2O} \tag{1.7}$$

$$\sum_{i,2} = 2v_{i,O_2} + v_{i,O} + v_{i,OH} + v_{i,H_2O}. \tag{1.8}$$

In other words, the first sum takes twice the stoichiometric coefficient for H_2 plus that for H plus that for OH and twice that for H_2O. For each reaction (row) this combination is exactly zero.

Chemists will recognize this summation in a different way: as the principle of conservation of atomic species. The total number of hydrogen atoms in the reactor must remain constant even though the exact chemical form—as free H or bound up in OH, H_2, and H_2O—may change. As there are two hydrogen atoms in H_2 and H_2O but only one in each of H and OH, we require

$$2[H_2] + [H] + [OH] + 2[H_2O] = \text{constant}. \tag{1.9}$$

This equation is completely equivalent to the statement $\sum_{i,1} = 0$ for all i (i.e. there is no net change in the number of H atoms in any of the reaction steps i) in eqn (1.7). Similarly, the sum $\sum_{i,2}$ represents the change in the number of O atoms in the ith step and again the conservation law gives $\sum_{i,2} = 0$ for all i. This is equivalent to the chemists' version

$$2[O_2] + [O] + [OH] + [H_2O] = \text{constant}. \tag{1.10}$$

If we start a reaction with a mixture of the pure reactants H_2 and O_2 then the constants in eqns (1.9) and (1.10) are simply given by twice the initial concentrations, $2[H_2]_0$ and $2[O_2]_0$.

In this rather roundabout way then, we have arrived at the conclusion that, as there are two different types of chemical building blocks (atoms), there will be two conservation equations associated with this kinetic scheme. We can use these two additional relationships to eliminate two of the variables, i.e. to reduce the number of governing rate equations which need to be considered. In the present case, for instance, we could choose to eliminate $[O_2]$ and $[H_2O]$. From eqn (1.9), we then have

$$[H_2O] = [H_2]_0 - \tfrac{1}{2}\{2[H_2] + [H] + [OH]\} \tag{1.11}$$

and from this and eqn (1.10)

$$[O_2] = [O_2]_0 - \tfrac{1}{2}\{[O] + \tfrac{1}{2}[OH] + [H_2]_0 - [H_2] - \tfrac{1}{2}[H]\}. \tag{1.12}$$

The full behaviour of the scheme can now be determined from the four rate equations $d[H_2]/dt$, $d/[H]/dt$, $d[OH]/dt$, and $d[O]/dt$, substituting for $[O_2]$ wherever it occurs in these equations from (1.12).

This all having been said, it is often easier not to attempt to find, or at least use, all of the possible conservation conditions but to compute with a slightly larger set of rate equations than is strictly necessary.

In many 'real' situations, a kinetic mechanism involving only true elementary steps is frequently not attainable—or always needed. Often a number of elementary steps will be so intimately coupled together that it is only possible to observe their combined effect: we refer then to an *overall process*. In these situations, empirical rate laws may become particularly useful. For instance, in the Belousov–Zhabotinskii reaction which we will discuss in some detail in a later chapter the following processes are important:

(5) $\qquad\qquad BrO_3^- + Br^- + 2H^+ \rightarrow HBrO_2 + HOBr$

(6) $\qquad\qquad HBrO_2 + Br^- + H^+ \rightarrow 2HOBr$

(7) $\qquad\qquad HOBr + Br^- + H^+ \rightarrow Br_2 + H_2O.$

None of these processes are necessarily believed to be elementary even though they have each been written in the same stoichiometric form as the elementary steps seen above in the $H_2 + O_2$ mechanism. Here, the form of reactions (5)–(7) is partly to identify the species which disappear in each process (the 'reactants') and those which are formed (the 'products'). In some cases, this form is also partly intended to convey information about the empirical rate laws which have been determined for each particular process. Thus for process (5), the rate of conversion of bromate and bromide ions into the species $HBrO_2$ and $HOBr$ is found to be first order in $[BrO_3^-]$ and $[Br^-]$ and second-order in the concentration of H^+: thus we can write for process (5)

$$-d[BrO_3^-]/dt = -d[Br^-]/dt = -\tfrac{1}{2}d[H^+]/dt = +d[HBrO_2]/dt$$

$$= +d[HOBr]/dt = k_5[BrO_3^-][Br^-][H^+]^2. \tag{1.13}$$

There is no suggestion, however, that this conversion is brought about by a single reactive collision between the four species on the left-hand side of reaction (5). We may also notice the appearance of stoichiometric factors (with appropriate signs) in eqn (1.13). If we were attending more carefully to the niceties of the mathematics, we would wish to make it clear that the term in eqn (1.13) is not a true rate: rather, with the appropriate stoichiometric factors, it gives the contribution from process (5) to the full reaction rate equations derived from the full mechanism. This distinction is not likely to cause much confusion here, however. Thus processes (6) and (7) also make the following contributions:

$$-d[HBrO_2]/dt = -d[Br^-]/dt = -d[H^+]/dt$$
$$= +\tfrac{1}{2}d[HOBr]/dt = k_6[HBrO_2][Br^-][H^+] \quad (1.14)$$

and

$$-d[HOBr]/dt = -d[Br^-]/dt = -d[H^+]/dt$$
$$= +d[Br_2]/dt = k_5[HOBr][Br^-][H^+] \quad (1.15)$$

respectively.

Another important part of the Belousov–Zhabotinskii mechanism is the pair of reactions

(8) $\qquad\qquad BrO_3^- + HBrO_2 + H^+ \rightarrow 2BrO_2{}^\cdot + H_2O$

(9) $\qquad\qquad BrO_2{}^\cdot + Ce^{3+} + H^+ \rightarrow HBrO_2 + Ce^{4+}.$

As well as providing the route by which the cerium ion catalyst is oxidized, this also gives rise to the crucial chemical feedback or *autocatalysis*. At the expense of a single $HBrO_2$ two molecules of the radical species $BrO_2{}^\cdot$ are produced in (8): these can then react relatively quickly via process (9), each reforming an $HBrO_2$ molecule. Thus there is an increase in the concentration of $HBrO_2$ through this cycle. What is particularly important is that, under typical conditions within the Belousov–Zhabotinskii mixture, reaction (8) is the slower of the two: thus the rate of reaction (8) determines the rate at which there is a net production of $HBrO_2$ (it is the *rate-determining step*). If we express this in rate equation terms, we would say that reactions (8) and (9) combine to produce an empirical rate of the form

$$d[HBrO_2]/dt = +k_8[HBrO_2][BrO_3^-][H^+]. \quad (1.16)$$

We may contrast this with the discussion above of elementary steps: here a term contributing to the rate of change of $[HBrO_2]$ has a *positive* coefficient, thus increasing the $HBrO_2$ concentration increases the rate at which it is being formed rather than the rate at which it is being removed. We can say that $HBrO_2$ is *catalysing* it own production: the process is *autocatalytic*.

The formal procedure for obtaining the autocatalytic form in eqn (1.16) is to write the two equations obtained from (8) and (9) for $[HBrO_2]$ and $[BrO_2{}^\cdot]$:

$$d[HBrO_2]/dt = -k_8[HBrO_2][H^+] + 2k_9[BrO_2{}^\cdot][Ce^{3+}][H^+] \quad (1.17)$$

$$d[BrO_2{}^\cdot]/dt = k_8[HBrO_2][Br^-][H^+] - k_9[BrO_2{}^\cdot][Ce^{3+}][H^+]. \quad (1.18)$$

If $BrO_2^.$ is a particularly reactive species we may apply the *stationary-state approximation*: this implies that $d[BrO_2^.]/dt \approx 0$, thus:

$$k_9[BrO_2^.][Ce^{3+}][H^+] \approx k_8[HBrO_2][Br^-][H^+]. \tag{1.19}$$

Substituting for the term $k_9[BrO_2^.][Ce^{3+}][H^+]$ from (1.19) into (1.17) yields eqn (1.16).

Autocatalysis or *chain branching* is also to be found in the hydrogen–oxygen reaction mechanism (0)–(4). The three steps (1)–(3) constitute the branching cycle for this system. Again this interpretation relies on the stationary-state approximation (SSA). In this case the SSA is applied to the two most reactive radicals: O and OH. If $d[OH]/dt$ and $d[O]/dt$ are set equal to zero in eqns (1.5) and (1.6), and the resulting relationships for [OH] and [O] are substituted into eqn (1.3), the rate equation for H atoms becomes

$$\frac{d[H]}{dt} = 2k_i[H_2][O_2] + (2k_2[O_2] - k_4)[H]. \tag{1.20}$$

The coefficient of [H] in the last term on the right-hand side is known as the *net branching factor*, $\phi = 2k_2[O_2] - k_4$. This may be positive or negative depending on the values of k_2 and $[O_2]$, which relate to the rate of increase in the total radical concentration through the branching cycle, and of k_4 which relates to the removal of radicals via the termination step (4). If branching exceeds termination, ϕ is positive: the radical concentration increases with time during a period of autocatalytic growth. If termination wins, ϕ is negative and there is no net autocatalytis.

In both the Belousov–Zhabotinskii and the hydrogen–oxygen reactions, the autocatalysis has arisen through a term which has been first order in the catalytic species ($HBrO_2$ and H respectively). Other concentration dependences can arise. These are not always simple, but sometimes can be split into suitable empirical forms. The reaction of iodate and iodide ions in acidic solutions produces iodine, I_2 in the presence of a suitable reducing agent, such as arsenite ions, this iodine is converted back to iodide: thus the process overall produces iodide from iodate. In this case the rate-determining step is the initial reaction between iodate and iodide. The rate of iodide formation in this system is well approximated by an empirical law which combines two forms of autocatalysis: one which is first order in iodide and one which is second order:

$$d[I^-]/dt = k_0[I^-][IO_3^-][H^+]^2 + k_c[I^-]^2[IO_3^-][H^+]. \tag{1.21}$$

Many experiments are carried out in buffered solutions at constant pH, so the concentration of H^+ is not a variable. Under such conditions, the two terms in eqn (1.21) can be said to have quadratic and cubic character respectively: the first term depends on the product of two varying concentrations, the second on three.

Specific experimental systems have played on important part in the oscillatory chemical reaction field, but so too has the analysis of simplified models. The latter seek to summarize or isolate one or two particular aspects of, say, chemical feedback in terms of prototype reaction rate equations which are suitable for algebraic manipulation—but not so abstracted as to lose immediate physical interpretation. In this way unifying features and general principles can emerge most clearly.

One set of models which has been particularly widely studied is that based on the simplest representations of autocatalysis. For the Belousov–Zhabotinskii or hydrogen–oxygen reactions, we have seen *quadratic autocatalysis*: such a process can be well mimicked by considering a general conversion of a 'reactant' species A to the autocatalyst B in a non-elementary 'step':

$$A + B \rightarrow 2B \qquad \text{rate} = k_q ab \qquad (1.22)$$

where a and b are the two concentrations. A process with *cubic autocatalysis* would be similar represented as

$$A + 2B \rightarrow 3B \qquad \text{rate} = k_c ab^2. \qquad (1.23)$$

We will make much use of such model schemes at various stages in this book.

It was mentioned earlier that most spontaneous chemical reactions are to some extent exothermic. If heat is released by a reaction with non-zero rate and if the transfer of heat from the reactor to the surroundings is not infinitely fast, we can expect that there will be some self-heating of the system. The exact magnitude of the temperature excursions will depend on such factors as the thermal capacity of the reactor and the reacting mixture, which may for instance be a dilute aqueous solution, the actual reaction rate and the heat transfer coefficient. If the self-heating becomes significant, and as we have seen chemical reaction rates are often quite sensitive functions of the temperature, we will have an extra variable to consider in our governing rate equations. Thus, we may have to augment our system of reaction rate equations, with their temperature dependence implicit in the rate constants, by adding an energy-balance equation expressing the rate of change of temperature with time. Typically, this will have the form of a difference between a chemical heat release rate and a physical heat transfer rate. The first of these will be a function of the instantaneous concentrations, the temperature itself, and the exothermicities of the various reactions: the heat transfer process may be essentially through Newtonian cooling and hence be governed by the temperature difference between the reacting mixture and the surroundings and such features as the surface-to-volume ratio.

We need not, however, consider ourselves confined to well-stirred systems governed by ordinary differential equations. In many situations of interest the diffusion of reactant species of the conduction of heat through porous media may be important. The governing equations will then have partial differential forms based on Fick's law or the Fourier heat equation, each augmented by reaction terms.

We must consider further the behaviour of chemical systems on the basis of the mechanisms and models introduced above. Once we have decided on a given set of governing rate equations (either ordinary or partial differential equations, perhaps with energy conservation included) a particular experimental situation is described by choosing suitable values for the various parameters such as activation energies, pre-exponential factors, ambient temperature, etc. To specify the problem fully we need then to select our initial conditions—the values of the concentrations of all species at time $t = 0$. We follow the same procedure if we are performing an experiment.

With this specification, the given chemical system now has a unique 'solution'. This means that the evolution of the concentrations of all species is completely fixed once the initial conditions have been chosen. (If we are dealing with partial differential equations, the solution involves the concentrations as functions of position as well as time.) We may represent this evolution of the reacting mixture as a path in a diagram such as Fig. 1.1. There we have taken for simplicity a reaction with two components, so the concentration axes form a plane: the evolution during the reaction then corresponds to a line in three dimensions, with time as the third axis. The reaction path, also known as a *trajectory* or *flow*, originates from a point in the concentration plane at $t = 0$ corresponding to the initial conditions and then moves through the *concentration–time space*.

Many reactions are conventionally carried out in a thermodynamically closed system. Such a system does not exchange mass with its surroundings. A characteristic of chemical reactions in closed systems is that for a given set of initial conditions there exists one and only one corresponding state of chemical equilibrium. As time increases, so the reaction trajectory moves from the initial point towards the equilibrium state, approaching it as $t \to \infty$. We can consider, as an example, the two consecutive reversible reaction sequence

(10) $$A \rightleftharpoons B \rightleftharpoons C.$$

Let us assume that the concentrations at time $t = 0$ are a_0, b_0, and 0 respectively. This is a two-variable system: once the concentrations of A and B are known at any time t, the concentration of C is uniquely fixed by mass conservation. If the equilibrium constants for the first and second steps are K_1 and K_2 respectively, then the chemical equilibrium state (for which $da/dt = db/dt = 0$) is given by

$$a_{\text{eq}} = \frac{(a_0 + b_0)}{K_1(1 + K_2) + 1} \qquad b_{\text{eq}} = \frac{K_1(a_0 + b_0)}{K_1(1 + K_2) + 1}. \qquad (1.24)$$

Thus all initial points with the same total initial concentration of reactants,

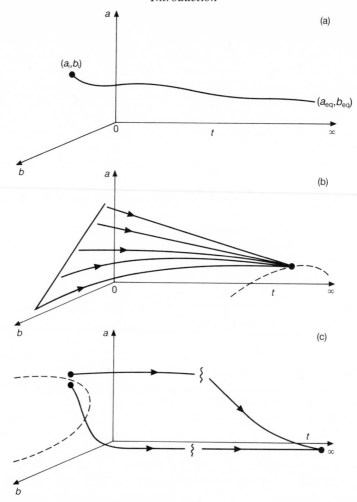

FIG. 1.1. The evolution of concentrations in time. (a) A 'trajectory' corresponding
to the development from an initial composition (a_i, b_i) to a final equilibrium state
(a_{eq}, b_{eq}). (b) The 'contraction of state space': a line of initial compositions evolve to
the same equilibrium point; the two dimensional initial concentration plane is reduced
to a line at infinite time. (c) 'Parametric sensitivity' to initial conditions. The two
systems shown differ only slightly initially, and ultimately attain the same equilibrium
composition, but evolve in significantly different ways at short times.

$a_0 + b_0$, are 'attracted' to the same equilibrium point as time tends to infinity:
a line on the concentration plane at $t = 0$ is 'mapped' to a point as $t \to \infty$,
as shown in Fig. 1.1(b). Going one stage further, the whole concentration
plane at $t = 0$ is mapped on to a line as $t \to \infty$. This 'contraction' in area
(or more generally of volume) is typical of 'dissipative dynamical systems',

of which chemical reactions are exemplary. In the limit of irreversible reactions steps, $K_1 \to \infty$ and $K_2 \to \infty$, all initial points are attracted to the same equilibrium point: $a_{eq} \to b_{eq} \to 0$ as $t \to 0$.

This contraction also underlies 'normal' chemical behaviour. Trajectories which start from similar initial points tend to move closer together as time increases. The implication here is that experiments which are set up with similar initial conditions tend to behave more and more similarly as time increases. This is the fundamental basis for scientific methodology itself: experiments are repeatable, by and large, even though we know that two experiments can never quite have identical initial conditions to all degrees of accuracy. This is, of course, not always so. Some chemical reactions exhibit 'ignition limits'. A classic example is the hydrogen–oxygen reaction at low pressures ($p \lesssim 30 \, \text{kN m}^{-2} \approx \frac{1}{4}$ atmosphere) and high ambient temperatures ($T \gtrsim 700 \, \text{K}$). At some initial composition, the spontaneous reaction between H_2 and O_2 is almost undetectably slow: the reaction trajectory departs only slowly from the initial point although it will eventually find its way to the equilibrium state as $t \to \infty$. A minute change in the initial composition, e.g. replacing a small amount of H_2 by more O_2, can change the character of the reaction completely. An ignition may occur, with essentially complete conversion of the reactants to product on a millisecond timescale. The departure from the vicinity of the initial point in the concentration–time space is rapid. Thus two close initial points may have strongly divergent trajectories at short times, as shown in Fig. 1.1(c). At long times, however, they all approach the same (or at least very similar) equilibrium state and so become close again.

We can draw a line or boundary on the initial concentration plane, separating initial states which diverge initially. This boundary is equivalent, in this case, to an ignition limit such as that shown in Fig. 1.2 for the $H_2 + O_2$ reaction.

There are some additional restrictions on the behaviour of chemical reactions in thermodynamically closed systems: these are not as severe as they once believed to be, but do have significant consequences for experiments and their interpretation. Once a given set of initial conditions has been chosen we have, in effect, specified a unique corresponding equilibrium state to which the reaction trajectory will move as t tends to infinity. The final approach of the trajectory to this equilibrium point is relatively simple, no matter how complex the kinetic mechanism: we always see a monotonic exponential decay to the equilibrium point. Chemical reactions cannot oscillate about equilibrium. These restrictions also apply 'close to' the equilibrium state, i.e. at long but still finite times in the reaction. Unfortunately, there is no universal, precise definition of 'close to' or 'long times' in this context. Also importantly, the rules make no qualification on the range of behaviour possible 'far from equilibrium', i.e. at short times. The reaction path may show quite complex wanderings in its early journey which will be

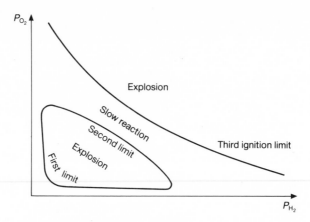

FIG. 1.2. A representative 'ignition diagram' typical of the hydrogen–oxygen reaction (see Chapter 9) showing loci of diverging trajectories, i.e. the 'ignition limits' that separate exploding from slow reaction systems.

reflected in the observed experimental behaviour (assuming we have some suitable method of detection). Again, it is not possible to produce much in the way of a precise definition of 'early on'. The 'long time' argument does, however, mean that complex responses such as oscillations or ignitions must eventually die out in a thermodynamically closed system: this is usually referred to as the 'effects of reactant consumption' by chemists. Any oscillatory behaviour will also take place against a continuously changing background of chemical composition.

Some of these considerations can be removed by operating the reaction in a thermodynamically open system. Open systems are those which do exchange mass with their surroundings. The simplest realization of this for a chemical system is the continuous-flow reactor: a supply of fresh reactants is continuously pumped into the reactor and an outflow maintains a constant volume. Such reactors are also commonly stirred as efficiently as possible to remove spatial gradients (the models ideally then involve ordinary rather than partial differential equations) and are known by the acronym CSTR (Continuous-flow Stirred Tank Reactor) or C*.

The continuous inflow and outflow mean that molecules do not spend infinite time in the reactor. The average time in the reactor, or mean residence time, can be calculated as the quotient of the volume divided by the total volumetric flow rate: the actual lifetimes are then distributed about this mean value. We can think of this as holding the reaction at some distance away from the chemical equilibrium state. The shorter the mean residence time, the further the system is 'constrained away from equilibrium'. The restrictive comments above relating to the final approach close to equilibrium are no longer relevant. We will have cause to examine models of reactions in a

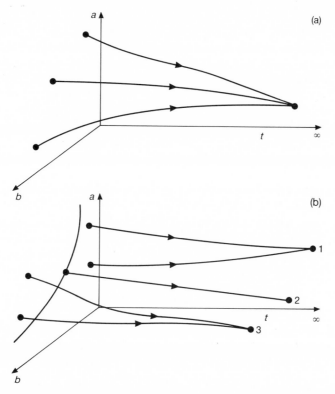

FIG. 1.3. The evolution of concentrations in an open reactor such as a CSTR. (a) A unique stable stationary state 'attracts' all initial conditions. (b) Multiple stationary states, with two attractors and a separating unstable state.

CSTR in later chapters. There we will see that the role of the chemical equilibrium state as an attractor for trajectories as time tends to infinity is taken over (at least in the simplest cases) by *stationary states* at which the rates of chemical change exactly match the net inflow or outflow of each reacting species.

We can again discuss some aspects of the behaviour in terms of trajectories through concentration–time space. In the simplest situation, a reaction may have a single stable stationary state. Then, all initial conditions (all points on the concentration plane at $t = 0$) move towards (are attracted) to this state as time increases (Fig. 1.3(a)). As we adjust the operating conditions, e.g. by varying the flow rate or the inflow concentrations of one of the reactants, so the stationary-state solution will change quantitatively, moving around on the t = ∞ plane. As well as quantitative effects, we may also find qualitative changes. For instance, new stationary-state solutions may appear: typically a new pair will emerge together, so we go from one stationary state to three. We now have *multiplicity*. Figure 1.3(b) shows these three solutions on the $t = \infty$ plane. The initial plane at $t = 0$ also has some additional significant features. Whereas all initial points were attracted to the single stable stationary state in Fig. 1.3(a), they now have three final states to choose between. We can again draw a boundary on the initial concentration plane separating initial points which move to one stationary state from those which go to another. In this way we divide the plane into two parts. On one side of the boundary the trajectories move to stationary-state 1: on the other, they move to stationary-state 3. But what about stationary-state 2? Only those points exactly on the boundary move to state 2, and this line has zero 'measure' (i.e. no area) on the initial plane. In practice, therefore, no trajectories approach this 'middle' state. It is not a stable stationary state: rather it is *unstable*. This is typical. When two new solutions appear, one of them (at least) will be unstable: this unstable state serves to identify a boundary on the initial concentration plane which separates those initial conditions which are attracted to one stable state from those attracted to the other. We may also note that, although this boundary or *separatrix* has been shown with a fairly simple form in Fig. 1.3(b), more complicated windings across the initial concentration plane are possible: the separatrix cannot in general, however, cross itself at any point.

Rather than presenting three-dimensional concentration–time space representations of the trajectories, the flows can be projected on to the concentration *phase plane*. For the cases shown in Fig. 1.3, these planes show one or three stationary-state solutions as *singular points* (Fig. 1.4(a) and (b) respectively). Any initial condition is also represented by a point on this plane. The concentration–time path now corresponds to a trajectory across the plane, starting from the initial point and moving to a singular (stationary-

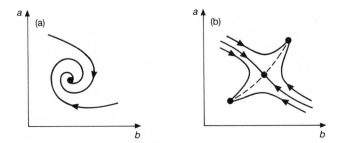

FIG. 1.4. Phase-plane portraits for stationary-state solutions. (a) A stable attractor ('sink') showing the spiralling-in as the concentrations vary: for an unstable 'source', the arrows are reversed and the concentrations move away across the plane. (b) Multiple stationary states with two sinks and a saddle point: only two trajectories approach the saddle—the insets, that divide the paramerer plane into two regions of attraction, one for each of the sinks.

state) point. In Fig. 1.4(a), where there is a unique stable singularity, all trajectories are attracted to the same point. In Fig. 1.4(b) we see the separatrix from Fig. 1.3(b) dividing the phase plane: the unstable middle stationary state lies on this separatrix and initial points along the separatrix are attracted to this state. All other points, however, are attracted to the stationary-state singularity on the appropriate side of the separatrix. Trajectories do not cross this boundary, nor do they intersect with themselves or each other: trajectories only meet at singular points.

Two other features of the phase-plane behaviour are worth mentioning at this point. We have seen one form of unstable stationary state: that associated with a separatrix. A second form of instability can also arise, even when there is a single stationary state. Although chemical systems are forbidden to oscillate about their (chemical) equilibrium state, there is no such restriction for stationary states. Thus the trajectories may continuously wind around an unstable stationary state, even as $t \to \infty$. In the concentration–time space, this corresponds to a trajectory winding on to a cylinder as shown in Fig. 1.5(a). If there are no other stationary-state (or other attracting) solutions, all initial points will have trajectories which move on to this cylinder, either winding in from outside or out from inside. If we project this

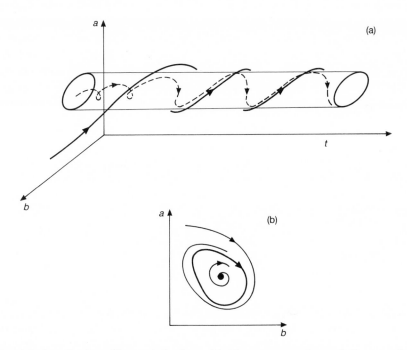

FIG. 1.5. Sustained oscillatory behaviour. (a) The concentrations wind on to the surface of a cylinder, giving rise to a limit cycle (b) in the phase plane.

behaviour on to the concentration phase plane, the cylinder becomes a closed loop or *limit cycle* which surrounds the (unstable) stationary state. The trajectories from given initial points again wind across the plane on to the limit cycle without crossing themselves or each other (Fig. 1.5(b)).

We have restricted our comment so far to so-called two-variable systems—those described by two independent concentrations (or, perhaps, a concentration and the temperature). This has been especially helpful in that the concentration–time space is only three-dimensional and hence relatively easily envisaged. The phase-plane projection is yet simpler—a two-dimensional picture. The requirement that paths on this surface should not cross means that only relatively simple responses such as stationary-state or peroiodic oscillations are possible. If we consider systems with three (or more) independent variables, greater complexity is allowed. The phase space is three-dimensional (or higher). The trajectories are still lines in this space: although different paths can still not cross, they may pass over or under each other, so if this higher-dimensional behaviour is projected on to any plane formed by two of the components such crossing will appear to arise. It is this greater degree of freedom that allows complex oscillations and even aperiodic responses and hence will be of much concern in this book.

We may note as a summary to the preceding discussions that virtually all of even the most highly complex patterns exhibited by chemical reactions can be approached and interpreted in terms of the behaviour exhibited by models of almost deceptive simplicity. There is generally no need for intricate, multi-reaction kinetic mechanisms in order to capture the essential features, even if these are ultimately required for the finest levels of quantitative detail. If the chemical system shows nothing more exotic than oscillations, it is basically behaving as a two-dimensional system: complex oscillations will require three variables. There are no extra qualitative complexities introduced by increasing a model from three to four variables. In addition to these minimum degrees of freedom, and perhaps more importantly, there are two other reqirements. There must be at least one mechanism for *feedback* within the chemical model: we have seen two examples already—chemical feedback through autocatalysis and thermal feedback through the temperature dependence of reaction rate constants. Secondly, such feedback must be sufficiently *non-linear*; we have seen both quadratic and cubic autocatalysis associated with the chemical feedback, and an exponential dependence on the inverse temperature from the Arrhenius rate law. These are all non-linear. If we only allow two variables, a quadratic non-linearity on its own will not be sufficient to produce oscillatory behaviour or multiple stationary states: it may do so if coupled with other sources of non-linearity or if there are three variables. Cubic and exponential non-linearities can support oscillations in two-variable schemes.

Before proceeding with a detailed analysis of a number of model chemical reaction schemes we will examine the behaviour of some simple 'mappings'.

At first sight these may appear rather abstract as chemical models. In fact they do have application in many physical situations, including chemical kinetics, but they also have an important role in themselves. Many of the (mathematical) analytical techniques which are in the current armoury of the non-linear scientist have their basis in mappings and many of the responses characteristic of 'flows' are revealed more clearly and computed more easily in maps.

References

Specific references are listed where appropriate at the end of chapters. Here it is convenient to list some relatively general texts of interest.

Babloyantz, A. (1986). *Molecules, dynamics and life*. Wiley, New York.

Berge, P., Pomeau, Y., and Vidal, C. (1984). *Order within chaos*. Wiley, Chichester.

Cvitanovic, P. (1984). *Universality in chaos*. Adam Hilger, Bristol.

Erdi, P. and Toth, J. (1989). *Mathematical models of chemical reactions*. Manchester University Press.

Field, R. J. and Burger, M. (1985). *Oscillations and traveling waves in chemical systems*. Wiley, New York.

Gray, P. and Scott, S. K. (1990). *Chemical oscillations and instabilities: non-linear chemical kinetics*. Oxford University Press.

Gray, P., Nicolis, G., Baras, F., Borckmans, P., and Scott, S. K. (1990). *Spatial inhomogeneity and transient behaviour in chemical kinetics*. Manchester University Press.

Holden, A. (1986). *Chaos*. Manchester University Press.

Kuramoto, Y. (1984). *Chemical oscillations, waves and turbulence*. Springer, Berlin.

Nicolis, G. and Prigogine, I. (1989). *Exploring complexity: an introduction*. Freeman, New York.

Pacault, A. and Vidal, C. (1979). *Synergetics, far from equilibrium*. Springer, Berlin.

Pippard, A. B. (1985). *Response and stability*. Cambridge University Press.

Rensing, L. and Jaeger, N. I. (1984). *Temporal order*. Springer, Berlin.

Sarkar, S. (1986). *Nonlinear phenomena and chaos*. Adam Hilger, Bristol.

Stewart, I. (1989). *Does God play dice?* Blackwells, Oxford.

Tabor, M. (1989). *Chaos and integrability in nonlinear dynamics*. Wiley, New York.

The Royal Society (1987). *Dynamical chaos*, (ed. M. V. Berry, I. C. Percival, and N. O. Weiss). Cambridge University Press.

Thompson, J. M. T. (1982). *Instabilities and catastrophes in science and engineering*. Wiley, Chichester.

Thompson, J. M. T. and Stewart, H. B. (1986). *Nonlinear dynamics and chaos*. Wiley, Chichester.

Vidal, C. and Pacault, A. (1981). *Nonlinear phenomena in chemical dynamics*. Springer, Berlin.

Vidal, C. and Pacault, A. (1984). *Non-equilibrium dynamics in chemical systems*. Springer, Berlin.

Warnatz, J. and Jäger, W. (1987) *Complex chemical reaction systems*. Springer, Berlin.

2

Mappings

The idea of a mapping in a dynamical system can be treated at various levels. At its simplest it can be accepted as a 'game'. Even in this form it is instructive and has the advantage of requiring nothing more than a non-programmable calculator for the reader to participate. Later, we will see that the lessons learnt are of surprising relevance to the behaviour in chemical reactions.

The game involves a variable x, perhaps a concentration. We cannot follow x as a continuous function of time, but can only observe its value at a series of times, $t = t_0, t_1, t_2, \ldots, t_n, \ldots$. (In the simplest case this could correspond to discrete sampling at regular time intervals $t = 0, \Delta t, 2\Delta t, \ldots, n\Delta t, \ldots$ etc., but in general there may be irregular time intervals between successive readings.) Thus we would obtain a series of measurements $x_0, x_1, x_2, \ldots, x_n, \ldots$ corresponding to these times. In order to model this we use a *map* (May 1976; Feigenbaum 1980*a*)—also known as an *iterative, recursive,* or *logistic map*—which allows us to predict the next value x_{n+1} from the current measurement x_n. In other words we want to find a formula that expresses x_{n+1} as some function of x_n. This mapping $f(x_n)$ forms the 'rule' by which the game progresses. We can add some qualifying by-laws: for instance, we can ask that the value of x should be limited to the range $0 \leqslant x \leqslant 1$.

To start the game, we choose a particular starting value x_0 for our variable x (lying in the range $0 \leqslant x \leqslant 1$). Now from x_0 we calculate a new value x_1 and from x_1 we find an x_2, from x_2 an x_3, and so on, according to the particular rule, i.e. the form of the mapping $f(x_n)$. As a particular example, we will consider the cubic map

$$x_{n+1} = Ax_n(1 - x_n)^2 \tag{2.1}$$

where A is some constant.

The map described by eqn (2.1) can be represented graphically, as shown in Fig. 2.1. The cubic curve has a maximum of $\frac{4}{27}A$ at $x_n = \frac{1}{3}$: if x_{n+1} is to remain always less than or equal to unity we thus require $A \leqslant \frac{27}{4}$. The map also has a minimum $f(x_n) = 0$ at $x_n = 1$.

The qualitative form of $f(x_n)$ is the same for all A. The specific quantitative values in Fig. 2.1 correspond to the choice $A = 3$. We can follow the game for this particular value of A. Let us start at $x_0 = \frac{1}{2}$: the resulting sequence from eqn (2.1) is then $x_1 = 0.375$, $x_2 = 0.439\,453\,1$, $x_3 = 0.414\,245\,3$, $x_4 = 0.426\,393\,2$, etc. For $n \geqslant 10$ we find $x_n = 0.422\,649\,7$. The sequence is

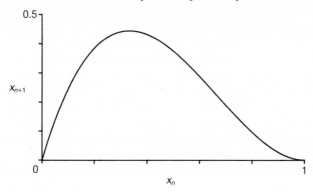

FIG. 2.1. The cubic map $x_{n+1} = Ax_n(1 - x_n)^2$ with $A = 3$, showing a maximum as $x_n = \frac{1}{3}$. The shape is general for all A.

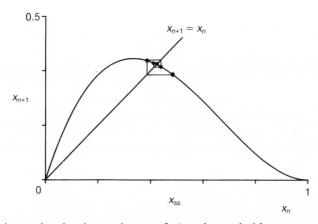

FIG. 2.2. A mapping showing attainment of a 'steady state' with $x_{n+1} = x_n$ for $A = 3$.

converging to a steady solution. The same 'stationary-state' value is obtained from any initial condition within the range specified above. The iterations of the map can also be seen in Fig. 2.2, which is simply Fig. 2.1 augmented by the additional line $x_{n+1} = x_n$. (Note that the 'path' followed, marked by the horizontal and perpendiculars between $f(x_n)$ and the line, is not a trajectory in the sense of the previous chapter (nor is there really any 'following'). Intermediate values of x for $n\Delta t < t < (n + 1)\Delta t$ are not defined in any sense.)

2.1. Stationary states of the map

The straight line $x_{n+1} = x_n$ also helps indicate the location of the stationary state appropriate to a given A as its intersection with the mapping curve. The stationary-state values can also be determined analytically for the present map. We require simply that $x_{n+1} = x_n$. If we denote such a value by x_{ss}, then this is given by the solution of

$$x_{ss} = Ax_{ss}(1 - x_{ss})^2. \tag{2.2}$$

One solution is $x_{ss} = 0$ as there is always an intersection at the origin. Non-zero solutions are also possible. These are given by

$$x_{ss} = 1 \pm A^{-1/2}. \tag{2.3}$$

For x_{ss} to lie between 0 and 1, we require $A \geqslant 1$ and take the lower root of eqn (2.3). For our example above, $A = 3$, so $x_{ss} = 1 - 1/\sqrt{3} = 0.422\,649\,7$, as we discovered by evaluating the sequence.

2.2. Parameter variation

The restriction that x must remain in the range between 0 and 1 has led us to allow A to have values only in the range

$$1 \leqslant A \leqslant \tfrac{27}{4}. \tag{2.4}$$

We have seen that for $A = 3$, the system evolves to the stationary state given by eqn (2.3). How does the response change as we vary A? The stationary-state solution exists across the whole range, but is it always stable?

As the parameter A is varied, so the point of intersection of the mapping curve and the straight line changes. The gradient of the cubic curve at the point of intersection also varies during this process. It turns out that the stationary state represented by the intersection is stable provided the value of the gradient at that point has a magnitude less than unity (May 1974; Li and Yorke; Guckenheimer *et al.* 1977). The gradient $\partial f/\partial x_n$ at any point is given by

$$\partial f/\partial x_n = A(1 - x_n)(1 - 3x_n). \tag{2.5}$$

Substituting the lower root from eqn (2.3), the gradient at the stationary-state solution is given in terms of the parameter A by

$$\partial f(x_{ss})/\partial x_n = 3 - 2\sqrt{A}. \tag{2.6}$$

For $A = 1$, the stationary-state solution is $x_{ss} = 0$ and the gradient is $+1$. This point has neutral stability. If A is increased beyond unity, the gradient decreases, so $|\partial f(x_{ss})/\partial x_n| < 1$, and the corresponding non-zero intersection is stable. There is a special case at $A = \frac{9}{4}$: here the intersection coincides with the maximum in the cubic curve and so $|\partial f(x_{ss})/\partial x_n| = 0$. Such points are called *superstable*: the iteration of the map converges most rapidly to the stationary state for such points, faster than the normal exponential approach. For $A > \frac{9}{4}$ the gradient at the stationary-state intersection is negative. The solutions will remain stable provided $|\partial f(x_{ss})/\partial x_n| < 1$, but will lose stability when

$$\partial f(x_{ss})/\partial x_n = -1 \qquad \text{at } A = A^* = 4 \tag{2.7}$$

which corresponds to $x_{ss} = \frac{1}{2}$. For $A > A^*$, we do not expect the map to iterate on to the solution given by eqn (2.3).

2.3. Period-doubling cascades

As an example, let us consider $A = 4.5$. The stationary-state solution given by eqn (2.3) is $x_{ss} = 0.5286$. If we start with any other initial value for x in the range between zero and unity, however, we soon find that the sequence settles to an alternation between $x = \frac{1}{3}$ and $\frac{2}{3}$. With this simple pattern of alternation, the value of x is repeated every second iteration: i.e. we have that

$$x_{n+2} = f[f(x_n)]. \tag{2.8}$$

Using our simple cubic form for $f(x_n)$, a double application of the mapping is equivalent to a higher-order polynomial form

$$x_{n+2} = A[Ax_n(1 - x_n)^2][1 - Ax_n(1 - x_n)^2]^2. \tag{2.9}$$

Figure 2.3 shows this double mapping, $x_{n+2}(x_n)$, with $A = 4.5$. This is typical for any A in the range $4 < A < 5$. There are three intersections, with $x_{n+2} = x_n$, over the range of x between zero and unity. One is at $x = 0.5286$, the (unstable) stationary-state solution, and the other two are at $x = \frac{1}{3}$ and $x = \frac{2}{3}$, the 'limit cycle' solutions between which x_n oscillates.

For the present value of the parameter, $A = 4.5$, the gradient of the double mapping described by eqn (2.9) is greater than -1 (with magnitude less than 1) at the two 'oscillatory' intersections, reflecting the stability of the simple alternating solution. As A increases, however, so the gradient at these points decreases, and may also reach -1. The system then bifurcates again, to a solution which repeats every four iterations. This occurs at $A = 5$. With $A = 5.121\,122$, the eventual oscillatory sequence is $x_n = x_{n+4} = 0.758\,685$, $x_{n+1} = 0.226\,254$, $x_{n+2} = 0.693\,678$, $x_{n+3} = 0.333\,33$.

The period-4 oscillations become unstable and give way, in turn, to period-8 at $A = 5.235\,09$. There are further period doublings, as listed in Table 2.1. Examples of the oscillatory time series for some of the different periodicities are given in Fig. 2.4.

The successive period-doubling bifurcations get closer and closer together as the number of iterations per complete cycle increases. For the highest

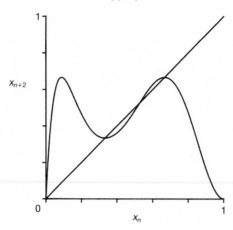

1

x_{n+2}

0

x_n

1

FIG. 2.3. The double mapping $x_{n+2}(x_n)$ for the cubic map, specifically for $A = 4.5$, showing an unstable intersection corresponding to the 'steady state' and the two stable intersections for the period-2 state.

periodicities, the range of A over which they are stable becomes vanishingly small. The final column of Table 2.1 shows the inverse of the ratio of the ranges for successive periodicities. As well as revealing the steady decrease, there is another point to this list. The value of the entry in this column is tending to the value 4.669 20 ... (Feigenbaum 1979, 1980a, b). We would find the same value for any particular form we might take for the function $f(x_n)$—at least, provided the map has a simple single maximum. It is a universal constant for such mappings, known as the Feigenbaum number.

We could continue the table with periods of 2^n for any n beyond 8. The bifurcations would occur as A increases, but will also tend to a finite limit as n tends to infinity. For the above model this limit is 5.300 506, so all of the periodicities 2^9, 2^{10}, ..., etc. must be squeezed in between 5.300 500 and this value.

TABLE 2.1
Location of period-doubling bifurcations for cubic map, from Tomlin (1990)

Period	A	$(A_n - A_{n-1})/(A_{n+1} - A_n)$
1	4.000 0	
2	5.000 0	4.253 7
4	5.235 09	4.577 39
8	5.286 449	4.649 14
16	5.297 496	4.673 01
32	5.299 86	4.653 54
64	5.300 368	4.668 29
128	5.300 477	4.669 01
256	5.300 500	

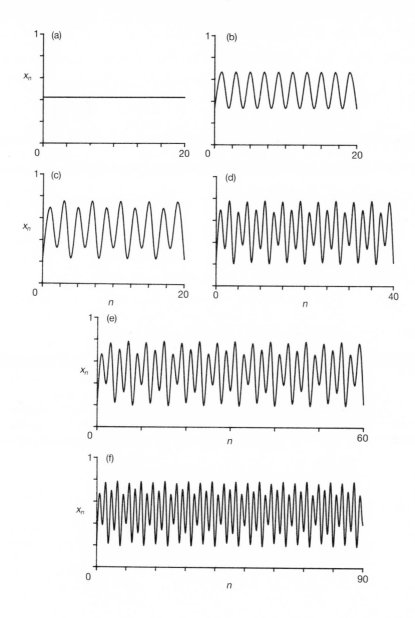

FIG. 2.4. Various periodicities in the period-doubling sequence as A is increased in the cubic map: (a) $A = 3$, period-1; (b) $A = 4.5$, period-2; (c) $A = 5.1$, period-4; (d) $A = 5.25$, period-8; (e) $A = 5.29$, period-16; (f) $A = 5.298$, period-32.

2.4. Aperiodicity and periodic windows

What happens for values of A above the limit of the period-doubling sequence? We have an upper limit on A of $\frac{27}{4}$, so we still have the range $5.300\,506 < A < 6.75$ to investigate. One particularly important form of behaviour, found just above the convergence limit, is that of aperiodicity or 'chaos'. Now the sequence x_n gives a different value at each step, never repeating itself no matter how many iterations we make.

Figure 2.5 gives an indication of how the behaviour varies with A across the whole range $1 < A < \frac{27}{4}$. It shows the values of $x_n(A)$ for 100 iterations (after initial transients have died away). For any given value in the range of stationary-state stability, $1 < A < 4$, we see that each x_n is the same for all n, so the curve is single valued. At $A = 4$, the response bifurcates, splitting into two branches: across the range $4 < A < 5$, the iterations alternate between the two branches for odd and even n. Further period doublings then ensue.

For $A > 5.300\,506$, x_n is different at each stage of the iteration and we could plot an infinite number of points for each A without repetition. Even so we would not fill the whole x_n range between 0 and 1—there are systematic gaps in the 'line' formed by the infinite number of points.

At larger A, there are occasional ranges where the behaviour becomes quite simple again. For instance at $A = 6.055\,678$, a period-3 pattern emerges. This is followed by a period-doubling sequence, with period-6, period-12, etc. This sequence also converges, to $A \approx 6.088$, reproducing the Feigenbaum

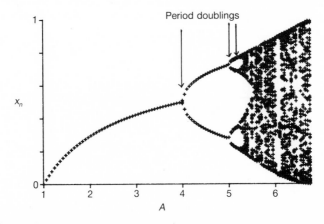

FIG. 2.5. The behaviour of the cubic map over the entire parameter range $1 \leqslant A \leqslant \frac{27}{4}$. 100 iterates are plotted for each A after transients have died out. For any $A \leqslant 4$ each iterate gives the same value of x_{n+1} as a period-1 steady state has been achieved, beyond this the period-2 solution gives two points etc. and the period-doubling cascades to chaos and subsequent periodic windows can be clearly seen.

constant as it does so. Other periodic 'windows' can be detected—one based on period-5 exists for $A \approx 6.28$. The range of A for such periodicities can become extremely small but not only do they exist, they occur in a well-defined order—a *universal sequence* which holds for all 'single-hump' maps. It is difficult to list the exact order of all of the (infinite number of) periodic windows, but restricting ourselves to the lower periodicities all maps of the above form will have the windows appearing in the following order (Metropolis *et al.* 1973; May 1976) as the bifurcation parameter A is increased: 6, 5, 3, 5, 6, 6, 5, 6. Notice that there are three different period-5 solutions and four distinct period-6 states, each distinguished by the order in which the x_n members of the series are visited and each giving rise to its own period-doubling cascade.

This universality has also given rise to a handy 'rule' often quoted (Li and Yorke 1975) as 'period-3 implies chaos'. Thus if for some experimental conditions we observe period-1 responses and a period-doubling sequence and elsewhere we find period-3, we can assume that a region of aperiodicity occurs somewhere between them.

2.5. Sensitivity to initial conditions: stretching and folding

For the simple stationary-state behaviour or with regular periodicity, the mapping converges on to the appropriate 'attractor' at least exponentially. Any information about the particular initial conditions is quite quickly lost as all initial x_0 move to the same repeating values. Such a response is therefore completely insensitive to the initial conditions.

The same is not true for the mapping when A is in a range corresponding to aperiodic responses. True, the value of x_n evolves according to completely defined rules—one value explicitly determines the next, with 100 per cent certainty. There is no randomness, no element of chance uncertainty or irregularity. If we know the rules and can measure a given starting condition exactly even a 'chaotic' pattern can be predicted exactly. The same initial condition (exactly) will always evolve in the same way (at least on the same computer with the same tolerances, etc.). However, such solutions are extremely sensitive to initial conditions. Unlike the periodic responses, two aperiodic sequences started from similar, but not identical, conditions will not stay 'close together'. In fact they will diverge exponentially, becoming completely uncorrelated. This may not be a particularly fast divergence, or it may be rapid. Figure 2.6 shows the evolution of two series with (a) $x_0 = 0.5000$ and (b) $x_0 = 0.5001$ for $A = 5.50$. The two traces are almost indistinguishable for 20 iterations or so, but then begin to go their own separate ways—we cannot predict the nth iterate of one map from that of the other. Figure 2.6(c) shows how this difference develops with n.

The origin of this divergence can be seen from the form of the mapping

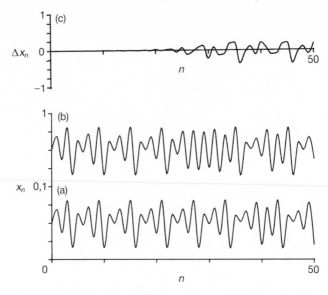

FIG. 2.6. The sensitivity to initial conditions under chaotic behaviour is revealed by the divergence of mappings from different x_0: (a) $x_0 = 0.5000$; (b) $x_0 = 0.5001$; (c) the difference between the two mappings.

curve itself. Figure 2.7(a) and (b) show two parts of the cubic curve in close-up. In (a) the slope of the map is particularly steep, i.e. $|\partial f/\partial x| \gg 1$. The behaviour of two iterated trajectories starting close together (separated by δx_n) reveals an increasing separation or stretching, with $\delta x_{n+1} > \delta x_n$: the greater the slope, the greater the degree of stretching. Thus divergence is associated with parts of the map with steep gradients. The trajectories cannot continue to diverge for ever, as we have required x to remain bounded in the range between 0 and 1. There must, therefore, be some compression or folding to counteract the stretching: folding will cause points initially far apart to be brought close together. This can be seen in the vicinity of the maximum in the curve, as shown in Fig. 2.7(b): the two trajectories on either side, separated by δx_n, are both mapped to very similar next points, so $\delta x_{n+1} < \delta x_n$. The distinction between periodic and aperiodic responses is thus related to this competition between stretching and folding: if the stretching wins 'on average', the trajectory will have an overall divergence corresponding to aperiodicity. We must be able to quantify 'on average' here: not only does the gradient vary with x_n but we must also remember that different parts of the map may be visited more frequently than others—some parts are not visited at all and hence make no contribution to the competition. For instance, with stationary-state behaviour only one point is visited (after transients have died out) so it is only the magnitude of the gradient at x_{ss} which is relevant.

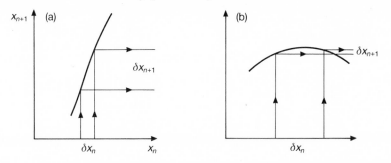

FIG. 2.7. Schematic representation of the origin of (a) divergence and (b) folding or reinjection by different sections of the mapping curve.

2.6. Lyapounov exponents

A quantitative measure of the divergence rate is the *Lyapounov exponent* λ. There are various recipes for calculating λ, some of which are discussed in Chapter 7. An important point is that these exponents play a similar role for flows, i.e. for the results from models based on differential equations and experimental results. Here we will use a simple route (Lauwerier 1986) which is suitable for mappings. We take

$$\lambda = \lim_{n \to \infty} \frac{1}{n} \sum_{i=0}^{n} \log_2 \left| \frac{\partial f}{\partial x} \right|. \tag{2.10}$$

We can thus compute our map for a large number n of iterations with various values of A: the derivative at each step is given by eqn (2.5). An alternative representation of eqn (2.10) is

$$\lambda = \int_0^1 p(x) \log_2 \left| \frac{\partial f}{\partial x} \right| dx \tag{2.12}$$

where $p(x)$ is a weighting factor given by the probability that a trajectory contains an iterate between x and $x + dx$. This weighting is carried out for us by the map in (2.10) as we iterate for sufficiently large n.

The variation of λ with A is shown in Fig. 2.8 for $5 \leqslant A \leqslant \frac{27}{4}$. For small A, $A < 5$, the map settles on to either a stationary-state or a regular periodicity. Such behaviour is typified by a negative Lyapounov exponent: the greater the magnitude the faster the convergence to the appropriate attractor. We have seen that there is a period-doubling bifurcation at $A = 5$ (Table 2.1). The attracting state has neutral stability and this is reflected by $\lambda = 0$. There are similar zeros at $A = 4$ and $A = 5.235\,09$, etc., i.e. at each period-doubling point. In between the zeros, λ becomes negative again (it does not pass through zero to become positive as computed here). In fact,

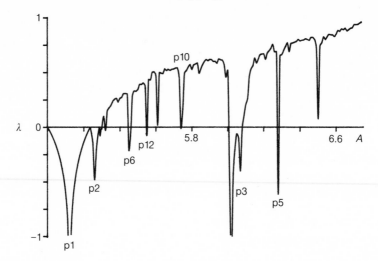

FIG. 2.8. Variation of the Lyapounov exponent λ with the bifurcation parameter A: $\lambda > 0$ indicates experimental divergence of initial points and hence corresponds to chaotic regimes.

between each zero there is a value of A for which $\lambda \to -\infty$: these are the superstable points mentioned above for which the convergence is faster than exponential.

The Lyapounov exponent does eventually pass through zero and become positive: for $A \gtrsim 5.300\,506$, the convergence limit of the first period-doubling cascade. Above this limit, the positive value of λ indicates the exponential divergence of neighbouring points in the aperiodic regime. As A is increased further, λ occasionally dips back down below the axis (going off to $-\infty$ in fact) as the periodic windows are encountered. The most obvious of these ranges come for $6.01 < A < 6.08$ and contain the period-doubling sequence beginning from period-3, whilst there is a period-5 sequence at $A \approx 6.28$.

The Lyapounov exponent has units of 'bits of information per iteration'. We can interpret the quantitative value of λ in another way. Imagine that our initial reading x_0 is made from an experimental system using an 8 bit analogue-to-digital converter and that we then try to predict the subsequent evolution from our numerical map. If λ has the value $+\frac{1}{2}$, then we lose $\frac{1}{2}$ of a bit per iteration: after eight iterations only 4 bits of our predicted x_8 will be significant: there is no significance to our prediction for x_{16}.

2.7. Experimental significance

The consequences and limitations on predictability arising from inexact knowledge of initial conditions have been mentioned. We should also recognize that in practice we will also have imprecise knowledge of the values of our parameters. Thus, on a good day, we may be able to measure our ambient temperature to ± 0.5 K and the reactant gas pressure to ± 0.1 per cent. Small uncertainties such as these will lead to smaller or larger uncertainties in our parameters. Furthermore, if we cannot measure these factors more precisely than this, we probably cannot control them to maintain absolutely constant values to any greater extent. Now, if a particular type of response is found over a wide range of parameter values, this may not be too important a limitation: there may be no qualitative change and the quantitative knock-on effect may be smaller than the precision of our measurements.

For our cubic map, the parameter A can run from 1 to 6.75. Simple stationary-state behaviour, period-2, and, to some extent, period-3 are likely to be 'stable' in that they survive over relatively wide parameter ranges. Only for these will the system be 'predictable' to any extent. The response is basically aperiodic for $A > 5.3$, i.e. for approximately 25 per cent of this range. With the various period-doubling sequences and periodic windows seen above, however, the parameter ranges over which some of the higher-order periodicities exist can become very small. If we cannot control our conditions within these ranges, we may not be able to sustain these patterns.

2.8. Interpretations of simple maps

As we have already commented, mappings of the type discussed above are not in any way easily related to a given set of reaction rate equations. Such mappings have, however, been used for chemical systems in a slightly different way. A quadratic map has been used to help interpret the oscillatory behaviour observed in the Belousov–Zhabotinskii reaction in a CSTR. There, the variable x_n is not a concentration, but the amplitude of a given oscillation. Thus the map correlates the amplitude of one peak in terms of the previous excursion.

With this identification, the stable 'stationary-state' behaviour (found for the cubic model with $1 < A < 4$) corresponds to oscillations for which each amplitude is exactly the same as the previous, i.e. to period-1 oscillatory behaviour. The first bifurcation ($A = 4$ above) would then give an oscillation with one large and the one smaller peak, i.e. a period-2 waveform. The period doubling could then continue in the same general way as described above.

More generally, a technique known as the Poincaré map will be of great importance when examining the behaviour of models based on ordinary differential equations, to which we turn our attention in the next chapter.

References

Feigenbaum, M. J. (1979). The universal metric properties of nonlinear transformations. *J. Stat. Phys.*, **21**, 669–706.

Feigenbaum, M. J. (1980*a*). Universal behavior in nonlinear systems. *Los Alamos
Sci.*, **1**, 4–27.

Feigenbaum, M. J. (1980*b*). The transition to aperiodic behavior in turbulent systems.
Commun. Math. Phys., **77**, 65–86.

Guckenheimer, J., Oster, G. F., and Ipatchki, A. (1977). The dynamics of density-
dependent population models. *J. Math. Biol.*, **4**, 101–47.

Lauwerier, H. A. (1986). One-dimensional iterative maps. In *Chaos*, (ed. A. V.
Holden), pp. 39–57. Manchester University Press.

Li, T.-Y. and Yorke, J. A. (1975). Period three implies chaos. *Am. Math. Mon.*, **82**,
985–92.

May, R. M. (1974). Biological populations with nonoverlapping generations: stable
points, stable cycles and chaos. *Science*, **186**, 645–7.

May, R. M. (1976). Simple mathematical models with very complicated dynamics.
Nature, **261**, 459–67.

Metropolis, N., Stein, M. L., and Stein, P. R. (1973). On finite limit sets for
transformations on the unit interval. *J. Comb. Theory*, **15(A)**, 25–44.

Tomlin, A. S. (1990). *Ph.D. thesis*. University of Leeds.

3

Flows I. Two-variable systems

In the previous chapter we investigated the behaviour of simple mappings. These showed 'stationary states' and bifurcations to periodicity, aperiodicity, and periodic windows. Even if we cannot claim to have completely understood these responses, a number of patterns with some definite structure have clearly emerged. Now we turn to models described by differential equations. In order to bring out the relevant features most clearly, we will employ the simplest of the model chemical schemes: the cubic autocatalator (Merkin *et al.* 1986, 1987*a,b*; Gray and Scott 1986, 1990*a,b*). This can be written in terms of mock stoichiometric reaction processes as

(0) $\qquad\qquad\qquad$ P $\to$ A $\qquad$ rate $= k_0 p$

(1) $\qquad\qquad\qquad$ A $\to$ B $\qquad$ rate $= k_u a$

(2) $\qquad\qquad$ A + 2B $\to$ 3B $\qquad$ rate $= k_c ab^2$

(3) $\qquad\qquad\qquad$ B $\to$ C $\qquad$ rate $= k_d b$.

Thus a reactant or precursor species P is converted to a final product C via a sequence of reactions involving two intermediates A and B. One of the reaction steps, (2), is autocatalytic with an overall cubic dependence on the concentrations. For now we will concentrate on the isothermal form of the model, i.e. we assume that there is no significant heat evolution, so we do not have to worry about the temperature dependence of the reaction rate constants. This aspect will be included later in this chapter.

A similar scheme, known as the Brusselator, has been widely studied and will be mentioned at various places throughout this book: it is formally related to the above model by swapping the A and the B in steps (2) and (3).

3.1. Governing reaction rate equations: pool chemical approximation

The scheme above involves four different chemical species: P, A, B, and C. The last of these is not involved in the kinetics as the reaction steps are assumed to be irreversible. A common approximation with such models of thermodynamically closed systems is that the initial concentration of the reactant is large compared with the maximum concentrations of the intermediates and that the 'initiation' step (0) has a relatively small rate constant. Then, the concentration of P will not vary much over a reasonable timescale and can be treated as a constant (we 'ignore reactant consumption' and assume $p = p_0$ for all times). This *pool chemical approximation* has been explained and justified in detail elsewhere (Merkin *et al.* 1986) and will be adopted here.

The reaction rate equations for the concentrations of A and B can be written in a particularly tidy form by introducing dimensionless parameters and variables

$$\frac{d\alpha}{d\tau} = \mu - \kappa_u \alpha - \alpha\beta^2 = f(\alpha, \beta; \mu, \kappa_u) \tag{3.1}$$

$$\frac{d\beta}{d\tau} = \kappa_u \alpha + \alpha\beta^2 - \beta = g(\alpha, \beta; \mu, \kappa_u). \tag{3.2}$$

Here $\mu = (k_0^2 k_c / k_d^3)^{1/2} p_0$ is the dimensionless (initial) concentration of the reactant; $\alpha = (k_c/k_d)^{1/2} a$ and $\beta = (k_c/k_d)^{1/2} b$ are the dimensionless concentrations of two intermediates; $\tau = k_0 t$ is the dimensionless time; and κ_u is the dimensionless rate constant for the uncatalysed step (1).

This transformation has no effect on the dynamics of the model but does reduce the number of apparent parameters from five (four rate constants and the initial concentration) to two: κ_u and μ.

3.2. Stationary states

The starting point for the analysis of a given model is nearly always the determination of the *stationary states*. These are specified by the condition that all the rates of change should vanish simultaneously. In the present case, therefore, we require

$$\mu - \alpha_{ss}\beta_{ss}^2 - \kappa_u \alpha_{ss} = 0 \tag{3.3}$$

$$\alpha_{ss}\beta_{ss}^2 + \kappa_u \alpha_{ss} - \beta_{ss} = 0. \tag{3.4}$$

These have the unique solution

$$\beta_{ss} = \mu, \qquad \alpha_{ss} = \frac{\mu}{\mu^2 + \kappa_u}. \tag{3.5}$$

The dependence of α_{ss} and β_{ss} on μ is shown in Fig. 3.1. The autocatalyst concentration is simply an increasing linear function of the reactant concentration: the locus for intermediate A shows a maximum, located at $\mu = \kappa_u^{-1/2}$ with $\alpha_{ss} = \alpha_{max} = \frac{1}{2}\kappa_u^{-1/2}$. The $\alpha_{ss}(\mu)$ and $\beta_{ss}(\mu)$ loci also cross, with $\alpha_{ss} = \beta_{ss} = \mu = (1 - \kappa_u)^{1/2}$ provided $\kappa_u \leqslant 1$.

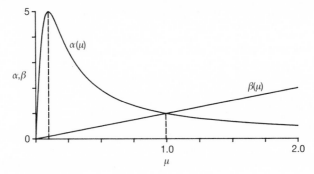

FIG. 3.1. The dependences of α_{ss} and β_{ss}, the dimensionless stationary-state concentrations of the intermediates A and B, on the reactant concentration μ for the autocatalator model, showing the maximum and crossing points.

The next test is that of stability: not all stationary states are stable, and a system cannot sit at an unstable state. More accurately, we wish to determine the *local stability* of a given stationary-state solution. As the stationary state above depends on the system parameters μ and κ_u, we can expect that the local stability will also vary with these.

To determine the local stability, we imagine a small (infinitesimal) perturbation $\Delta\alpha_0$, $\Delta\beta_0$ from the given stationary state at some time τ_0. The evolution of these perturbations will also be governed by eqns (3.1) and (3.2):

$$\frac{d\Delta\alpha}{d\tau} = f(\Delta\alpha, \Delta\beta; \mu, \kappa_u) \tag{3.6}$$

$$\frac{d\Delta\beta}{d\tau} = g(\Delta\alpha, \Delta\beta; \mu, \kappa_u) \tag{3.7}$$

where $f(\Delta\alpha, \Delta\beta; \mu, \kappa_u)$ and $g(\Delta\alpha, \Delta\beta; \mu, \kappa_u)$ are the appropriate forms of the right-hand sides of eqns (3.1) and (3.2). However, because of the small magnitude of the perturbations, non-linear terms such as $(\Delta\alpha)^2$, etc. will be negligible (at least initially). We can thus apply the linearized equations

$$\frac{d\Delta\alpha}{d\tau} = f(\alpha_{ss}, \beta_{ss}; \mu, \kappa_u) + (\partial f/\partial\alpha)_{ss}\Delta\alpha + (\partial f/\partial\beta)_{ss}\Delta\beta + \cdots \tag{3.8}$$

$$\frac{d\Delta\beta}{d\tau} = g(\alpha_{ss}, \beta_{ss}; \mu, \kappa_u) + (\partial g/\partial\alpha)_{ss}\Delta\alpha + (\partial g/\partial\beta)_{ss}\Delta\beta + \cdots \tag{3.9}$$

Here $f(\alpha_{ss}, \beta_{ss}; \mu, \kappa_u)$ and $g(\alpha_{ss}, \beta_{ss}; \mu, \kappa_u)$ correspond to the right-hand sides—the *rate functions*—of eqns (3.1) and (3.2) evaluated with the stationary-state solutions. This last point is especially useful as $g(\alpha_{ss}, \beta_{ss}; \mu, \kappa_u) = f(\alpha_{ss}, \beta_{ss}; \mu, \kappa_u) = 0$, by definition of a stationary state. The terms $(\partial f/\partial\alpha)_{ss}$, etc. are partial derivatives of the rate functions, again evaluated by substituting in for the stationary states: as an example for the present model

$$(\partial f/\partial \alpha)_{ss} = -(\kappa_u + \beta_{ss}^2) = -(\kappa_u + \mu^2). \tag{3.10}$$

The present model has two variables and hence we have two equations governing the evolution of the perturbations and four partial derivatives. We can easily generalize this to n variables, writing the resulting equations in matrix form for compactness: if the variables are $x_1, \ldots, x_n$ and the rate equations have the form $dx_i/d\tau = f_i(x_1, \ldots, x_n)$ then we will have

$$\frac{d\Delta \mathbf{x}}{d\tau} = \mathbf{J}(\mathbf{x}_{ss})\Delta \mathbf{x} \tag{3.11}$$

where $\Delta \mathbf{x} = (\Delta x_1, \ldots, \Delta x_n)^{\mathrm{T}}$ and $\mathbf{J}(\mathbf{x}_{ss})$ is the $n \times n$ *Jacobian matrix*

$$\mathbf{J} = \begin{pmatrix} \dfrac{\partial f_1}{\partial x_1} & \dfrac{\partial f_1}{\partial x_2} & \cdots & \dfrac{\partial f_1}{\partial x_n} \\ \vdots & & \ddots & \vdots \\ \dfrac{\partial f_n}{\partial x_1} & \dfrac{\partial f_n}{\partial x_2} & \cdots & \dfrac{\partial f_n}{\partial x_n} \end{pmatrix}. \tag{3.12}$$

The various partial derivatives are then evaluated using the stationary-state solution $(x_{1,ss}, x_{2,ss}, \ldots, x_{n,ss})$.

For the present model, the Jacobian has the form

$$\mathbf{J} = \begin{pmatrix} -(\kappa_u + \beta_{ss}^2) & -2\alpha_{ss}\beta_{ss} \\ (\kappa_u + \beta_{ss}^2) & 2\alpha_{ss}\beta_S - 1 \end{pmatrix} = \begin{pmatrix} -(\kappa_u + \mu^2) & \dfrac{-2\mu^2}{(\mu^2 + \kappa_u)} \\ (\kappa_u + \mu^2) & \dfrac{(\mu^2 - \kappa_u)}{(\mu^2 + \kappa_u)} \end{pmatrix}. \tag{3.13}$$

Because the rate equations governing the evolution of the perturbations are linear, their solutions will involve the sum of n exponential terms. Thus for our two-variable model, we will express $\Delta \alpha(\tau)$ and $\Delta \beta(\tau)$ in the form

$$\Delta \alpha = c_1 \exp(\lambda_1 \tau) + c_2 \exp(\lambda_2 \tau) \tag{3.14}$$

$$\Delta \beta = c_3 \exp(\lambda_1 \tau) + c_4 \exp(\lambda_2 \tau). \tag{3.15}$$

The pre-exponential terms c_1–c_4 are constants determined by the initial perturbations. More important for the qualitative behaviour are the exponents λ_1 and λ_2. These are obtained from the Jacobian matrix as its eigenvalues (Davis 1962; Arnol'd 1973; Jordan and Smith 1977), i.e. by $\mathbf{J}(\mathbf{x}_{ss}) - \lambda \mathbf{I} = 0$. For our 2×2 matrix, the eigenvalues are given by the roots of the quadratic equation

$$\lambda^2 - \mathrm{tr}(\mathbf{J})\lambda + \det(\mathbf{J}) = 0 \tag{3.16}$$

where tr(**J**) and det(**J**) are the trace and determinant of the Jacobian

$$\text{tr}(\mathbf{J}) = (\partial f/\partial \alpha) + (\partial g/\partial \beta)$$

$$\text{det}(\mathbf{J}) = (\partial f/\partial \alpha)(\partial g/\partial \beta) - (\partial f/\partial \beta)(\partial g/\partial \alpha).$$

(3.17a)

For now we need only consider three situations. If the determinant is negative, the roots of eqn (3.16) will be real and have opposite sign, i.e. one will be positive and the other negative. The exponential term in eqns (3.14) and (3.15) associated with the negative eigenvalue will decrease in magnitude as time increases, so this part of the perturbation will decay back to the stationary state. However, the term associated with the positive eigenvalue will increase (exponentially) with time. The net effect is that the perturbation grows and the system moves away from what is an unstable stationary-state solution. This form of behaviour is characterized on a *saddle*. We may note that there is a limited sense of stability, but only those perturbations for which the coefficients c_i associated with the positive eigenvalue are exactly zero will decay.

If the determinant of the Jacobian is positive, the sign of the eigenvalues is determined by the trace. For tr(**J**) < 0, the two roots of eqn (3.16) will be negative or will be complex numbers with negative real parts. The exponential terms in eqns (3.14) and (3.15) will then decrease in magnitude as time increases, so the perturbation decays to zero and the system returns back to the same, stable stationary state. This behaviour is characteristic of a *sink*. A distinction is sometimes worthwhile between sinks with real eigenvalues and those with a complex pair: in the first case the return is monotonic and the stationary state is termed a stable node; with complex eigenvalues, there is a damped oscillatory approach to the stationary state, characteristic of a stable focus.

If the trace of the Jacobian is positive, the roots of eqn (3.16) will be positive or a complex pair with positive real parts. The exponential terms now increase in magnitude with time—growing in an oscillatory manner if the eigenvalues are complex. The stationary state is unstable: a *source* (or an unstable node or focus).

For the present model, the form of the Jacobian matrix is such that

$$\text{tr}(\mathbf{J}) = -(\kappa_u + \mu^2) + \frac{(\mu^2 - \kappa_u)}{(\mu^2 + \kappa_u)} \qquad \text{det}(\mathbf{J}) = \mu^2 + \kappa_u. \qquad (3.17b)$$

Clearly, as μ and κ_u are positive, so the determinant is always positive for this system. The stationary-state solution, eqn (3.5), is never a saddle point. Turning now to the trace, we wish to identify the parameter values, if any, at which the stationary state changes from a sink (stable, with tr(**J**) < 0) to a source (unstable, with tr(**J**) > 0). This change in stability is associated with the trace changing sign, i.e. with tr(**J**) = 0. The latter condition is

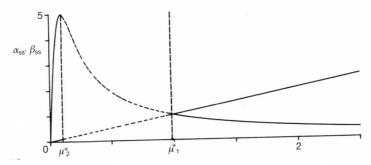

FIG. 3.2. Changes in local stability of the autocatalator stationary state with μ showing the two Hopf bifurcation points $\mu_{1,2}^*$ relative to the maximum and crossing point. The stationary state is unstable for $\mu_2^* < \mu < \mu_1^*$.

satisfied when

$$\mu^4 - (1 - 2\kappa_u)\mu^2 + \kappa_u(1 + \kappa_u) = 0. \tag{3.18}$$

This is a biquadratic equation, which has roots μ_1^* and μ_2^* given by

$$(\mu_{1,2}^*)^2 = \tfrac{1}{2}[1 - 2\kappa_u \pm (1 - 8\kappa_u)^{1/2}]. \tag{3.19}$$

This equation has two real, positive roots provided the discriminant is positive, i.e. provided

$$\kappa_u < \tfrac{1}{8}. \tag{3.20}$$

If condition (3.20) is satisfied, the roots μ_1^* and μ_2^* mark the end points of a range of the reactant concentration over which the stationary state is unstable. This situation is illustrated in Fig. 3.2.

3.3. Hopf bifurcation and limit cycles

The preceding discussion has introduced an important concept: the loss of stability of a stationary state. For a two-variable system, this is associated with the trace of the Jacobian matrix becoming zero, provided also that the determinant is positive. More fundamentally, this corresponds to the eigenvalues $\lambda_{1,2}$ changing in character from a complex pair with negative real parts when $\mathrm{tr}(\mathbf{J}) < 0$ to a pair with positive real parts for $\mathrm{tr}(\mathbf{J}) > 0$. At the point of changing stability, the *Hopf bifurcation point*, the eigenvalues have the form

$$\lambda_{1,2} = \mathrm{Re}(\lambda) \pm i\,\mathrm{Im}(\lambda) \qquad \text{with } \mathrm{Re}(\lambda) = 0 \tag{3.21}$$

i.e. the eigenvalues form an imaginary pair. There are a number of other qualifications we must make to ensure that the stability of the stationary state does indeed change when eqn (3.21) is satisfied (Hopf 1942; Hassard *et al.* 1981). For two-variable systems we also require that the imaginary part should be non-zero: this is equivalent, in this case, to requiring a positive determinant. Also, the real part must actually pass through zero transversally. If eqn (3.21) is satisfied by a particular value μ^* of some parameter μ, these conditions can be expressed in the form

$$\text{Im}(\lambda) > 0 \quad \text{and} \quad \text{d Re}(\lambda)/\text{d}\mu \neq 0 \quad \text{for } \mu = \mu^*. \quad (3.22)$$

For general systems with n variables ($n > 2$), conditions (3.21) and (3.22) are augmented by an extra requirement on the remaining $n - 2$ eigenvalues, $\lambda_3 – \lambda_n$. These must all either be negative or have negative real parts at μ^*:

$$\text{Re}(\lambda_i) < 0 \quad \text{for } i = 3, n \quad (3.23)$$

To see the implications of a Hopf bifurcation, we can return to our two-variable model and examine the *phase plane*. First we must check the various Hopf conditions. Equation (3.21) is satisfied for $\mu = \mu^*_{1,2}$ given by eqn (3.19). The imaginary part is given by $[\det(\mathbf{J})]^{1/2}$ and is non-zero for non-zero μ and κ_u from eqn (3.17b). The derivative of the trace with respect to μ, which gives d $\text{Re}(\lambda_{1,2})/\text{d}\mu$ is given by

$$\frac{\text{d tr}(\mathbf{J})}{\text{d}\mu} = \frac{\mu}{(\mu^2 + \kappa_u)^2} [1 - 8\kappa_u \pm (1 - 8\kappa_u)^{1/2}] \quad (3.24)$$

and so is non-zero for $\kappa_u < \frac{1}{8}$, the same condition on the uncatalysed reaction rate constant as that given in (3.20) for the existence of Hopf points.

For $\mu < \mu^*_2$ and for $\mu > \mu^*_1$, the eigenvalues have negative real parts and so perturbations decay back to the stationary state in a damped oscillatory manner. If we project this behaviour on to the α–β phase plane, the stationary state is represented by a point whose coordinates are given by eqn (3.5). The damped oscillatory approach to this *singular point* from conditions close by corresponds to a spiral trajectory on the phase plane, as shown in Fig. 3.3(a): the origin of the term stable focus can clearly be seen from this representation.

In the opposite case, with $\mu^*_2 < \mu < \mu^*_1$, the trace of $\mathbf{J}$ is positive and so is the real part of the eigenvalues. The oscillatory departure from the stationary state gives rise to a spiralling out from the singular point in the phase plane, as shown in Fig. 3.3(b). For the moment we leave the question of where the trajectory winds to.

At the two Hopf points themselves, $\mu = \mu^*_2$ or $\mu = \mu^*_1$, the linear analysis is inconclusive. Here, the small non-linear terms which are neglected in the formulation of eqns (3.8) and (3.9) play an important role. Provided the various conditions mentioned above are met, the non-linear terms lead to the formation of a single periodic trajectory. The sustained oscillatory evolution projects on to the phase plane to give a closed curve or *limit cycle* around which the trajectory moves. At the Hopf points, this cycle is born but has zero size and sits 'on' the singular point. As the parameter μ is

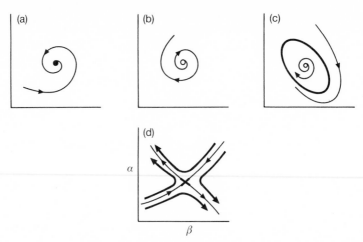

FIG. 3.3. Phase portraits for stationary states: (a) stable sink; (b) unstable source; (c) unstable source + stable limit cycle (reversing the arrows gives a sink + unstable limit cycle); (d) the local flow at a saddle point showing the inset and outset, and the division of the parameter plane (this singularity does not occur for the autocatalator.

varied, however, the limit cycle grows to surround the stationary-state singularity. Figure 3.3(c) shows a typical limit cycle.

Limit cycles are also *attractors* (or *repellors*) in the phase plane. Like stationary states, they can also have stability or instability. A general route to the determination of stability of a limit cycle will be presented shortly: this involves numerical methods, but close to the Hopf points it is possible (at least in theory) to assess the stability of the emerging cycle analytically. There are two possible scenarios (each of which has a mirror image). If we consider a situation like that found for μ_2^*, the trace of $\mathbf{J}$ and real part of λ change from negative to positive as the *bifurcation parameter* μ is increased. Thus the stationary state is stable for $\mu < \mu_2^*$ and unstable for $\mu > \mu_2^*$. Now, if the limit cycle grows as μ is increased beyond μ_2^* it will take the stability lost by the stationary state: a stable limit cycle emerges and surrounds the unstable singular point. Figure 3.4(a) represents this case by showing the 'size' (amplitude) of the emerging cycle as a function of μ. Conversely, if the limit cycle born at μ_2^* grows as μ is decreased, it will be unstable (Fig. 3.4(c)). The mirror images, which represent the possibilities at the upper Hopf point μ_1^* where the stationary state regains stability as μ is increased through the Hopf bifurcation, are shown in Fig. 3.4(b) and (d).

The recipe for determining which of the four possibilities arises in a given model with a given set of parameter values has been presented in a relatively simple-to-use form elsewhere (Gray and Scott 1990*a*, Chapter 5) and so will not be reproduced here. As well as distinguishing the nature of the Hopf

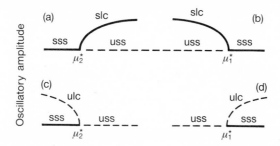

FIG. 3.4. Schematic representations of the four possible arrangements at a Hopf bifurcation: the diagrams show a stationary state that loses stability (——, stable; – – –, unstable) at a Hopf bifurcation (μ_2^* or μ_1^*). A limit cycle emerges (slc, stable limit cycle; ulc, unstable limit cycle) whose amplitude increases as the system moves away from the bifurcation point. (a) and (b) represent supercritical Hopf bifurcations with a stable limit cycle growing around the unstable stationary state; (c) and (d) are subcritical with an unstable limit cycle emerging around the stable stationary state.

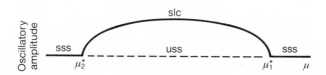

FIG. 3.5. The Hopf bifurcation scenario for the autocatalator, with two supercritical Hopf bifurcation points and a stable limit cycle existing over the range of stationary-state instability between them.

bifurcation, such analyses can also predict the rate at which the amplitude and period of the limit cycle changes with the bifurcation parameter close to the Hopf point.

With the present model we find only the situations in Fig. 3.4(a) at μ_2^* and Fig. 3.4(b) at μ_1^*. A stable limit cycle emerges at each Hopf point and grows as we adjust the concentration of the reactant μ so as to move into the region of stationary-state instability. In fact, it is the same limit cycle in each case, and the two loci of amplitudes joint up between the Hopf points as shown in Fig. 3.5. This form of response is the simplest situation we can have, but cannot be confirmed by the Hopf analysis which only gives information about behaviour 'local' to $\mu_{1,2}^*$. We can now also answer the question as to where the trajectories winding away from an unstable stationary state go to. They move towards the stable limit cycle and so the system settles into sustained oscillatory motion.

3.4. Development of oscillations

Figure 3.5 shows how the amplitude of the limit cycle or the corresponding oscillations varies with the reactant concentration μ: this form is typical of the response for all $\kappa_u < \frac{1}{8}$. The actual form of the oscillations is not conveyed in such a representation, but we can integrate the governing equations for a series of μ across the range $\mu_2^* < \mu < \mu_1^*$. In each case it is important to compute for sufficient time such that any transient influence of the initial conditions dies away, thus ensuring that the final attractor is revealed. Figure 3.6 shows four images each for a selection of reactant concentrations: the first two give the actual time series $\alpha(\tau)$ and $\beta(\tau)$, then the corresponding

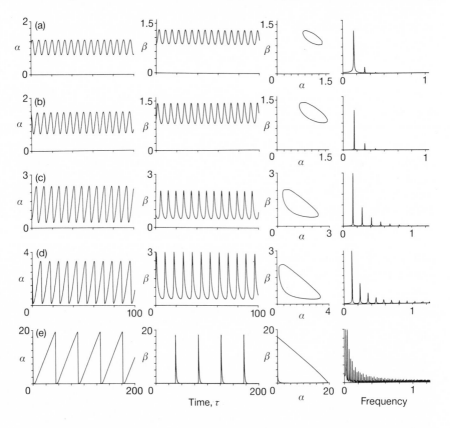

FIG. 3.6. The development of oscillations and the corresponding limit cycles and Fourier power spectra over the range of stationary-state instability for the auto-catalator: (a) small-amplitude, almost sinusoidal oscillations close to the upper Hopf bifurcation, $\mu = 0.995$ (and $\kappa_u = 0$, so $\mu_1^* = 1$); (b) $\mu = 0.99$; (c) $\mu = 0.95$; (d) $\mu = 0.5$, $\kappa_u = 0.001$ showing the typical 'sawtooth' waveform.

limit cycle in the α–β phase plane. Finally we show the *power spectrum* for each oscillatory record, obtained by Fourier transforming the time series (usually $\alpha(\tau)$).

Close to the Hopf bifurcation point μ_1^* (in Fig. 3.6(a) $\mu = 0.995$ whilst the corresponding Hopf bifurcation with $\kappa_u = 0$ is at $\mu_1^* = 1$) the oscillations have small amplitude and basically a sinusoidal waveform. The latter point is reflected in the power spectrum which shows a fundamental peak with a frequency $\omega = 0.157$ and a small contribution at the first overtone only. The limit cycle has the form of an ellipse.

As μ is decreased, the amplitude and period increase, the limit cycles become markedly more 'triangular', and more overtones appear as contributions in the power spectrum. With $\mu = 0.5$ and $\kappa_u = 10^{-3}$, the system is close to the middle of the oscillatory region. The oscillations now have developed to a characteristic sawtooth in the $\alpha(\tau)$ time trace and a pulse form for $\beta(\tau)$. The power spectrum contains significant contributions from a great number of overtones.

3.5. The Salnikov model: subcritical Hopf bifurcations

The simple cubic autocatalator examined above gives rise only to *supercritical Hopf bifurcations*, i.e. those at which a stable limit cycle is born at the Hopf point and grows towards the side on which the stationary state is unstable (Fig. 3.4(a) and (b)). The other form of the Hopf bifurcation, shown in Fig. 3.4(c) and (d), is termed *subcritical*: then an unstable limit cycle emerges. A simple scheme which shows both supercritical and subcritical Hopf bifurcations is the Salnikov scheme (Salnikov 1948, 1949; Gray *et al.* 1988a; Gray and Roberts 1988; Kay and Scott 1988). This is closely related to the cubic autocatalator, but the chemical feedback is replaced by a thermal effect. The scheme requires only two steps and both of these are first order:

(4) $P \rightarrow A$ rate = $k_4 p$

(5) $A \rightarrow B + \text{heat}$ rate = $k_5(T)a$.

The first step again involves a pool chemical reactant whose concentration will be regarded as constant over the timescale of an experiment. Now the second step is exothermic and its rate constant has an Arrhenius form

$$k_5 = A \exp(-E/RT) \qquad (3.25)$$

where T is the temperature within the reactor. The reaction takes place in a volume V with surface area S across which heat is transferred by Newtonian cooling with a heat transfer coefficient χ to a reservoir at an ambient temperature T_a. The reaction rate and heat-balance equations can be written in the following dimensionless forms:

$$\frac{d\alpha}{d\tau} = \mu - \kappa\alpha \exp[-\theta/(1 + \varepsilon\theta)] \tag{3.26}$$

$$\frac{d\theta}{d\tau} = \alpha \exp[-\theta/(1 + \varepsilon\theta)] - \theta \tag{3.27}$$

where α is again a measure of the concentration of A, θ is the dimensionless temperature difference $E(T - T_a)/RT_a^2$: μ is the (initial) concentration of the reactant P, κ the value of the reaction rate constant k_5 evaluated with $T = T_a$, and $\varepsilon = RT_a/E$.

The last parameter is likely to be small for typical chemical systems: if we take $\varepsilon = 0$ for simplicity, the behaviour of the above model is qualitatively no different from that of the cubic autocatalator. For $\kappa < e^{-2}$, the stationary-state locus, which even has the maximum in $\alpha_s^s(\mu)$ and a linear relationship between θ_{ss} and μ, shows two supercritical Hopf bifurcation points $\mu_{1,2}^*$ between which there exists a stable limit cycle. However, for non-zero ε in the range $0 < \varepsilon < \frac{2}{9}$ and for small values of κ, the character of the upper Hopf point changes. We can now have a subcritical Hopf bifurcation from which an unstable limit cycle emerges and grows as the reactant concentration increases, as shown in Fig. 3.7.

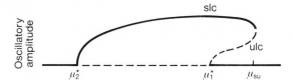

FIG. 3.7. The limit cycle development appropriate to the Salnikov model when the upper Hopf bifurcation is subcritical, shedding an unstable limit cycle from μ_1^*. The lower Hopf bifurcation at μ_2^* is supercritical and sheds a stable limit cycle. Both cycles grow as μ increases and they coalesce at μ_{su}.

As it turns out, this unstable limit cycle meets the stable limit cycle which has emerged from the lower, supercritical Hopf point μ_2^*. This merging of the limit cycles at the point μ_{su} somewhere above μ_1^* destroys both and appears as a turning point in the amplitude locus in Fig. 3.7. In the range $\mu_1^* < \mu < \mu_{su}$ the (stable) stationary-state solution is surrounded by two limit cycles: the first is unstable, the outer one stable—as shown in Fig. 3.8. The system cannot remain on an unstable limit cycle anymore than it can on an unstable stationary state, so there are no corresponding concentration–time oscillations. The unstable cycle does, however, play an important role in dividing the phase plane into two regions. An initial condition corresponds to a point on the phase plane. Those which lie inside the unstable limit cycle

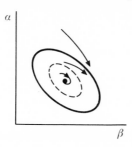

FIG. 3.8. A stable stationary state surrounded by an unstable limit cycle (broken curve) and then a stable limit cycle (solid curve). Trajectories cannot cross the unstable limit cycle so any system starting within the broken cycle must approach the stable stationary state; those starting outside the unstable limit cycle tend to the stable cycle.

will give rise to trajectories which wind towards the stable stationary-state solution. Trajectories cannot cross a limit cycle, stable or unstable, so any point which starts outside the unstable cycle will not be able to reach the stationary state: instead it must wind on to the other attractor—the stable limit cycle. Thus the unstable limit cycle marks the edge of the *basin of attraction* of the two competing attractors. We see that for identical experimental conditions (i.e. identical parameters) there are two coexisting stable forms of behaviour—stationary state or sustained oscillation: which will be observed in any given run depends on the initial conditions.

The Salnikov scheme is one for which the full Hopf analysis can be carried out analytically. The condition for Hopf bifurcation for a system with fixed ε is given parametrically by

$$\kappa = \left(\frac{\theta}{(1 + \varepsilon\theta)^2} - 1 \right) \exp\left(-\frac{\theta}{(1 + \varepsilon\theta)} \right) \tag{3.28}$$

$$\mu^* = \kappa\theta \tag{3.29}$$

with $\theta_- \leqslant \theta \leqslant \theta_+$ where

$$\theta_\pm = \frac{1}{2\varepsilon^2}[1 - 2\varepsilon \pm (1 - 4\varepsilon)^{1/2}]. \tag{3.30}$$

As an example, if $\varepsilon = 0.1$, Hopf bifurcations can have temperature excesses in the range $1.27 < \theta < 78.8$: the parameter values for which the dimensionless temperature excess at Hopf bifurcation is $\theta = 2$ are $\kappa = 7.345 \times 10^{-2}$ and $\mu = 0.147$. The qualitative form of the Hopf bifurcation locus for and given ε in the range $0 \leqslant \varepsilon < \frac{1}{4}$ is shown in Fig. 3.9: for $\kappa < (1 - 4\varepsilon)\,\mathrm{e}^{-2}$ there are two Hopf points between which the stationary-state solution is stable. Also marked on the right-hand boundary (the upper Hopf bifurcation in terms of μ) is a special point κ_{deg}: for $\kappa > \kappa_{\mathrm{deg}}$ both Hopf bifurcations are supercritical; for $\kappa < \kappa_{\mathrm{deg}}$ the upper Hopf is subcritical.

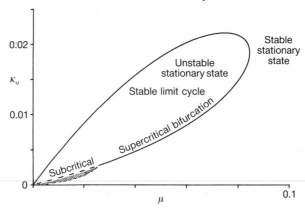

FIG. 3.9. The Hopf bifurcation locus for the Salnikov model with $\varepsilon = 0.21$, showing the change from subcritical to supercritical bifurcation at the upper Hopf point for low κ_u. Within the region of unstable stationary-state behaviour the system has a stable limit cycle. In the shaded region stable and unstable limit cycles coexist with the stable stationary state.

One of the additional quantities which is evaluated during this process is a *Floquet exponent*, β_2. The sign of this term at the Hopf bifurcation point indicates the stability of the emerging limit cycle: $\beta_2 < 0$ corresponds to a stable limit cycle; $\beta_2 > 0$ to instability. The special or *degenerate* point on the Hopf locus in Fig. 3.9 corresponds to a change between these two patterns and hence can be located from the condition $\beta_2 = 0$. The parametric form for β_2 is

$$\beta_2 = \frac{1 - \theta - 2\varepsilon(3 - 2\theta) - 3\varepsilon^2\theta(4 - \theta) - 6\varepsilon^3\theta^2}{8(1 + \varepsilon\theta)^6}. \tag{3.31}$$

Setting $\beta_2 = 0$ and rearranging the resulting equation gives the temperature excess at the degenerate point:

$$\theta = \frac{(1 - 4\varepsilon + 12\varepsilon^2) + (1 - 8\varepsilon + 28\varepsilon^2)}{6\varepsilon^2(1 - 2\varepsilon)} \tag{3.32}$$

and this has solutions corresponding to positive values for the parameters κ and μ^* provided $\varepsilon < \frac{2}{9}$ (Hopf bifurcations exist beyond this, for $\varepsilon < \frac{1}{4}$, but do not show this degeneracy).

The procedure of locating parameter combinations corresponding to degeneracies is a powerful technique (Golubitsky and Langford 1981; Guckenheimer and Holmes 1983; Golubitsky and Schaeffer 1985): the mathematical criteria can be written explicitly and often suitable numerical codes exist for solving the resulting systems of equations. Once a given type of degenerate point has been located it is often possible to predict qualitatively almost all of the behaviour which the model may show for parameter values close by, or to use path-following techniques to trace the degeneracy through higher-order parameter spaces.

So far we have seen two ways in which limit cycle solutions can appear in or disappear from the phase plane: they may grow from Hopf bifurcation points or be born as a stable–unstable pair of cycles (the reverse of the coalescence described above as μ increases through μ_{su}). This completes the list for systems which have only two variables and a unique stationary state. Soon we will examine the behaviour of two-variable models which can show multiple stationary states, where another route to the formation of limit cycles will be important. The much wider range of responses for three or more variable systems will be introduced in the next chapter.

3.6. The stability of time-dependent solutions: Poincaré sections and Floquet multipliers

An important technique for locating limit cycle and other periodic attractors involves the idea of a Poincaré section. For a two-variable system this section corresponds to a line in the phase plane and the procedure is easily visualized. Figure 3.10 shows an x–y phase plane and the evolution of a particular

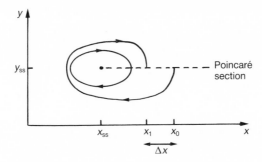

FIG. 3.10. The construction of a Poincaré plane and application of the shooting method to find periodic solutions.

trajectory from some given initial condition (x_0, y_0). Also shown are the stationary-state singularity $(x_{\mathrm{ss}}, y_{\mathrm{ss}})$ and a Poincaré line: the latter is simply the line $y = y_{\mathrm{ss}}$ for $x > x_{\mathrm{ss}}$. For convenience we have chosen (x_0, y_0) on this line, i.e. $y_0 = y_{\mathrm{ss}}$. As time evolves, so the trajectory winds around the phase plane and after one circuit it reaches the Poincaré line again—at some point (x_1, y_{ss}). In general we expect that the initial and return points will not be identical as the trajectory will be moving the system towards some form of attractor. There will thus be some distance $\Delta x = x_1 - x_0$ separating the two intersections of the trajectory and the Poincaré line. If, however, by some chance we had started our trajectory from a point exactly on a limit cycle solution, then the return point would correspond exactly to the initial point, i.e. $\Delta x = 0$. We can use this last condition as a prescription for efficiently locating the point (or points) on the Poincaré line which lies (lie) on the limit cycle(s)—assuming such a cycle or cycles exist. The procedure is to use a Newton–Raphson iteration process, so an improved estimate for the x coordinates of the point on the limit cycle passing through the Poincaré line x_{n+1} is obtained from the current trial value x_n from

$$x_{n+1} = x_n - (\Delta x_n)/(\mathrm{d}\Delta x/\mathrm{d}x)_{x_n} \tag{3.33}$$

where $(\mathrm{d}\Delta x/\mathrm{d}x)_{x_n}$ is the derivative of the distance Δx_n evaluated with the current trial value of x_n. This latter quantity must be evaluated numerically by computing a second trajectory from a point $x_n + \delta x$ along with that from x_n.

This shooting method has a number of advantages. First, the convergence on to a solution will generally be faster than the exponential approach given by simply integrating the differential equations for a long time—provided a reasonable first estimate can be chosen. Secondly, this process will also converge on to unstable limit cycles (and we do not need to know about the stability in advance). In theory we could find an unstable limit cycle for a two-variable system by integrating backwards in time: then we do need to know in advance that it is unstable, and even this will generally not work for three of more variables. Thirdly, the intermediate output from the iteration procedure can be used to determine the *Floquet multipliers* of the final limit cycle (Coddington and Levinson 1955): these provide information about the stability of the periodic solution and warn of and allow us to categorize any oncoming bifurcations of that solution.

Also, we may notice that the repeated intersection of the limit cycle with the Poincaré line as the trajectory evolves gives us a *mapping* with similar features to those discussed in the previous chapter (although we will not, in general, be able to find a simple analytical formula determining the mapping). The various iterates will, if the process is successful, converge on to a particular point along the Poincaré line: this is then a stationary state or *invariant point* of the map. In this way, the limit cycle—a one-dimensional attractor (or repellor)—has been identified with (or 'reduced to') a zero-dimensional point in the Poincaré line. With a higher number of variables, the phase space would be n-dimensional, the Poincaré plane $(n-1)$-dimensional, but a limit cycle is still reduced from one to zero dimension.

Floquet analysis comes from the theory of differential equations with time-dependent (and, specifically, periodic) coefficients. Just as we have examined the local stability to infinitesimal perturbations of stationary-state solutions of the governing differential equations in Section 3.2, so we can ask the same questions of limit cycles or, in particular, the corresponding invariant points in the Poincaré plane (line). The invariant points can again be identified as sinks, sources, or saddles: with more than two variables we will also see an analogy to Hopf bifurcation of the invariant point in the Poincaré plane, leading to *quasiperiodicity*, and other more complex responses.

In order to assess the stability of a limit cycle solution, we must first find our limit cycle. This will generally be done numerically, so the stability analysis must be a numerical process. We must also note that the system does not move round the limit cycle with a constant (angular) velocity: on some sections of the curve the motion is particularly slow; elsewhere there is fast motion. Because of this, we can only decide if a perturbation is growing or decaying if we compare successive measurements at the same point around the orbit. An appropriate point is the intersection with the Poincaré line (we can, of course, choose any line in the phase plane that is cut transversally by the limit cycle as our Poincaré line). First we determine the limit cycle solution (e.g. by the shooting method above) and, in particular, its invariant point on the Poincaré line and its period τ_p. We can then impose an initial perturbation Δx_0 and observe its evolution in time, i.e. the iterations of the Poincaré map by integrating the governing equations from this new starting point.

At the end of the first cycle, the new position of the perturbed solutions can be written as

$$\Delta x(\tau_p) = J_{lc}(\tau_p)\Delta x_0 \qquad (3.34)$$

where $J_{lc}(\tau_p)$ is the (time-dependent) Jacobian matrix appropriate to the limit cycle, evaluated at the end of the period (when the trajectory has returned to the Poincaré line). In fact, J_{lc} has the form

$$J_{lc}(\tau_p) = \exp(B\tau_p) \qquad (3.35)$$

where B is also an $n \times n$ matrix.

If this relationship between successive iterations is applied recursively, the perturbation after q periods will be given by

$$\Delta x(m\tau_p) = [J_{lc}(\tau_p)]^q\Delta x_0 = \exp(qB\tau_p)\Delta x_0. \qquad (3.36)$$

If the 'magnitude' of the Jacobian matrix is 'less than unity', then the perturbation will decrease on repeated iteration, corresponding to a stable invariant point (limit cycle).

Thus the local stability of a given invariant point can be determined from the eigenvalues of the matrix J_{lc} (or of B). The former are known as the Floquet multipliers m_i and are generally easier to obtain then the Floquet exponents β_i, the eigenvalues of B. An n-variable system (with an $(n-1)$-dimensional Poincaré plane) will have n Floquet multipliers. One of these, m_1 say, will always have the value $+1$ (corresponding to $\beta_1 = 0$): this is because an initial perturbation exactly along the limit cycle solution in the phase plane will retain its magnitude exactly after each period, i.e. will neither grow nor decay. The remaining multipliers are determined from the iterations of the Poincaré map. For a two-variable system we have just m_2 (or β_2) which must be real. If $|m_2| < 1$, and thus $\beta_2 < 0$, the limit cycle is stable and the invariant point is a sink: with $|m_2| > 1$ and, hence, $\beta_2 > 0$, we have an unstable limit cycle and a source or saddle-type invariant point. The critical case is clearly $|m_2| = 1$ or $\beta_2 = 0$: this situation corresponds to the merging or bifurcation of stable and unstable limit cycles as an μ_{su} in Fig. 3.7. If the bifurcation with $|m_2|$ occurs in coincidence with the Hopf bifurcation itself we have the degeneracy discussed towards the end of Section 3.5.

With three or more variables, the Floquet multipliers can occur as complex pairs as well as real numbers and a wider range of bifurcations is possible. Before going on to such systems (Chapter 4) there is another significant development of two-variable schemes which has important consequences: that of *multistability* or *multiplicity of stationary-state solutions*.

3.7. Autocatalysis in flow reactors

Multiple stationary-state phenomena can be investigated by means of a particularly simple cubic autocatalytic model (Gray and Scott 1983, 1984). Imagine a reaction converting reactant A to product C via an autocatalytic species B:

(2) $$A + 2B \rightarrow 3B \qquad \text{rate} = k_c ab^2$$

(3) $$B \rightarrow C \qquad \text{rate} = k_d b$$

occurring in a CSTR. As indicated, this pair of reactions has been abstracted from the cubic autocatalator model discussed earlier: the production of A from the precursor P can be replaced in a flow system by a constant inflow, and we have also omitted the uncatalysed reaction (1) whose role can be taken over by an inflow of B.

The governing mass-balance equations for the rates of change of concentration include both chemical reaction and inflow or outflow terms. With a suitable choice of dimensionless variables and parameters, these become

$$\frac{d\alpha}{d\tau} = \frac{(1 - \alpha)}{\tau_{res}} - \alpha\beta^2 \qquad (3.37)$$

$$\frac{d\beta}{d\tau} = \frac{(\beta_0 - \beta)}{\tau_{res}} + \alpha\beta^2 - \kappa_d\beta \qquad (3.38)$$

where $\alpha = a/a_0$, $\beta = b/a_0$, and $\beta_0 = b_0/a_0$: a_0 and b_0 are the inflow concentrations of A and B; $\tau_{res} = k_c a_0^2 t_{res}$ is the mean residence time (inverse of the flow rate) and $\kappa_d = k_d/k_c a_0^2$.

The stationary-state solutions of this model are given by

$$\beta_{ss} = \frac{(1 + \beta_0 - \alpha_{ss})}{(1 + \kappa_d \tau_{res})} \qquad (3.39)$$

and the roots of the cubic equation

$$\frac{(1 + \kappa_d \tau_{res})^2}{\tau_{res}} (1 - \alpha_{ss}) - \alpha_{ss}(1 + \beta_0 - \alpha_{ss})^2. \qquad (3.40)$$

There are a number of special cases for which the cubic is easy to solve. If there is no autocatalyst inflow and no decay ($\beta_0 = \kappa_d = 0$), both terms on the left-hand side have the factor $(1 - \alpha_{ss})$: thus $\alpha_{ss} = 1$ is a solution for all residence times, corresponding to a state of no reaction. The quadratic equation remaining after this solution has been factored out has roots

$$(1 - \alpha_{ss})_\pm = \tfrac{1}{2}[1 \pm (1 - 4\tau_{res}^{-1})^{1/2}]. \qquad (3.41)$$

Thus for $\tau_{res} > 4$, there are two extra stationary-state solutions. This situation continues to hold with low inflow concentrations of autocatalyst, $\beta_0 < \tfrac{1}{8}$, and the dependence of the stationary-state *extent of reaction* $(1 - \alpha_{ss})$ on the residence time shows a typical S-shape hysteresis loop (Fig. 3.11(a)). For some range $\tau_{res}^- < \tau_{res} < \tau_{res}^+$, the system has three different stationary-state solutions for any given residence time. As β_0 increases, the hysteresis loop *unfolds* and for $\beta_0 > \tfrac{1}{8}$ there is a unique stationary state for each residence time: for the special (degenerate) case $\beta_0 = \tfrac{1}{8}$ the stationary-state locus shows a vertical inflexion point.

If the autocatalyst decay reaction has a small, but non-zero, rate ($0 < \kappa_d \ll 1$), the stationary-state curves show significantly different behaviour at long residence times. If there is no catalyst inflow ($\beta_0 = 0$), the zero reaction solution $(1 - \alpha_{ss}) = 0$ again satisfies the stationary-state condition for all residence times and can be factored out. The remaining quadratic has roots

$$(1 - \alpha_{ss})_\pm = \tfrac{1}{2}\{1 \pm [1 - 4(1 + \kappa_d \tau_{res})^2 \tau_{res}^{-1}]^{1/2}\}. \qquad (3.42)$$

These will be real provided the discriminant is positive, i.e. over the range $\tau_{res}^- < \tau_{res} < \tau_{res}^+$ given by

$$\tau_{res}^\pm = \frac{1 - 8\kappa_d \pm (1 - 16\kappa_d)^{1/2}}{8\kappa_d^2}. \qquad (3.43)$$

Thus, multiple stationary states are possible provided $\kappa_d < \tfrac{1}{16}$: the two non-zero solutions form a closed curve or *isola* as shown in Fig. 3.11(b). As κ_d is increased, the size of the isola decreases, shrinking to a point when $\kappa_d = \tfrac{1}{16}$: for higher values of the decay rate constant, there is only the zero conversion state at each τ_{res}. The variation of $\tau_{res}^\pm$ with the autocatalyst decay rate κ_d is shown in Fig. 3.12: this has a typical cusp shape. Parameter values within the cusp correspond to combinations of decay rate and residence time for which there exist multiple stationary states.

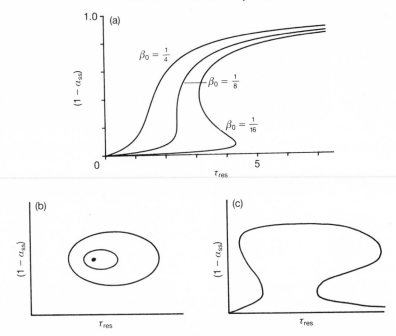

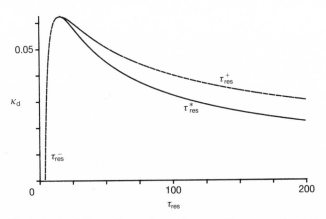

Fig. 3.11. Different possible stationary-state bifurcation loci for cubic autocatalysis in a CSTR. (a) The unfolding of a single hysteresis loop for the reaction without decay as the inflow concentration of the autocatalyst is increased. (b) Isola patterns for the model with decay but no autocatalyst inflow: as κ_d increases the size of the isola decreases, shrinking to a point for $\kappa_d = \frac{1}{16}$. (c) A mushroom pattern possible with non-zero autocatalyst inflow showing two hysteresis loops.

Fig. 3.12. The loci of saddle–node turning points marking the ends of the isola and the Hopf bifurcation point for cubic autocatalysis in a CSTR with no inflow of B.

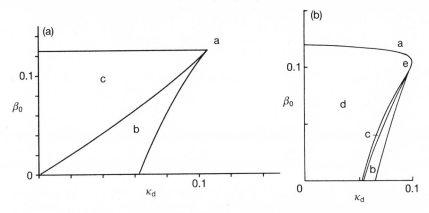

FIG. 3.13. The regions in the κ_d–β_0 parameter plane for the different stationary-state patterns of Fig. 3.11: (a) no uncatalysed reaction step with a (unique), b (isola), and c (mushroom); (b) inclusion of a slow uncatalysed reaction A → B with dimensionless rate constant $\kappa_u = 0.002$, for which the single hysteresis response d and an isola + hysteresis loop pattern e also exist.

With both β_0 and κ_d non-zero, unique or isola responses are still possible for different combinations of these parameters. There is another stationary-state pattern: a mushroom (Fig. 3.11(c)), which has two separate ranges of multiplicity (S-shaped and Z-shaped hysteresis loops). The different stationary-state patterns in this model are typical of many chemical systems. Figure 3.13(a) shows the different combinations of β_0 and κ_d for which each occurs for the present model: allowing for a non-zero rate of the uncatalysed reaction step (1) slightly changes the appearance of this *parameter plane*, enhancing the region of a single hysteresis loop (Fig. 3.13(b)). The details of the analysis leading to the loci in these figures are given in Gray and Scott (1990*a,b*).

3.8. Local stability analysis for flow reactor

The idea of local stability of a given stationary-state solution is quite general and the analysis of Section 3.2 applies directly to the present model. In the case $\kappa_d = 0$ (no autocatalyst decay) we really have just a one-variable model, as the concentrations of A and B are linked uniquely by the stoichiometry: $\beta = (1 + \beta_0 - \alpha)$ at all times, not just under stationary-state conditions. The single mass-balance equations can then be written as

$$\frac{d\alpha}{d\tau} = \frac{(1 - \alpha)}{\tau_{\text{res}}} - \alpha(1 + \beta_0 - \alpha)^2. \tag{3.44}$$

The stability is determined by the sign of the single eigenvalue $\lambda = \partial(d\alpha/d\tau)/\partial\alpha$

evaluated at the stationary state of interest. Thus

$$\lambda = -\tau_{res}^{-1} - (1 + \beta_0 - \alpha_{ss})(1 + \beta_0 - 3\alpha_{ss}). \tag{3.45}$$

It turns out that λ is always negative if the stationary-state solution is unique, i.e. the unique state is stable. If there are three stationary states for a given residence time, then the eigenvalue corresponding to the highest and lowest of states will be negative (stability) whilst that for the middle solution is positive (unstable). This alternation of stability is typical of such systems: in particular, the middle solution is a saddle point, separating the two nodal states. The *ignition* and *extinction* points marking the ends of the range of multiple stationary states correspond to the merging of a saddle and a node. As λ is generally negative for a node and positive for a saddle, the condition for a turning point, a *saddle–node bifurcation*, is clearly $\lambda = 0$.

The saddle–node condition is appropriate for models with more than one variable, such as the present scheme with $\kappa_d > 0$. The discussion of Section 3.2 has prepared us for the existence of saddle point solutions, even though none were found with the simple model studied there. In the present case, we again find that the middle branch is one of saddles which have one positive and one negative eigenvalue. As we approach the turning point of an isola, mushroom, or hysteresis loop, so one of the eigenvalues tends to zero. If the positive eigenvalue changes sign as we go around the saddle–node point on to the uppermost or lowest branch, the stationary state becomes a stable node: if, however, it is the negative eigenvalue passing through zero, the emerging node is unstable.

Still, in general, saddle–node bifurcations are parametrized by one eigenvalue becoming zero or, equivalently, the determinant of the Jacobian matrix vanishing. This gives a readily applicable mathematical criterion for locating ignition and extinction points in chemical systems.

As our two-variable system has greater than quadratic non-linearity, we also expect that Hopf bifurcation phenomena should be a feature. The Jacobian matrix has the form

$$\mathbf{J} = \begin{pmatrix} -(\tau_{res}^{-1} + \beta_{ss}^2) & -2\alpha_{ss}\beta_{ss} \\ \beta_{ss} & 2\alpha_{ss}\beta_{ss} - (\tau_{res}^{-1} + \kappa_d) \end{pmatrix} \tag{3.46}$$

evaluated with the appropriate forms for the stationary-state solution. Thus, the trade and determinant are

$$\text{tr}(\mathbf{J}) = 2\alpha_{ss}\beta_{ss} - (\beta_{ss}^2 + 2\tau_{res}^{-1} + \kappa_d) \tag{3.47}$$

$$\det(\mathbf{J}) = \tau_{res}^{-1}(\tau_{res}^{-1} + \kappa_d - 2\alpha_{ss}\beta_{ss}) + \beta_{ss}^2(\tau_{res}^{-1} + \kappa_d). \tag{3.48}$$

For the system with no autocatalyst inflow, $\beta_0 = 0$, the Hopf condition $\text{tr}(\mathbf{J}) = 0$ can be solved analytically, giving

$$\tau_{res}^* = \frac{1 - 2\kappa_d^{1/2} + (1 - 4\kappa_d^{1/2})^{1/2}}{2\kappa_d^{3/2}}. \tag{3.49}$$

Equation (3.49) shows that for slow catalyst decay, $\kappa_d < \frac{1}{16}$, there is one Hopf bifurcation point. This condition on the decay rate is the same as that for the existence of isola patterns (cf. eqn (3.45)): the Hopf point always lies on the upper shore of the isola—that corresponding to the highest extent of reaction, as shown in Fig. 3.14. This is something of a change from the behaviour shown by the pool chemical model of Section 3.2 for which Hopf points always appeared in pairs.

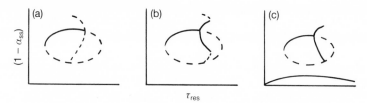

FIG. 3.14. Hopf bifurcation phenomena along the upper shore of an isola for cubic autocatalysis in a CSTR: (a) subcritical bifurcation with an unstable limit cycle growing and terminating in a homoclinic orbit as τ_{res} decreases; (b) supercritical bifurcation shedding a stable limit cycle that grow as τ_{res} increases and coalescing with an unstable limit cycle born from a homoclinic orbit; (c) as with (b) but now the stable limit cycle forms the homoclinic orbit.

The Floquet exponent analysis reveals further detail (Farr and Scott 1988): for $\frac{9}{256} < \kappa_d < \frac{1}{16}$, the recipe gives that β_2 is positive, so the emerging limit cycle is unstable and grows as the residence time decreases below τ_{res}^*, where the upper stationary state is stable. For the smallest values of the decay rate constant, $\kappa_d < \frac{9}{256}$, the Hopf bifurcation is supercritical with a stable limit cycle emerging as τ_{res} increases.

3.9. Homoclinic orbits

What happens to the limit cycle as it grows away from the Hopf bifurcation on an isola? As there is no other Hopf point, it cannot shrink back to zero amplitude. Figure 3.13(a) appears to show the (unstable) limit cycle growing until it hits the lower branch of saddle points. In fact this is a reasonable summary, for which we can provide more detail. The appropriate phase-plane portraits are shown in Fig. 3.15(a)–(c). In each portrait there are three stationary-state points, each lying on the straight line specified by eqn (3.39), and with one corresponding to no reaction ($\alpha_{ss} = 1$, $\beta_{ss} = 0$). Of the other two, the middle singular point is a saddle through which pass two special trajectories: the *separatrices*. The separatrices (each consisting of two branches) can be further distinguished: along one, the evolution proceeds in the direction of the saddle, as indicated by the arrows. This is the saddle *inset*.

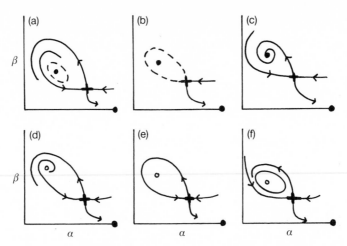

FIG. 3.15. Creation or death of limit cycles via a homoclinic bifurcation: (a)–(c) disappearance of an unstable limit cycle; (d)–(f) birth of a stable limit cycle.

The other separatrix is the *outset* and provides motion away from the saddle. The highest stationary state is a stable focus.

For residence times just shorter than the Hopf point (Fig. 3.15(a)), the stable focus is surrounded by the (unstable) limit cycle. The upper inset to the saddle emerges from the unstable limit cycle. The left outset from the saddle point spirals around this inset and eventually tends towards the stable node of no reaction. Only trajectories starting within the unstable limit cycle move towards the upper focal state.

As the residence time is decreased, the limit cycle increases in size. It remains cradled within the upper branch of the inset, but eventually this gets very close to the outset. The culmination of this process sees the inset and outset joining up, and hence both becoming part of the limit cycle (Fig. 3.15(b)). The closed loop which contains the saddle is called a *homoclinic orbit*: a closed curve which leaves and returns to the same stationary state. For shorter residence times there is no closed trajectory: the limit cycle has been destroyed and the outset from the saddle now spirals within the inset, approaching the upper focal point (Fig. 3.15(c)).

Viewed in reverse, we can see the creation of a limit cycle solution through a homoclinic orbit as the inset and outset join up. In the next chapters we will see other examples of homoclinicity, leading to the creation of orbits which have complex periodicity or even aperiodicity, for systems with more than two variables.

Homoclinic orbits cannot be observed from any change in the eigenvalues of the stationary state around which the limit cycle exists. In this context it is a non-local phenomenon. However, the formation of such an orbit is local to the saddle point, as this lies on the orbit. We can tell whether the limit cycle is stable or unstable at the homoclinic bifurcation by comparing the two eigenvalues of the saddle point. If the positive one has greater magnitude, the limit cycle is unstable: if the negative eigenvalue has greater magnitude the limit cycle will be stable.

Figure 3.14(b) shows a stable limit cycle growing from a supercritical Hopf point. At some point to the right of τ_{res} (higher residence time) a homoclinic orbit is formed and sheds an unstable limit cycle. At still higher τ_{res}, the stable and unstable cycles collide and annihilate each other. A slightly simpler scenario is shown in Fig. 3.14(c): now the stable limit cycle emerging from τ_{res}^* terminates in a homoclinic orbit itself, giving rise to the sequence of phase-plane portraits in Fig. 3.15(d)–(f). Such behaviour only occurs for the present model if there is a non-zero inflow of autocatalyst.

With $\beta_0 > 0$, mushroom patterns can exist: again there may be Hopf points along either the highest or lowest branches (but not along the middle, saddle point branch). If there is only one Hopf point, this will lie on the highest branch (highest extent of reaction): the emerging limit cycle must terminate either in a homoclinic orbit itself or by collision with another limit cycle of opposite stability and which will have been born from a homoclinic orbit in much the same way as the responses on an isola just described. If, however, there are two Hopf points the emerging limit cycles have all the options discussed so far.

If both Hopf points lie on the upper branch then both will be supercritical (for the present model). The limit cycle emerging at one can simply vanish at the other (Fig. 3.16(g)), much like the behaviour seen around the unique stationary state for the pool chemical model. However, the existence of the saddle point branch can become important and the locus of the limit cycle solution become broken (Fig. 3.16(h)). For this response, there are oscillatory solutions close to both Hopf points, but not over the whole range between them. If the Hopf points are on separate branches then that on the lowest branch is a subcritical point (for the present model) and sheds an unstable limit cycle as the residence time increases. Again there are (at least) two possibilities: either the unstable limit cycle can grow until it meets the stable limit cycle from the upper Hopf point (Fig. 3.16(i)) or both can die at homoclinic orbits (Fig. 3.16(j)).

Even this list does not complete the variety of responses possible in this or other two-variable schemes: all the known 'bifurcation diagrams for the present model are shown in Fig. 3.16(a)–(u), where up to three limit cycle solutions can coexist with one or three stationary states in various ways. It is important to stress that the number of limit cycle solutions is in no way restricted by the number of variables: the determining features are the degree of non-linearity in the governing reaction rate equations and the number of parameters we have under our control.

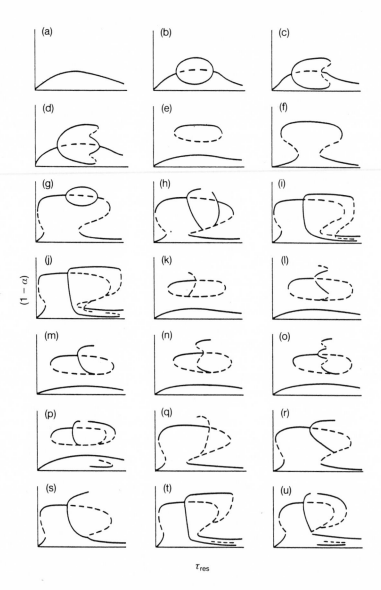

Fig. 3.16. Different combinations of stationary-state multiplicity and Hopf bifurcation behaviour for cubic autocatalysis in a CSTR (generic behaviour for two-variable systems): solid lines represent stable stationary states or limit cycles, dashed curves correspond to unstable states and cycles.

3.10. Degeneracies and double zero eigenvalues

We can locate the parameter values for which there is a change in the number of Hopf bifurcation points. Two routes for this are possible. One sees two Hopf points merging together, as seen previously when $\kappa_u = \frac{1}{8}$ for the pool chemical model. This degeneracy is parametrized by the condition that the trace of the Jacobian matrix becomes zero at some point (the eigenvalues have zero real part) but that it should not pass through zero (or it crosses tangentially): i.e. we require

$$\text{tr}(\mathbf{J}) = 0 \qquad d\,\text{tr}(\mathbf{J})/d\tau_{\text{res}} = 0 \qquad \text{for a two-variable system} \quad (3.50)$$

or

$$\text{Re}(\lambda) = 0 \qquad d\,\text{Re}(\lambda)/d\tau_{\text{res}} = 0 \qquad \text{in general.} \quad (3.51)$$

This condition gives the line H in Fig. 3.17: to the right of this line, the decay rate and autocatalyst inflow concentrations are such that there are no Hopf bifurcations as the residence time is varied. To the left, there are two Hopf points. The line H does not continue to the κ_d axis but stops at a special point with coordinates $\kappa_d = \frac{9}{2}(2 - \sqrt{3})^3$, $\beta_0 = \frac{1}{2}(3^{3/2} - 5)$.

The change from two Hopf points to one, or from one to none, is associated with a different degeneracy—a *double zero eigenvalue* (DZE). At a normal Hopf point, the eigenvalues have zero real parts but non-zero imaginary

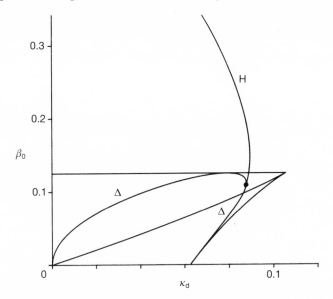

FIG. 3.17. Loci of degenerate Hopf bifurcations: the curves Δ and H are superimposed on the degenerate stationary state loci from Fig. 3.13(a).

parts. At a saddle–node turning point one of the (real) eigenvalues becomes zero. Now imagine the special case that a Hopf point occurs exactly on the turning point. The eigenvalues at such a point will have both zero real and imaginary parts $\lambda_1 = \lambda_2 = 0$. For a two-variable system this is also parametrized by

$$\text{tr}(\mathbf{J}) = \det(\mathbf{J}) = 0. \tag{3.52}$$

For the present model this condition is satisfied by

$$\beta_0 = x(1 - 2x) \qquad \kappa_d = 4x^2(1 - x)^4 \tag{3.53}$$

with $0 < x < \frac{1}{2}$, giving rise to the curve Δ in Fig. 3.17. As we cross this curve a Hopf point moves around the turning point at the end of an isola or mushroom on to the saddle point branch (where it ceases to correspond to the Hopf conditions and hence does not shed a limit cycle). Below Δ the model gives rise to a single Hopf point. The curve H described above emerges from Δ at the special point $\kappa_d = \frac{9}{2}(2 - \sqrt{3})^3$, $\beta_0 = \frac{1}{2}(3^{3/2} - 5)$. If we cross Δ below this, the existing Hopf point disappears. If we cross Δ but stay below H, a second Hopf point appears.

The DZE curve has yet more significance: close to parameter values which satisfy condition (3.52) as well as a Hopf point and a saddle–node bifurcation we can *guarantee* the existence of a homoclinic orbit (Arnol'd 1972; Takens 1974; Bogdanov 1975; Guckenheimer and Holmes 1983, pp. 364–76). Basically, the Hopf bifurcation occurs so close to the saddle solution that the emerging limit cycle must collide with the separatrices before it has grown to finite size.

3.11. Codimension and structural stability

In the preceding sections we have examined a hierarchy of different possible responses for chemical systems (which are also characteristic of dynamical systems in general). The simplest are stationary states and periodic solutions. All others, such as Hopf points, saddle nodes, or homoclinic orbits, are to some greater or lesser extent degenerate. It may seem rather a let-down after all this analysis to learn that only the non-degenerate forms will ever be realized in practice. While this is true, it is some comfort to know that the effects of the others are important and can be seen, even if they will not occur exactly themselves. Thus we cannot arrange for a system to sit exactly on a Hopf bifurcation point nor on a homoclinic orbit: we can keep an experimental system in limit cycle or stationary-state operation—although we cannot keep it exactly on a particular stationary-state or oscillatory value. The key to these statements is that we will not have perfect control over our experimental operating conditions. In mathematical terms, we must consider the structural stability of the different responses, with the related concept of *codimension* (Poston and Stewart 1978; Iooss and Joseph 1980; Golubitsky and Schaeffer 1985).

Chemists who know their phase rule (phases of matter rather than phase planes or spaces) will recognize the similarity of codimension and constraints. We can characterize a given type of solution in terms of the number of parameters that are constrained to achieve it. Thus for a stationary state there is no restriction. If the decay rate κ_d and the autocatalyst inflow concentration β_0 are chosen arbitrarily we still have complete freedom with our choice of the residence time, because at least one stationary state exists for all parameter combinations. There are no restrictions so a stationary state has codimension zero.

Limit cycles also have zero codimension. Such solutions do not exist for all parameter values, but once all but one have been fixed there is still a finite range of the remaining parameter corresponding to such a form: limit cycle solutions are found for a finite volume in the full $(\tau_{res} - \kappa_d - \beta_0)$ parameter space.

The first stage of degeneracy is that for saddle–node bifurcations, Hopf bifurcations, and homoclinic orbits. These are codimension-1 phenomena: once the parameters κ_d and β_0 have been specified, the attainment of one of these forms requires τ_{res} to take a specific value (or one of a number of discrete values). The final parameter is fixed, so there is one less degree of freedom. These solutions are found on surfaces in the parameter plane.

At the next stage come double zero eigenvalues and degenerate Hopf bifurcations or the unfolding of hysteresis loops and the loss of isola patterns either by shrinking to a point or opening up to a mushroom. These correspond to lines in the parameter plane. Once one of the parameters, β_0 say, has been chosen, the condition for these degeneracies specifies the remaining two. We have lost two degrees of freedom and so have a codimension-2 phenomenon. Finally come the codimension-3 points, such as that where the curves H and Δ meet in Fig. 3.17, for which all three parameters are specified.

If we deal with systems with more parameters it is possible that higher codimension phenomena may be encountered, but things get a bit esoteric by then.

The foregoing discussion now allows us to explain why only stationary-state or limit cycle behaviour can be directly observed experimentally: only these have *structural stability*. We will always have to settle for some inaccuracy in our control of each of our experimental parameters: that's life. Instead of each experimental situation corresponding to a point in parameter space, the uncertainties mean that we must expect a (hopefully) small volume. Any response which requires us to have more control than this is structurally unstable: a requirement that we at least sit on a surface—worse still a line—just cannot be met. Thus, codimension 1 and higher cannot be realized—although, as explained above, they do still retain their influence and importance.

3.12. Other two-variable models

3.12.1. Brusselator model

Perhaps the most widely used model for oscillations amongst chemists is the *Brusselator* scheme of Lefever, Prigogine, and Nicolis (Prigogine and Lefever 1968; Glansdorff and Prigogine 1971; Nicolis and Prigogine 1977; Lefever *et al.* 1988; see also Gray *et al.* 1988*b*). This is a pool chemical model, similar to that used in Sections 3.1–3.4 and thus most closely associated with thermodynamically closed systems. The model has the form

$$A \rightarrow X$$

$$B + X \rightarrow D + Y$$

$$Y + 2X \rightarrow 3X$$

$$X \rightarrow E$$

so two reactant species A and B are converted to products D and E via a sequence of irreversible reaction processes involving two intermediates X and Y. The concentrations of the latter are the two variables for which we write the reaction rate equations in the (dimensionless) form:

$$\frac{dx}{dt} = A - Bx + x^2 y - x \tag{3.54}$$

$$\frac{dy}{dt} = Bx - x^2 y. \tag{3.55}$$

There are two parameters A and B representing the concentrations of the pool chemical species. The stationary-state solutions are simply

$$x_{ss} = A \qquad y_{ss} = B/A. \tag{3.56}$$

From these and the rate laws, the Jacobian matrix has the following form

$$\mathbf{J}(x_{ss}, y_{ss}) = \begin{pmatrix} B - 1 & A^2 \\ -B & -A^2 \end{pmatrix}. \tag{3.57}$$

The determinant is always positive, $\det(\mathbf{J}) = A^2$, whilst the trace is given by

$$\text{tr}(\mathbf{J}) = B - 1 - A^2. \tag{3.58}$$

The stationary state is stable for low concentrations of the reactant B, $B < (1 + A^2)$, and unstable if the inequality is reversed. The Hopf condition is then

$$B^* = 1 + A^2. \tag{3.59}$$

Although the Brusselator was not the first model proposed which shows limit cycle oscillations, its tractability and relationship to closed vessels and its emergence at the same time as the first wide reports of the Belousov–Zhabotinskii reaction—an experimental example of an oscillatory system—have given it the highest position within the field.

3.12.2. FONI model

Another 'classical' system is that of a first-order exothermic reaction in a well-stirred flow reactor (termed FONI by Aris for first-order, non-isothermal) much studied by chemical engineers—see, for example, Liljenroth (1918), MacMullin and Weber (1935), Zeldovich (1941), Zeldovich and Zysin (1941), Denbigh (1944, 1947), van Heerden (1953), Bilous and Amundson (1955), Aris and Amundson (1958a, b, c, d), Aris (1969), Hlavacek *et al.* (1970), Uppal *et al.* (1974, 1976), Vaganov *et al.* (1978), Chang and Calo (1979), Golubitsky and Keyfitz (1980), Williams and Calo (1981), Balakotaiah and Luss (1981, 1982a, b, c, 1983, 1984), Kwong and Tsotsis (1983), Denbigh and Turner (1984), Farr and Aris (1986), Planeaux and Jensen (1986). This is closely related to the autocatalytic scheme of Sections 3.7–3.11. A single reaction converts reactant A to a product, releasing heat: the self-heating increases the value of the reaction rate constant (cf. Section 3.5). Cooling occurs by Newtonian heat transfer and by the outflow of heated reactant/product.

The mass- and heat-balance equations can be written as

$$\frac{d\alpha}{d\tau} = \frac{(1 - \alpha)}{\tau_{res}} - \alpha \exp[\theta/(1 + \varepsilon\theta)] \tag{3.60}$$

$$\frac{d\theta}{d\tau} = B\alpha \exp[\theta/(1 + \varepsilon\theta)] - (\tau_{res}^{-1} + \tau_N^{-1})\theta. \tag{3.61}$$

This model has four parameters: the residence time, τ_{res}; the reaction exothermicity, B; the Newtonian cooling time τ_N; and the activation energy ε. Again, the stationary-state curve can show isolas, mushrooms, or hysteresis loops and these can be combined with supercritical or subcritical Hopf bifurcations, homoclinic orbits, etc. Up to three coexisting limit cycles have been observed.

3.12.3. Oregonator model

The Belousov–Zhabotinskii (BZ) reaction will have a chapter to itself later on, but we can mention here one of the simplest models which has been derived from its mechanism. This is the 'two-variable' *Oregonator*: we will see various Oregonators with more variables later. The chemistry which drives the BZ oscillations can be condensed down to the following representation (Field and Noyes 1974):

$$A + Y \rightarrow X + P \qquad \text{rate} = k_3 A Y$$

$$X + Y \rightarrow 2P \qquad \text{rate} = k_2 X Y$$

$$A + X \rightarrow 2X + 2Z \qquad \text{rate} = k_5 A X$$

$$2X \rightarrow A + P \qquad \text{rate} = k_4 X^2$$

$$B + Z \rightarrow \tfrac{1}{2} f Y \qquad \text{rate} = k_0 B Z.$$

The different species can be identified in chemical terms as: $A = BrO_3^-$, B = all oxidizable organic species, $X = HBrO_2$, $Y = Br^-$, $Z = Ce(IV)$ the oxidized form of the catalyst, and $P = HOBr$. The pool chemical approximation is applied to the major reactants A and B, and the product P does not enter any of the forward reactions. Initially we have a three-variable scheme for the intermediate species X, Y, and Z. With suitable choices of dimensionless groups, the reaction rate equations for these can be written as (Tyson 1979, 1982)

$$\varepsilon_1 \frac{dx}{d\tau} = qy - xy + x(1 - x) \tag{3.62}$$

$$\varepsilon_2 \frac{dy}{d\tau} = -qy - xy + fz \tag{3.63}$$

$$\frac{dz}{d\tau} = x - z \tag{3.64}$$

where $x = 2k_4 X / k_5 A$, $y = k_2 Y / k_5 A$, $z = k_0 k_4 B Z / (k_5 A)^2$, $\tau = k_0 B t$, and the three dimensionless parameters are $q = 2k_3 k_4 / k_2 k_5$, $\varepsilon_1 = k_0 B / k_5 A$, and $\varepsilon_2 = 2k_0 k_4 B / k_2 k_5 A$. Using typical values for the various rate constants we can estimate ε_1 and ε_2: $\varepsilon_1 \sim 0.04$ and $\varepsilon_2 \sim 4 \times 10^{-4}$.

The small value of ε_2, in particular, is helpful. Because this parameter multiplies the differential, it will appear in the denominator of the right-hand side if we divide through. Then we can argue that either the numerator must also be small or $dy/d\tau$ will be large. In the latter case, the dimensionless concentration of bromide ion will adjust quickly to a value relative to the more slowly changing x and z such that the numerator becomes small. Thus we expect that 'for almost all of the time' the following relationship will hold:

$$-qy - xy + fz = 0 \qquad \text{or} \qquad y = \frac{fz}{q + x}. \tag{3.65}$$

With y now expressed in terms of z and x, we can reduce our system to one of two governing equations and two variables:

$$\varepsilon_1 \frac{dx}{d\tau} = x(1 - x) + \frac{f(q - x)z}{(q + x)} = g(x, z) \tag{3.64}$$

$$\frac{dz}{d\tau} = x - z = h(x, z). \tag{3.67}$$

We can now do a Hopf analysis and find the bifurcation value of the 'stoichiometric coefficient' f in terms of ε_1 and q.

For a stationary state, clearly $z_{ss} = x_{ss}$. One possible stationary state is $x_{ss} = 0$ (with then $z_{ss} = 0$) and this is a saddle point solution: the remaining possibilities are given by the roots of the quadratic

$$x_{ss}^2 - (1 - f - q)x_{ss} - q(1 + f) = 0 \tag{3.68}$$

i.e.

$$x_{ss}^{\pm} = \tfrac{1}{2}\{1 - f - q \pm [(1 + q)^2 - 2(1 - 3q)f + f^2]^{1/2}\}. \tag{3.69}$$

The lower root is negative (and hence physically unacceptable); the upper root is positive.

To find any Hopf bifurcation points, we look for the trace of the Jacobian for eqns (3.64) and (3.67) to vanish:

$$\varepsilon_1 \operatorname{tr}(\mathbf{J}) = 1 - 2x_{ss}\left(1 + \frac{fq}{(q + x_{ss})^2}\right) - \varepsilon_1 = 0. \tag{3.70}$$

This and the upper root of eqn (3.69) can be used to determine the Hopf bifurcation point for given values of q and ε_1.

There is another way of analysing the rate equations for oscillatory solutions—the *method of nullclines*. This makes use of the occurrence of a (second) small parameter ε_1 multiplying the differential $dx/d\tau$. Using the same argument as that rehearsed above, we expect the concentration x to adjust quickly to slow variations in z, so that the function $g(x, z)$ given by the right-hand side of eqn (3.66) remains as close to zero as possible.

Figure 3.18 shows the *nullcline* $g(x, z) = 0$ in the x–z phase plane. This is a cubic curve with a minimum and maximum in the range $q < x < 1$. Also shown is the nullcline $h(x, z)$: this is simply a straight line with unit gradient. The two nullclines intersect: this point is the positive, non-zero stationary state as $g(x, z) = h(x, z) = 0$.

In Fig. 3.18(a) and (b), the intersection occurs on a portion of the $g(x, z)$ nullcline with negative gradient (the right-hand and left-hand branches, respectively). If we examine the evolution from a given initial point, these two cases have similar behaviour. Consider Fig. 13.8(a) and the initial point x_1, z_1. This does not lie on the nullcline, so $dx/d\tau$ will be large. The first motion from this point is a rapid movement on to the right-hand branch of

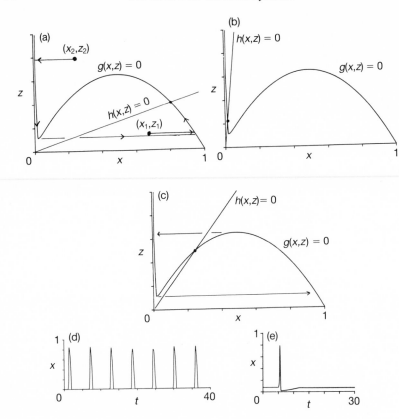

FIG. 3.18. The method of nullclines applied to the simple Oregonator: (a) $f < 0.5$, showing a stable stationary-state intersection at high x; (b) $f > 1 + \sqrt{2}$ showing a stable stationary-state intersection at low x; (c) an unstable stationary-state intersection for intermediate f lying between the extrema of the $g(x, z) = 0$ nullcline, showing the corresponding relaxation oscillation limit cycle; (d) typical relaxation oscillations in x for $f = 1$; (e) excitability, with a single large excursion from a perturbed stable intersection 'close' to the minimum in the $g(x, z) = 0$ nullcline.

the $g(x, z)$ nullcline. Subsequent motion will occur along the nullcline and the trajectory simply moves towards the stationary-state intersection (which also lies on this branch) as $\tau \to \infty$. If we start from a point such as x_2, z_2 the initial rapid motion takes us to the left-hand branch of the nullcline. The trajectory then moves along the curve until it reaches the minimum. At this point, with z still decreasing, the trajectory leaves the $g(x, z)$ nullcline, so $g(x, z)$ is no longer close to zero. We get another rapid jump, from the minimum to the right-hand branch. In the limit of vanishingly small ε_1 this transition will give an almost vertical line on the phase plane, i.e. z will not vary as x jumps: as ε_1 becomes finite, so the path becomes curved slightly attaining a minimum as it crosses the $h(x, z)$ nullcline. Once the trajectory hits the right-hand branch it can move to the stationary-state point.

A stationary-state intersection on the left-hand branch is also attained through much the same sequences in Fig. 3.18(b). For different values of the parameters, however, the intersection can occur on the middle branch, for which $g(x, z)$ has positive gradient. Now the stationary state is not attained (at least in the limit $\varepsilon_1 \to 0$). Instead the trajectory cycles around the $g(x, z)$ nullcline, anticlockwise in Fig. 3.18(c), with periods of relatively slow evolution along the outside branches separated by rapid motions as the system jumps at the maximum or minimum. Typical *relaxation oscillations* described by this process are shown in Fig. 3.18(d).

The condition for oscillations in this approach then is that the stationary state should lie between the maximum and minimum on the $g(x, z)$ nullcline. The turning points can be located. The defining condition for the $g(x, z)$ nullcline is

$$fz = \frac{x(1 - x)(q + x)}{(x - q)}. \tag{3.71}$$

For the extrema, $dz/dx = 0$,

$$2x^3 - (1 + 2q)x^2 + 2q(1 - q)x - q^2 = 0. \tag{3.72}$$

With $q \ll 1$ (and typically $q \sim 10^{-3}$) we have

$$x_{\min} = (1 + \sqrt{2})q \qquad \text{and} \qquad x_{\max} = \tfrac{1}{2} - q. \tag{3.73}$$

Thus for instability, we require $x_{\min} < x_{ss} < x_{\max}$, for the positive, non-zero stationary state. Rearranging the stationary-state condition (3.68) we have for f

$$f = \frac{(1 - x_{ss})(q + x_{ss})}{(x_{ss} - q)}. \tag{3.74}$$

Thus, substituting here for the ends of the oscillatory range $x_{ss} = x_{\min}$ and $x_{ss} = x_{\max}$ we have in terms of the stoichiometric factor

$$\tfrac{1}{2} - q < f < (1 + \sqrt{2})[1 - (1 + \sqrt{2})q]. \tag{3.75}$$

This condition is now independent of ε_1.

Returning to Fig. 3.18(a) or (b), this scheme can also be used to illustrate the phenomenon of *excitability*. With a stable intersection (one on the right-hand or left-hand branches of the $g(x, z)$ nullcline) that lies close to either the maximum or the minimum, very small perturbations simply decay back to the stationary state. However, if the system is given a larger perturbation a much larger excursion is made before the return. For example, if the system in Fig. 3.18(a) is perturbed such that the concentration z becomes larger than that corresponding to the maximum in $g(x, z)$, the initial fast motion takes the trajectory horizontally across the phase plane, to the left-hand branch. The system will then move along this branch to the minimum, from which it then jumps back to the right-hand branch and returns along that to the intersection. Thus we get a single, large excursion or pulse of excitation, as shown in Fig. 3.18(e).

More recently, Gaspar and Showalter (1988) have analysed the behaviour of this two-variable model in a CSTR context, using traditional Hopf bifurcation techniques and direct numerical integrations. They revealed responses of the form described previously, with multiple stationary states, supercritical and subcritical Hopf bifurcations, and homoclinic orbit formation.

3.12.4. *Heterogeneous catalysis*

Two other models of interest have been studied with regard to oxidation reactions involving heterogeneous catalysis. The first is similar to a form used by Eigenberger (1978*a*, *b*) and can be analysed in much the same way as the Oregonator scheme above. This scheme begins with the reversible adsorption of a reactant P on to a vacant surface site S of a catalyst. The adsorbed species can then either desorb or react to produce a desorbed product C: further, the model requires that this reaction step also involves two additional sites. The scheme then is

$$P + S \rightleftharpoons P - S \qquad \text{rate} = k_1 s - k_{-1}(ps)$$

$$P - S + 2S \rightarrow 3S + C \qquad \text{rate} = k_2(ps)s^2$$

$$Q + S \rightleftharpoons Q - S \qquad \text{rate} = k_3 s - k_{-3}(qs).$$

The rate law for the second step here is clearly of a similar form to the cubic autocatalysis of earlier sections. The rate of the first step will also be a function of the concentration (or partial pressure) of reactant P above the catalyst, but if this remains constant it can be included in the rate coefficient k_1. The third step is a reversible adsorption–desorption process involving a 'catalyst poison' Q.

If the total concentration of sites on the catalyst is s_0, then conservation of these sites requires $s + ps + qs = s_0$ or, in terms of the fractional surface coverages,

$$\theta_s + \theta_p + \theta_q = 1 \qquad (3.76)$$

where $\theta_s = s/s_0$, $\theta_p = ps/s_0$, and $\theta_q = qs/s_0$. With this constraint, the system has only two independent surface concentrations and the surface reaction

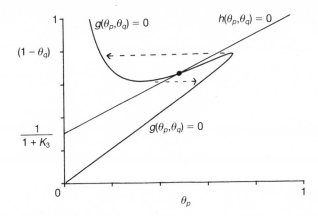

FIG. 3.19. The method of nullclines applied to the Eigenberger model for surface reactions showing an unstable stationary-state intersection on the middle branch of the $g(x, z) = 0$ nullcline.

rate equations can be written in the form

$$\frac{d\theta_p}{d\tau} = (1 - \theta_p - \theta_q) - K_{-1}\theta_p - \kappa_2\theta_p(1 - \theta_p - \theta_q)^2 = g(\theta_p, \theta_q) \quad (3.77)$$

$$\varepsilon \frac{d\theta_q}{d\tau} = K_3(1 - \theta_p - \theta_q) - \theta_q = h(\theta_p, \theta_q) \quad (3.78)$$

where $K_{-1} = k_{-1}/k_1$, $\kappa_2 = k_2 s_0^2/k_1$, $K_3 = k_3/k_{-3}$, $\varepsilon = k_1/k_{-3}$, and $\tau = k_1 t$.

Figure 3.19 shows the nullclines corresponding to $g(\theta_p, \theta_q) = 0$ and $h(\theta_p, \theta_q) = 0$ in the $\theta_p - (1 - \theta_q)$ phase plane. The latter is simply a straight line with non-zero intercept and gradient less than unity, given by

$$(1 - \theta_q) = \frac{1 + K_3\theta_p}{1 + K_3}. \quad (3.79)$$

The $g(\theta_p, \theta_q)$ nullcline is specified by

$$(1 - \theta_q) = \frac{1 + 2\kappa_2\theta_p^2 \pm (1 - 4\kappa_2 K_{-1}\theta_p^2)^{1/2}}{2\kappa_2\theta_p} \quad (3.80)$$

which has two branches for $\theta_p^2 < 1/(4\kappa_2 K_{-1})$. The lower root emerges from the origin and merges with the upper solution at the maximum in the nullcline. The upper branch has a minimum (if $K_{-1} < \frac{1}{8}$) before tending to infinity as $\theta_p \to 0$. For various values of the parameters the two nullclines have either three or one intersection as stationary-state multiplicity is possible. If there is a unique intersection and if this lies on the middle section (between the extrema) of the $g(\theta_p, \theta_q)$ nullcline, then the stationary state will be unstable (in the limit $\varepsilon \to 0$) allowing oscillatory solutions.

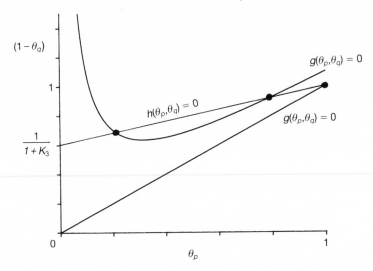

FIG. 3.20. The method of nullclines applied to the Eigenberger model for surface reactions in the absence of desorption $K_{-1} \to 0$, for which the $g(x, z) = 0$ nullcline has two disconnected branches. There is always one (stable) intersection at $\theta_p = (1 - \theta_q) = 1$. Two other intersections may exist, as shown, but oscillations now occur around the leftmost point rather than the middle intersection.

For the special case of irreversible adsorption of the reactant $(K_{-1} \to 0)$ the above analysis is not so obviously applicable. Now eqn (3.80) factorizes to give $(1 - \theta_q) = \theta_p$ and $(1 - \theta_q) = (1 + \kappa_2\theta_p^2)/(\kappa_2\theta_p)$. The latter of these is a curve with a minimum at $\theta_p = \kappa_2^{-1/2}$ and which asymptotes to the other solution, as shown in Fig. 3.20. There is always a stationary-state intersection at $\theta_p = 1 - \theta_q = 1$. For some parameter values there may also be two other intersections, corresponding to stationary-state multiplicity. Any oscillations now occur around the intersection furthest to the left in Fig. 3.20. Gray and Scott (1990a) have given a Hopf bifurcation analysis for general values of the parameters, showing that the oscillations emerge from a supercritical bifurcation as stable limit cycles around the upper stationary state of a hysteresis loop and terminate in a homoclinic orbit.

The second model, and one to which we shall return, is due to Takoudis, Schmidt, and Aris (Takoudis *et al.* 1981*a*, *b*, *c*). This allows for the competitive chemisorption of two reactants, still involving the participation of vacant sites in the reaction step but not requiring a catalyst poison. The model here is

$$P + S \rightleftharpoons P - S$$

$$R + S \rightleftharpoons R - S$$

$$P - S + R - S + 2S \rightarrow \text{products} + 4S.$$

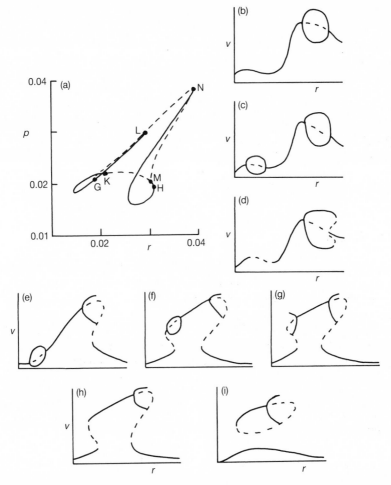

FIG. 3.21. (a) Hopf (——) and saddle–node (– – –) bifurcation loci for the Takoudis–Schmidt–Aris model: (b)–(i) different possible bifurcation diagrams with multiplicity and limit cycle solutions corresponding to different horizontal cuts across (a). (Reprinted with permission from McKarnin *et al.* (1988a). *Proc. R. Soc.*, **A415**, 363–87. © The Royal Society, London.) (See next page for textual reference.)

Again we have a two-variable system, for which the dimensionless rate equations can be written in the form

$$\frac{\mathrm{d}\theta_p}{\mathrm{d}\tau} = p(1 - \theta_p - \theta_r) - \kappa_1\theta_p - \theta_p\theta_r(1 - \theta_p - \theta_r)^2 \qquad (3.81)$$

$$\frac{\mathrm{d}\theta_r}{\mathrm{d}\tau} = r(1 - \theta_p - \theta_r) - \kappa_2\theta_r - \theta_p\theta_r(1 - \theta_p - \theta_r)^2. \qquad (3.82)$$

We will follow a path already charted (McKarnin *et al.* 1988*a, b*) and consider two of the parameters, κ_1 and κ_2, as fixed with $\kappa_1 = 10^{-3}$ and $\kappa_2 = 2 \times 10^{-3}$.

Figure 3.21(a) shows the *p–r* parameter plane, with the boundaries corresponding to saddle–node and Hopf bifurcation points. Within the roughly triangular region, the system exhibits three stationary states. If the dimensionless partial pressure of reactant R is varied for constant *p*, corresponding to a horizontal traverse of the figure, we find one of four stationary-state patterns: unique, single hysteresis loop, mushroom, or isola depending on how the region of multiplicity is cut. These categories are further subdivided by the possibility of up to four Hopf bifurcations. Eight of the possible responses are shown in Fig. 3.21(b)–(i): the system can also exhibit a unique stationary-state locus with no Hopf bifurcation.

3.13. Summary

We are now appraised of the typical behaviour of two-variable chemical systems, and have been introduced to a number of specific models that will reappear in an augmented form at various stages through the remaining text. The basic responses of stationary-state solutions, Hopf bifurcation, limit cycle oscillation, and homoclinic orbit formation, along with the associated degeneracies formed from these, will all be important as first stages in the analysis and understanding of three (or more) variable models and, later, of real experimental systems.

References

Aris, R. (1969). *Elementary chemical reactor analysis*. Prentice-Hall, Englewood Cliffs, NJ.

Aris, R. and Amundson, N. R. (1958*a*). An analysis of a chemical reactor stability and control I. *Chem. Eng. Sci.*, **7**, 121–31.

Aris, R. and Amundson, N. R. (1958*b*). An analysis of a chemical reactor stability and control II. *Chem. Eng. Sci.*, **7**, 132–47.

Aris, R. and Amundson, N. R. (1958*c*). An analysis of a chemical reactor stability and control III. *Chem. Eng. Sci.*, **7**, 148–55.

Aris, R. and Amundson, N. R. (1958*d*). Stability of some chemical systems under control. *Chem. Eng. Prog.*, **54**, 227–36.

Arnol'd, V. I. (1972). Lectures on bifurcations in versal families. *Russ. Math. Surv.*, **27**, 54–123.

Arnol'd, V. I. (1973). *Ordinary differential equations*. MIT Press, Cambridge, MA.

Balakotaiah, V. and Luss, D. (1981). Analysis of multiplicity patterns of a CSTR. *Chem. Eng. Commun.*, **13**, 111–32.

Balakotaiah, V. and Luss, D. (1982*a*). Analysis of multiplicity patterns of a CSTR. *Chem. Eng. Commun.*, **19**, 185–9.

Balakotaiah, V. and Luss, D. (1982*b*). Exact steady-state multiplicity criteria for two consecutive or parallel reactions in lumped-parameter systems. *Chem. Eng. Sci.*, **37**, 433–45.

Balakotaiah, V. and Luss, D. (1982*c*). Structure of the steady-state solutions of lumped parameter chemically reacting systems. *Chem. Eng. Sci.*, **37**, 1611–23.

Balakotaiah, V. and Luss, D. (1983). Dependence of the steady-states of a CSTR on the residence time. *Chem. Eng. Sci.*, **38**, 1709–29.

Balakotaiah, V. and Luss, D. (1984). Global analysis of the multiplicity features of multi-reaction lumped-parameter systems. *Chem. Eng. Sci.*, **39**, 865–81.

Bilous, O. and Amundson, N. R. (1955). Chemical reactor stability and sensitivity. *AIChE J.*, **1**, 513–21.

Bogdanov, R. I. (1975). Versal deformations of a singular point on the plane in the case of zero eigenvalues. *Funct. Anal. Appl.*, **9**, 144–5.

Chang, H.-C. and Calo, J. M. (1979). Exact criteria for uniqueness and multiplicity of an *n*th order chemical reaction via a catastrophe theory approach. *Chem. Eng. Sci.*, **34**, 285–99.

Coddington, E. A. and Levinson, N. (1955). *Theory of ordinary differential equations.* McGraw-Hill, New York.

Davis, H. T. (1962). *Introduction to nonlinear differential and integral equations.* Dover, New York.

Denbigh, K. G. (1944). Velocity and yield in continuous reactions I. *Trans. Faraday Soc.*, **40**, 352–73.

Denbigh, K. G. (1947). Velocity and yield in continuous reactions II. *Trans. Faraday Soc.*, **43**, 648–60.

Denbigh, K. G. and Turner, J. C. R. (1984). *Chemical reactor theory.* Cambridge University Press.

Eigenberger, G. (1978*a*). Kinetic instabilities in heterogeneously catalyzed reactions I. *Chem. Eng. Sci.*, **33**, 1255–61.

Eigenberger, G. (1978*b*). Kinetic instabilities in heterogeneously catalyzed reactions II. *Chem. Eng. Sci.*, **33**, 1263–8.

Farr, W. W. and Aris, R. (1986). 'Yet who would have thought the old man to have had so much blood in him?'—reflections on the multiplicity of steady-states of the stirred tank reactor. *Chem. Eng. Sci.*, **41**, 1385–402.

Farr, W. W. and Scott, S. K. (1988). Dynamic fine structure in the cubic autocatalator. *Chem. Eng. Sci.*, **43**, 1708–10.

Field, R. J. and Noyes, R. M. (1974). Oscillations in chemical systems, part 4. Limit cycle behaviour in a model of a real chemical reaction. *J. Chem. Phys.*, **60**, 1877–84.

Gaspar, V. and Showalter, K. (1988). Period lengthening and associated bifurcations in a two-variable, flow Oregonator. *J. Chem. Phys.*, **88**, 778–91.

Glansdorff, P. and Prigogine, I. (1971). *Thermodynamics of structure, stability and fluctuations.* Wiley, New York.

Golubitsky, M. and Keyfitz, B. (1980). A qualitative study of the steady-state solutions for a continuous flow stirred tank reactor. *SIAM J. Math. Anal.*, **11**, 316.

Golubitsky, M. and Langford, W. F. (1981). Classification and unfoldings of degenerate Hopf bifurcations. *J. Differ. Equations*, **41**, 375–415.

Golubitsky, M. and Schaeffer, D. (1985). *Singularities and groups in bifurcation theory.* Springer, New York.

Gray, B. F. and Roberts, M. J. (1988). Analysis of chemical kinetic systems over the entire parameter space. *Proc. R. Soc.*, **A416**, 391–402.

Gray, P. and Scott, S. K. (1983). Autocatalytic reactions in the isothermal, continuous stirred tank reactor. Isolas and other forms of multistability. *Chem. Eng. Sci.*, **38**, 29–43.

Gray, P. and Scott, S. K. (1984). Autocatalytic reactions in the isothermal, continuous stirred tank reactor. Oscillations and instabilities in the system A + 2B → 3B; B → C. *Chem. Eng. Sci.*, **39**, 1087–97.

Gray, P. and Scott, S. K. (1986). A new model for oscillatory behaviour in closed systems: the autocatalator. *Ber. Bunsenges. Phys. Chem.*, **90**, 985–96.

Gray, P. and Scott, S. K. (1990a). *Chemical oscillations and instabilities: non-linear chemical kinetics.* Oxford University Press.

Gray, P. and Scott, S. K. (1990b). Archetypal response patterns for open chemical systems with two components. *Philos. Trans. R. Soc.*, **A332**, 69–87.

Gray, P., Kay, S. R., and Scott, S. K. (1988a). Oscillations of an exothermic reaction in a closed system I. *Proc. R. Soc.*, **A416**, 321–41.

Gray, P., Scott, S. K., and Merkin, J. H. (1988b). The Brusselator model of oscillatory reactions. *J. Chem. Soc. Faraday Trans. 1*, **84**, 993–1012.

Guckenheimer, J. and Holmes, P. (1983). *Nonlinear oscillations, dynamical systems and bifurcations of vector fields.* Springer, New York.

Hassard, B. D., Kazarinoff, N. D., and Wan, Y.-H. (1981). *Theory and applications of Hopf bifurcation*, London Mathematical Society Lecture Note Series 41. Cambridge University Press.

Hlavacek, V., Kubicek, M., and Jelinek, J. (1970). Modelling of chemical reactors—XVIII. *Chem. Eng. Sci.*, **25**, 1441–61.

Hopf, E. (1942). Abzweigung einer periodischen Lösung von einer stationären Lösung eines Differentialsystems. *Ber. Verh. Sachs. Akad. Wiss. Leipzig Math.-Nat.*, **94**, 3–22.

Iooss, G. and Joseph, D. D. (1980). *Elementary stability and bifurcation theory.* Springer, New York.

Jordan, D. W. and Smith, P. (1977). *Nonlinear ordinary differential equations.* Oxford University Press.

Kay, S. R. and Scott, S. K. (1988). Oscillations of an exothermic reaction in a closed system II. *Proc. R. Soc.*, **A416**, 343–59.

Kwong, V. K. and Tsotsis, T. T. (1983). Fine structure of the CSTR parameter plane. *AIChE J.*, **29**, 343–7.

Lefever, R., Nicolis, G., and Borckmans, P. (1988). The Brusselator: it does oscillate all the same. *J. Chem. Soc. Faraday Trans. 1*, **84**, 1013–23.

Liljenroth, F. G. (1918). Starting and stability phenomena of ammonia oxidation and similar reactions. *Chem. Metall. Eng.*, **19**, 287–97.

McKarnin, M. A., Aris, R., and Schmidt, L. A. (1988a). Autonomous bifurcations of a simple bimolecular surface-reaction model. *Proc. R. Soc.*, **A415**, 363–87.

McKarnin, M. A., Aris, R., and Schmidt, L. A. (1988b). Forced oscillations of a self-oscillating bimolecular surface reaction model. *Proc. R. Soc.*, **A417**, 363–88.

MacMullin, R. B. and Weber, M. (1935). The theory of short-circuiting in continuous-flow mixing vessels in series and the kinetics of chemical reactions in such systems. *Trans. Amer. Inst. Chem. Engrs.*, **31**, 409–56.

Merkin, J. H., Needham, D. J., and Scott, S. K. (1986). Oscillatory chemical reactions in closed vessels. *Proc. R. Soc.*, **A406**, 299–323.

Merkin, J. H., Needham, D. J., and Scott, S. K. (1987a). On the structural stability of a simple pooled chemical system. *J. Eng. Math.*, **21**, 115–27.

Merkin, J. H., Needham, D. J., and Scott, S. K. (1987b). On the creation, growth and extinction of oscillatory solutions for a simple pooled chemical reaction scheme. *SIAM J. Appl. Math.*, **47**, 1040–60.

Nicolis, G. and Prigogine, I. (1977). *Self-organization in nonequilibrium systems. From dissipative structures to order through fluctuations.* Wiley, New York.

Planeaux, J. B. and Jensen, K. F. (1986). Bifurcation phenomena in CSTR dynamics: a system with extraneous thermal capacitance. *Chem. Eng. Sci.*, **41**, 1497–523.

Poston, T. and Stewart, I. (1978). *Catastrophe theory and its applications.* Pitman, London.

Prigogine, I. and Lefever, R. (1968). Symmetry breaking instabilities in dissipative systems II. *J. Chem. Phys.*, **48**, 1695–700.

Salnikov, I. Ye. (1948). Thermokinetic model of a homogeneous periodic reaction. *Dokl. Akad. Nauk SSSR*, **60**, 405–8.

Salnikov, I. Ye. (1949). Thermokinetic model of a homogeneous periodic reaction. *Zh. Fiz. Khim.*, **23**, 258–60.

Takens, F. (1974). Singularities of vector fields. *Publ. Math. IHES*, **43**, 47–100.

Takoudis, C. G., Schmidt, L. D., and Aris, R. (1981a). Multiple steady states in reaction controlled surface catalysed reactions. *Chem. Eng. Sci.*, **36**, 377–86.

Takoudis, C. G., Schmidt, L. D., and Aris, R. (1981b). Isothermal sustained oscillations in a very simple surface reaction. *Surf. Sci.*, **105**, 325–33.

Takoudis, C. G., Schmidt, L. D., and Aris, R. (1981c). Isothermal oscillations in surface reactions with coverage independent parameters. *Chem. Eng. Sci.*, **37**, 69–76.

Tyson, J. J. (1979). Oscillations, bistability and echo waves in models of the Belousov–Zhabotinskii reaction. *Ann. NY Acad. Sci.*, **316**, 279–95.

Tyson, J. J. (1982). Scaling and reducing the Field–Körös–Noyes mechanism of the Belousov–Zhabotinskii reaction. *J. Phys. Chem.*, **86**, 3006–12.

Uppal, A., Ray, W. H., and Poore, A. B. (1974). On the dynamic behavior of continuous stirred tank reactors. *Chem. Eng. Sci.*, **29**, 967–85.

Uppal, A., Ray, W. H., and Poore, A. B. (1976). The classification of the dynamical behavior of continuous stirred tank reactors—influence of reactor residence time. *Chem. Eng. Sci.*, **31**, 205–14.

Vaganov, D. A., Samoilenko, N. G., and Abramov, V. G. (1978). Periodic regimes of continuous stirred tank reactors. *Chem. Eng. Sci.*, **33**, 1133–40.

van Heerden, C. (1953). Autothermic processes. *Ind. Eng. Chem.*, **45**, 1242–7.

Williams, D. C. and Calo, J. M. (1981). 'Fine structure' of the CSTR parameter space. *AIChE J.*, **27**, 514–16.

Zeldovich, Ya. B. (1941). Towards the theory of combustion intensity. The evolution of an exothermic reaction in a flow, I. *Zh. Tekh. Fiz.*, **XI**, 493–500.

Zeldovich, Ya. B. and Zysin, Yu. A. (1941). Towards the theory of combustion intensity. The evolution of an exothermic reaction in a flow, II. *Zh. Tekh. Fiz.*, **XI**, 501–8.

4

Flows II: Three-variable systems

The previous chapter revealed a great deal about the behaviour of two-variable chemical systems. Despite the variety of bifurcation phenomena, only two structurally-stable responses were found: stationary-state operation or simple period-1 limit cycles. More complex patterns, such as oscillations that repeat every other excursion (period-2), are forbidden with only two variables by the restriction that trajectories may not, in general, cross on the phase plane (which is a two-dimensional surface). If we consider models or systems with more than two variables, the phase space becomes at least three dimensional. Trajectories are still lines through this space and so there is much greater freedom—actual crossing is still forbidden, but trajectories can now pass over or under (or around) each other. In this chapter we turn our attention to the simplest of these higher-dimensional schemes—three-variable models. We will find two additional, structurally-stable responses: a T^2-*torus* and a *strange attractor*. These open a Pandora's box of extra responses in the concentration–time traces.

4.1. The non-isothermal autocatalator

The model through which we will begin our investigation of three-variable schemes is a simple extension of the pool chemical, cubic autocatalator of the previous chapter. The physical motivation for increasing the complexity of our model is this. Most spontaneous chemical reactions are exothermic: in reactions involving radical or active species, such as an autocatalyst, the most significantly exothermic processes are the termination steps—reaction (3) of Chapter 3. The most highly temperature-sensitive reactions are those with the highest activation energies: in general these are the initiation such as reaction (0). With these principles in mind, the augmented model can be written in the form

(0)	$P \rightarrow A$	rate $= k_0(T)p$
(1)	$A \rightarrow B$	rate $= k_u a$
(2)	$A + 2B \rightarrow 3B$	rate $= k_c ab^2$
(3)	$B \rightarrow C + \text{heat}$	rate $= k_0 b$.

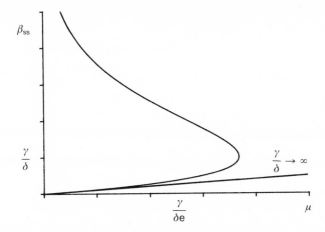

FIG. 4.1. The stationary-state behaviour $\beta_{ss}(\mu)$ for the three-variable non-isothermal autocatalator, showing a turning point (thermal explosion) at $\mu = \gamma/\delta e$, $\beta_{ss} = \gamma/\delta$. Also shown is the linear relationship appropriate to the isothermal model obtained in the limit $\gamma/\delta \to \infty$.

Using similar dimensionless groups to those from Chapter 3, the reaction rate and heat-balance equations can be written in the form

$$\frac{d\alpha}{d\tau} = \mu\, e^{\theta} - \alpha\beta^2 - \kappa_u\alpha \tag{4.1}$$

$$\frac{d\beta}{d\tau} = \alpha\beta^2 + \kappa_u\alpha - \beta \tag{4.2}$$

$$\frac{d\theta}{d\tau} = \delta\beta - \gamma\theta. \tag{4.3}$$

Here δ represents the exothermicity of reaction (3) whilst γ is a measure of the surface heat transfer coefficient. The full Arrhenius temperature dependence has been simplified by the 'exponential approximation' discussed briefly in Section 3.5, whereby $\exp[\theta/(1 + \varepsilon\theta)]$ is replaced by $\exp(\theta)$.

The stationary-state solutions satisfying $d\alpha/d\tau = d\beta/d\tau = d\theta/d\tau = 0$ are given by

$$\theta_{ss} = \delta\beta_{ss}/\gamma \qquad \alpha_{ss} = \beta_{SS}/(\kappa_u + \beta_{ss}^2) \tag{4.4}$$

where β_{ss} is a solution of the equation

$$\mu = \beta_{ss}\exp(-\delta\beta_{ss}/\gamma). \tag{4.5}$$

The qualitative form of the stationary-state locus is the same for all μ, δ, and γ, and is shown in Fig. 4.1. For $\mu < \gamma/\delta\, e$ there are two stationary-state solutions, one with $\beta_{ss} < \gamma/\delta$ and the other with $\beta_{ss} > \gamma/\delta$. These two solutions merge when $\mu = \gamma/\delta$ e and the system shows a *thermal explosion* for higher initial concentrations of the reactant.

The classic behaviour of the isothermal autocatalator is regained in either limit $\delta \to 0$ or $\gamma \to \infty$, with then the linear relationship $\beta_{ss} = \mu$ showing no maximum or runaway.

4.2. Hopf bifurcation for three-variable systems

The local stability of a given stationary state can be determined in the same way as that discussed in Section 3.2. The Jacobian is now a 3×3 matrix and so has three eigenvalues λ_{1-3}. These will be given as the roots of a cubic equation of the form

$$\lambda^3 + b\lambda^2 + c\lambda + d = 0. \tag{4.6}$$

For the present model, with the above stationary-state relationships, the Jacobian becomes

$$\mathbf{J} = \left\{ \begin{array}{ccc} -(\beta_{ss}^2 + \kappa_u) & \dfrac{-2\beta_{ss}^2}{(\beta_{ss}^2 + \kappa_u)} & \beta_{ss} \\[3mm] \beta_{ss}^2 + \kappa_u & \dfrac{(\beta_{ss}^2 - \kappa_h)}{(\beta_{ss}^2 + \kappa_u)} & 0 \\[3mm] 0 & \delta & -\gamma \end{array} \right\}. \tag{4.7}$$

Thus the coefficients in the cubic (4.6) are

$$b = 1 + \beta_{ss}^2 + \kappa_u + \gamma - 2\beta_{ss}^2(\beta_{ss}^2 + \kappa_u)^{-1} \tag{4.8}$$

$$c = (\beta_{ss}^2 + \kappa_u)(1 + \gamma) + \gamma[1 - 2\beta_{ss}^2(\beta_{ss}^2 + \kappa_u)^{-1}]. \tag{4.9}$$

$$d = (\beta_{ss}^2 + \kappa_u)(\gamma - \delta\beta_{ss}) \tag{4.10}$$

We may use the generalized results from Section 3.3 to establish the conditions for bifurcation. There is a saddle–node bifurcation when the determinant vanishes, i.e. for $d = 0$: this gives $\beta_{ss} = \gamma/\delta$ as shown above for the turning point in the stationary-state locus. For the upper branch, $\beta_{ss} > \gamma/\delta$ so the determinant d is negative: the stationary states here are saddles.

Any Hopf bifurcation point must occur on the lower branch in Fig. 4.1. The conditions on the eigenvalues represented by eqns (3.21)–(3.23) for a three-variable system can be expressed in terms of the coefficients b, c, and d as

$$bc = d \qquad \text{with } b, c, \text{ and } d > 0. \tag{4.11}$$

There may be two, one, or no Hopf points, depending on the values of the parameters κ_u, δ and γ. We will be particularly interested in the first of these

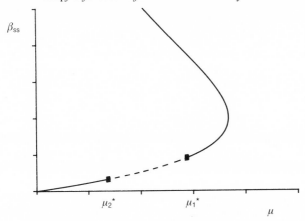

FIG. 4.2. Loss of local stability and Hopf bifurcation phenomena along the lower branch of stationary-state solutions for the non-isothermal autocatalator: the solution is unstable (---) between the Hopf points $\mu_2^* < \mu < \mu_1^*$. The upper branch is an unstable saddle point over all its range.

cases. The lower solution becomes unstable for some range of μ given by $\mu_2^* \leqslant \mu_1^*$, with $\mu_2^* > 0$ and $\mu_1^* < \gamma/\delta$ e as shown in Fig. 4.2.

In the limit of a thermoneutral reaction, $\delta = 0$, the temperature is decoupled from the concentrations. The cubic Hopf condition can be factorized to yield

$$(\lambda + \gamma)\left[\lambda^2 + \left(1 + \mu^2 + \kappa_u - \frac{2\mu^2}{\mu^2 + \kappa_u}\right)\lambda + \mu^2 + \kappa_u\right] = 0 \quad (4.12)$$

with $\beta_{ss} = \mu$. One eigenvalue is now simply related to the cooling rate, $\lambda = -\gamma$: the other two are given by the quadratic term. The conditions for Hopf bifurcation in this isothermal model is then

$$\mu^2 = \tfrac{1}{2}[1 - 2\kappa_u \pm (1 - 8\kappa_u)^{1/2}] \quad (4.13)$$

as found previously, requiring $\kappa_u < \tfrac{1}{8}$ and with $\mu \to 1$ and 0 as $\kappa_u \to 0$.

Returning to the fully coupled non-isothermal system, Hopf bifurcations require $c > 0$, i.e.

$$\frac{\gamma + 1}{\gamma} \geqslant \frac{\beta_{ss}^2 - \kappa_u}{(\beta_{ss}^2 + \kappa_u)^2}. \quad (4.14)$$

In the limit of no uncatalysed reactions, $\kappa_u \to 0$, this rearranges to the condition $\beta_{ss}^2 \geqslant \gamma/(1 + \gamma)$. Since we must also satisfy $d > 0$, i.e. $\beta_{ss} < \gamma/\delta$, the condition for Hopf bifurcation in this case becomes

$$\delta^2 \leqslant \gamma(1 + \gamma). \quad (4.15)$$

The dimensionless exothermicity of the reaction must, therefore, not be too large compared with the heat transfer coefficient of the system.

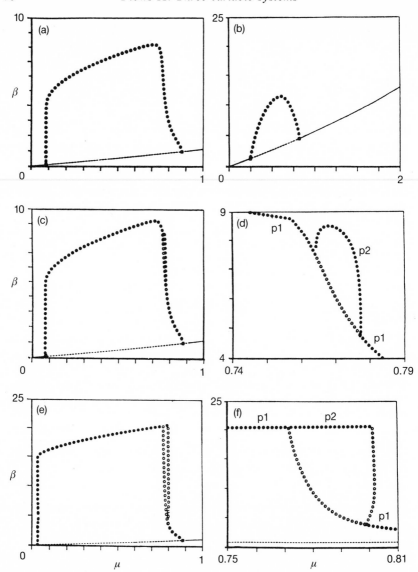

FIG. 4.3. Development of the oscillatory solutions between the Hopf bifurcation points for the non-isothermal autocatalator and the influence of the uncatalysed reaction rate constant κ_u: in each case $\gamma = 1.0$ and $\delta = 0.1$. (a) $\kappa_u = 7 \times 10^{-3}$, ther period-1 state is stable over the whole range between the Hopf bifurcations but shows a 'false bifurcation' or 'canard' for $\mu < \mu_1^*$. (b) $\kappa_u = 0.05$, the canard has disappeared, and the region of instability reduced. (c) With lower uncatalysed rate constants, $\kappa_u = 5.5 \times 10^{-3}$, the period-1 limit cycle loses stability over a range of reactant concentration at period-doubling bifurcations, leading to a stable period-2 oscillation, shown in more detail in (d).

4.3. Period-1 solutions

Figure 4.3(a) shows a section of the lower stationary-state branch for parameter values $\gamma = 1.0$, $\delta = 0.1$, and $\kappa_u = 7 \times 10^{-3}$. The behaviour of this system is essentially two dimensional. Both of the Hopf bifurcations are supercritical and the stable period-1 limit cycle born at one survives across the whole range of stationary-state instability, disappearing at the other. The variation in amplitude of the limit cycle with the reactant concentration for $\mu_2^* \leqslant \mu \leqslant \mu_1^*$ is marked on the figure. There are two sections of the curve for which the amplitude varies rapidly with μ. These rapid variations are known as *canards* (from the French for 'false' as these can falsely suggest a bifurcation in the limit cycle solutions). Their occurrence often indicates that more complex responses may appear close by. The canard near the upper Hopf point will be of particular interest.

If the uncatalysed reaction rate constant is increased, the Hopf bifurcation points move closer together, decreasing the range of stationary-state instability. (In the isothermal autocatalator model, Hopf bifurcations exist only if $\kappa_u < \frac{1}{8}$.) Figure 4.3(b) shows the variation in amplitude of the period-1 limit cycle for a system with $\kappa_u = 5 \times 10^{-2}$. The canard behaviour noted above has now been lost so the amplitude varies relatively smoothly with μ.

For both of the two loci just described, the corresponding phase-space behaviour is relatively simple. Outside the range of instability, the lower stationary state is a stable sink. As we pass through a Hopf point, the sink becomes a source and a stable limit cycle emerges to surround it. The limit cycle in this three-dimensional system is still just a one-dimensional closed curve. There is a second singularity in the phase space—the saddle point. This has little influence in the above cases: there is no homoclinic orbit formation; the limit cycle merely varies in size and then shrinks back to the lower stationary state as this regains stability.

4.4. Bifurcation of periodic solutions

For small values of κ_u we see behaviour which has no counterpart in two-variable systems. With $\kappa_u = 5.5 \times 10^{-3}$ things start out normally as shown in Fig. 4.3(c). There are two Hopf bifurcation points: each of these is supercritical, leading to the emergence of a stable period-1 limit cycle which grows as we move into the region of stationary-state instability. Furthermore, this period-1 cycle exists over the whole range between μ_2^* and μ_1^* as before. However, it is not always now a stable periodic solution. If we consider the scenario as we decrease μ from μ_1^*. At some point μ_1^{**} there is a *secondary bifurcation* at which the period-1 solution becomes unstable, as shown in detail in Fig. 4.3(d). We can characterize this bifurcation in terms of the

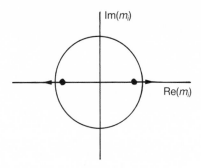

FIG. 4.4. Representation of the Floquet multiplier behaviour for an oscillatory three-variable system: one multiplier has the value $+1$; the two remaining multipliers, $m_{2,3}$, may be real or complex and remain within the unit circle for a stable solution. A bifurcation of the oscillatory state occurs when m_2 and/or m_3 cross the unit circle: if one exits through -1, there is a period-doubling bifurcation. Other bifurcations described within the text correspond to different crossings of the unit circle.

Floquet multipliers introduced in Section 3.6. Any periodic solutions for this three-variable system will have three Floquet multipliers. One of these, m_1, always has the value $+1$ for the reasons discussed earlier. The remaining two, m_2 and m_3, may either both be real or form a complex pair. The stability of a given periodic solution is determined by the magnitude of these multipliers. If they both have magnitude less than unity, the periodic solution is stable to perturbations: if either or both has magnitude which increases through unity then the associated periodic solution becomes unstable. Figure 4.4 shows how the Floquet multipliers can be represented as points in the complex plane. For a stable periodic solution, m_2 and m_3 must remain inside the unit circle: the bifurcation of the periodic orbit is associated with one or both of the multipliers crossing the unit circle. There are different ways in which this crossing can occur as we shall see in this chapter. Each of these has a different implication for the nature of the bifurcation which occurs.

Returning to our example, for $\mu_1^{**} < \mu < \mu_1^*$, the multipliers m_2 and m_3 for the period-1 solution are both real and lie within the unit circle. As μ approaches μ_1^{**} from above, the Floquet multiplier with largest magnitude, m_2, becomes more negative. At the bifurcation point, m_2 crosses the unit circle at -1 on the real axis. This is characteristic of a *period-doubling bifurcation*. We have seen period doublings in the mapping of Chapter 2. The change in the $\beta(\tau)$ time trace is from oscillations which repeat every maximum to a waveform which repeats every second peak, as shown in Fig. 4.5(a) and (b). The corresponding changes in the $\alpha-\beta-\theta$ phase space are slightly more intricate. The now unstable period-1 limit cycle does not disappear, but a new cycle emerges from it (much as it emerged from the stationary-state singular point at the primary Hopf bifurcation). The new period-2 cycle has the appearance of the (single) edge of a Möbius strip, with

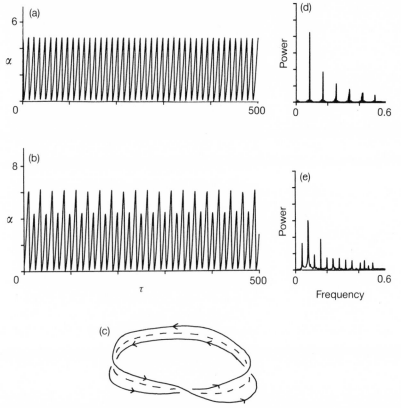

FIG. 4.5. Transition through a period-doubling bifurcation: the non-isothermal autocatalator model from Fig. 4.3(c), (d). Here, (a) $\mu = 0.7800$, period-1; (b) $\mu = 0.772$, period-2. The corresponding Fourier power spectra are shown in (d) and (e), the latter revealing the extra subharmonic peaks. (c) Geometrical representation of a period-2 solution on a Möbius band, with the unstable period-1 limit cycle (broken curve) at the centre of the band.

the unstable period-1 cycle lying on the strip 'inside' the former, as shown in Fig. 4.5(c). The trajectory can move around the strip edge in three dimensions without crossing itself. The change in waveform can also be seen in the Fourier power spectrum. Figure 4.5(d) and (e) compare the spectra for the period-1 and period-2 solutions from Fig. 4.5(a) and (b) respectively either side of the period-doubling bifurcation. A second set of contributing frequencies can cleary be seen in the latter, corresponding to a frequency half that of the original fundamental peak (twice the period) and its various overtones. This emergence of the subharmonic peak leads to the period doubling being referred to as a *subharmonic bifurcation*.

For the present example, the period-doubling bifurcation is supercritical and so the emerging period-2 oscillation is stable and grows as μ decreases below μ_1^{**} (where the period-1 solution is unstable). Initially, the difference in maxima between the first and second peaks in the waveform is not particularly pronounced, and both resemble the previous period-1 oscillations. As we move away from the bifurcation, however, this difference increases. The period-2 oscillation remains stable for a range of the reactant concentration, $\mu_2^{**} < \mu < \mu_1^{**}$. The Floquet multiplier m_2 for the period-2 solution has the value $+1$ at μ_1^{**} (and at μ_2^{**}). Inside this range both m_2 and m_3 for the period-2 are inside the unit circle, whilst m_2 for the period-1 cycle remains outside. Eventually there is a 'period-halving' bifurcation at which the period-2 disappears (m_2 passes through $+1$): the period-1 regains stability as its Floquet multiplier m_2 moves back into the unit circle through -1.

Subcritical period doubling is also possible. An example for the upper doubling bifurcation with the present model is shown in Fig. 4.3(e) and (f) for a system with a smaller uncatalysed rate constant $\kappa_u = 10^{-3}$. Now, the period-2 solution that emerges is unstable and grows in amplitude as μ increases. Later, in this particular instance, there is then a turning point in the period-2 locus and the period-2 limit cycle is stable along the upper branch. The 'period-halving' bifurcation is supercritical as before. Such a system will thus show 'hard excitation', jumping from period-1 into period-2 oscillations for which the two maxima are immediately very different, as μ is decreased below μ_1^{**}.

4.5. More period doubling and chaos

The system of Fig. 4.3(c) and (d) with $\delta = 0.1$, $\kappa_u = 5.5 \times 10^{-3}$, and $\gamma = 1.0$ shows two each of the supercritical Hopf bifurcation and period doublings. If the heat transfer parameter γ is decreased, corresponding to slower heat loss, further complexity is found. Between $\gamma = 0.7$ and 0.65, a second pair of period doublings appear, as shown in Fig. 4.6(a) and (b). Now stable period-4 oscillations exist over a narrow range of the precursor concentration μ. Yet more period doublings (bubbling) are rapidly introduced by further decreasing γ. At each, the Floquet multiplier m_2 for the current period-2^n solution passes out of the unit circle at -1 whilst a period-2^{n+1} appears with its multiplier m_2 moving in from $+1$. For $\gamma = 0.5$ a full (subharmonic) cascade to chaotic responses, reminiscent of the Feigenbaum sequence found in the simple mapping of Chapter 2, has been achieved (Fig. 4.6(c)).

Representative time series (showing the autocatalyst concentration β as a function of time) are presented in Fig. 4.7. Considering a sequence with decreasing μ, following the upper Hopf point a period-1 limit cycle emerges (Fig. 4.7(i)). This period doubles, and a typical period-2 trace for $\mu = 0.715$ is given in Fig. 4.7(h). By $\mu = 0.715$, a period-4 oscillation (Fig. 4.7(g)) is

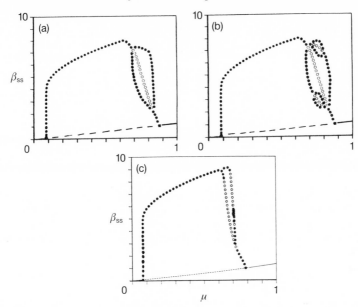

FIG. 4.6. 'Bubbles' of period-doubling bifurcation for different heat transfer coefficients with $\delta = 0.1$ and $\kappa_u = 5.5 \times 10^{-3}$: (a) $\gamma = 0.7$, with a range of period-2 on the period-1 locus; (b) $\gamma = 0.65$, the period-2 solution also becomes unstable for some μ giving a period-4 state; (c) with $\gamma = 0.5$ many (infinitely) more period doublings form a nested cascade into and out of chaos.

found. For $\mu = 0.708$, the time trace (Fig. 4.7(f)) is chaotic. A window of period-3 responses emerges at smaller μ (Fig. 4.7(e)) before another region of aperiodicity (Fig. 4.7(d)). For $\mu = 0.688\,70$, period-4 has returned: later period halvings occur as μ is decreased further.

Figure 4.8 shows a selection of the attractors in phase space corresponding to these time series. In the case of the periodic responses, the trajectory forms a closed curve which winds a more or less twisting path through the three-dimensional volume.

For aperiodic systems, the trajectory never closes up as there is no repeating unit. In infinite time, then, we can imagine the line densely covering some form of surface with finite area in the phase plane (there will, however, be holes in this surface not covered even as time tends to infinity). The resulting structure is known as a *strange attractor*: trajectories originating points which lie off this 'surface' are attracted on to it, but do not then settle to any normal periodicity. In fact, this attractor is more than a surface, i.e. it is more than two-dimensional. And yet it also has zero volume—it must have a lower dimension than the (three-dimensional) phase space as the flow is contracting. The strange attractor has a non-integer dimension between two and three: it is a *fractal* object.

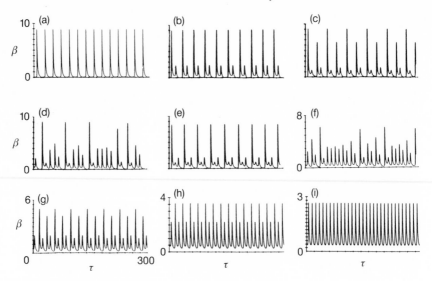

FIG. 4.7. Selected time series for the system with $\delta = 0.1$, $\kappa_u = 5.5 \times 10^{-3}$, and $\gamma = 0.5$: (a) $\mu = 0.6$, period-1; (b) $\mu = 0.65$, period-2; (c) $\mu = 0.687$, period-4; (d) $\mu = 0.6887$, aperiodic; (e) $\mu = 0.695$, period-3; (f) $\mu = 0.708$, aperiodic; (g) $\mu = 0.71$, period-4; (h) $\mu = 0.715$, period-2; (i) $\mu = 0.72$, period-1.

Much of the strange attractor appears quite regular, with neighbouring 'tracks' following similar courses. However, we know that chaotic behaviour is always associated with sensitivity to initial conditions and an overall divergence of trajectories (a positive Lyapounov exponent) so the tracks must separate at some finite rate. If the trajectories were to diverge on a finite surface, they would necessarily cross (infinitely many times). This is not allowed (not even once) for the reasons mentioned earlier. Any 'crossing' therefore must occur out of the plane. Once this has been achieved, the trajectories can be folded back on to the 'surface'—to satisfy the contractional requirement.

(We will see a truly two-dimensional attractor later—a torus—but this is associated with quasiperiodicity rather than divergence or aperiodicity.)

4.6. Next-amplitude maps, Poincaré sections, and next-return maps

The series of bifurcations above for the non-isothermal autocatalator provides a good example of how behaviour observed in mappings can also be exhibited by flows governed by differential equations, i.e. its universality. We have even seen again the relationships between period-3 and aperiodicity. Is it possible to find some formal connection between these two cases? We

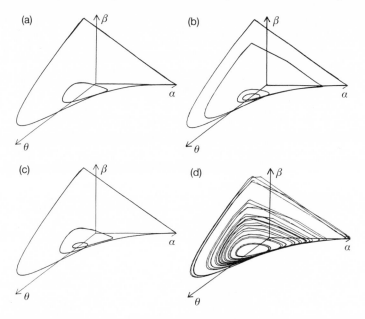

FIG. 4.8. Phase-space attractors for (a) $\mu = 0.65$, period-2; (b) $\mu = 0.687$, period-4; (c) $\mu = 0.695$, period-3; (d) $\mu = 0.708$ aperiodic behaviour from Fig. 4.7.

can do this by examining the *next-amplitude* or *next-maximum* maps for the time series. A next-amplitude (maximum) map plots the amplitude (maximum) of the $(n + 1)$th peak against the amplitude (maximum) of the nth. For regular periodic responses this process will reveal a finite number of discrete points (as always we must allow transients to die away).

For an aperiodic motion on a strange attractor, the next-amplitude map will consist of an infinite number of different points. Figure 4.9 shows the appropriate maps for chaotic traces for each of the two regons of aperiodicity in the above example. In both cases, the points appear to lie on some form of mapping curve. If we fit this curve to some analytic function, we can then iterate the map as in Chapter 2 to predict future amplitudes (but, of course, we remain limited by the positive Lyapounov exponent).

For the chaotic region at lower reactant concentrations, the next-amplitude maps have some similarity to the cubic curve of Chapter 2, with a relatively broad and flat maximum (Fig. 4.9(a)). Renormalization theory has shown that all maps with single, broad maxima will show basically the same sequences to chaos through period doubling and give rise to exactly the same value for the Feigenbaum number.

In the chaotic region for $\mu \approx 0.708$, the next-amplitude map has a sharp peak (Fig. 4.9(b)). A similar sharp maximum is found with another model, the Lorenz equations. The latter have no special chemical relevance, but we

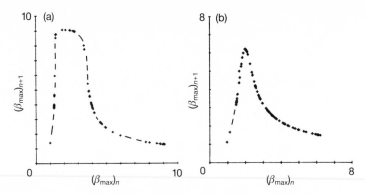

FIG. 4.9. Next-maximum maps for chaotic responses from Fig. 4.7: (a) for $\mu = 0.6887$ showing a flat maximum in the mapping curve; (b) $\mu = 0.708$ showing a sharper maximum.

will mention them briefly later because of their historical importance. This form of map can still support a period-doubling route to chaos (as we have indeed seen in the previous section) but may have different scaling properties and hence a different value for the Feigenbaum number.

Next-amplitude maps are special cases of a more general concept—the *next-return map*. Figure 4.10(a) reproduces one of the phase-space attractors— a strange attractor from an aperiodic trace in this instance. Also marked is a *Poincaré plane*: this is simply a plane which cuts a portion of the flow on this attractor transversally. Figure 4.10(b) shows the typical form of such interaction of the flow with this plane: the *Poincaré section*. This may appear to describe a simple curve, but that is not strictly the case (the attractor is not two dimensional). We could choose almost any plane for this procedure, and the form of resulting section will vary depending on whether we cut part of the attractor on which 'neighbouring' trajectories are diverging, roughly parallel or one where they are being folded back together.

The points on the Poincaré section represent repeated intersections at the end of each 'circuit' of the attractor. If we follow a give trajectory evolving in time for several circuits and observe where successive points appear on the section we will see that they do not move progressively along the curve in any sense: such a sequence is indicated in Fig. 4.10(b) by numbering 10 successive points. We can represent the order of return as *next-return map* by plotting the value of β, say, for the $(n + 1)$th intersection against that for the nth. Figure 4.10(c) shows the typical form of such a map. For the present system this would again have a broad maximum.

The next-maximum map is no more than a particular case of this procedure: the Poincaré plane is then defined by the condition $d\beta/d\tau = 0$. It is a particularly convenient choice for systems whose governing differential equations are known because only a small amount of extra program code is

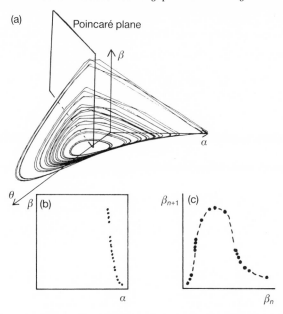

FIG. 4.10. Representation of the construction of a general Poincaré section and return map. (a) A plane is chosen that cuts the attractor transversally; (b) the intersections of the motion on the attractor with the plane form the Poincaré section; (c) plotting the value of one of the components for the $(n + 1)$th intersection against its value at the nth intersection forms a next-return map.

needed to monitor for turning points and we can check for transversality by testing that the second derivative is non-zero.

4.7. Other models showing period doubling

We have seen two examples of period-doubling sequences so far in this book. In fact this is a common scenario for the progression from simple periodicity to chaos in chemical models—and has been seen in some experiments as well. Here we briefly mention a number of other systems which support this behaviour.

4.7.1. C-feedback system

Comparing the model discussed above with its two-variable isothermal predecessor, we can see that the self-heating has introduced a second feedback mechanism through the 'Arrhenius' term $\mu\, e^{\theta}$. As well as coupling the temperature variable back into the chemistry, this temperature dependence of the reaction rate constant also introduces an extra non-linearity to the equations. Peng *et al.* (1990) have shown that this latter feature is not necessary, i.e. that all the extra three-dimensional behaviour arises merely because of the existence of the extra feedback step and does not rely on it being non-linear. These authors studied the model in which the second feedback route is played chemically by the species C—the final product in the two-variable scheme of Chapter 3. If this species can catalyse the 'initiation' process converting P to A via a step

(4) $P + C \rightarrow A + C$ rate $= k_4 pc$

and the scheme is also augmented by a decay step

(5) $C \rightarrow D$ rate $= k_5 c$

the governing differential equations for the three variables a, b, and c participating in reactions (0)–(5) can be written in the dimensionless form

$$\frac{d\alpha}{d\tau} = \mu + \delta\xi - \alpha\beta^2 - \kappa_u\alpha \tag{4.16}$$

$$\frac{d\beta}{d\tau} = \alpha\beta^2 + \kappa_u\alpha - \beta \tag{4.17}$$

$$\frac{d\xi}{d\tau} = \beta - \gamma\xi \tag{4.18}$$

with $\xi = (k_c/k_d)^{1/2}c$, $\delta = k_4 p/k_d$, and $\gamma = k_5/k_d$. The form of eqn (4.16) resembles that of (4.1) with a linear temperature dependence of the initiation process on the extra variable (ξ here, θ there). The extra decay step (5) then plays the role equivalent to that of the Newtonian heat transfer in the non-isothermal scheme.

Equations (4.16)–(4.18) also support period doubling, bubbling, mixed-mode oscillations, and chaos, closely following the qualitative scenario described above for the model with thermal feedback.

4.7.2. *Multiple exothermic reactions in a CSTR*

As an extension of the classic FONI system (Section 3.12.2), Jorgensen and Aris (1983) considered the dynamics of two consecutive exothermic reactions

$$A \rightarrow B: \quad \text{rate} = k_1(T)a, \quad Q_1 = -\Delta H_1$$

$$B \rightarrow C: \quad \text{rate} = k_2(T)b, \quad Q_2 = -\Delta H_2$$

in a well-stirred flow reactor. The mass- and energy-balance equations can be written as

$$\frac{\mathrm{d}a}{\mathrm{d}t} = \frac{a_0 - a}{t_{\text{res}}} - k_1(T)a \qquad (4.19)$$

$$\frac{\mathrm{d}b}{\mathrm{d}t} = \frac{b_0 - b}{t_{\text{res}}} + k_1(T)a - k_2(T)b \qquad (4.20)$$

$$c_p\sigma\frac{\mathrm{d}T}{\mathrm{d}t} = \frac{c_p\sigma(T - T_0)}{t_{\text{res}}} + Q_1k_1(T)a + Q_2k_2(T)b - \frac{\chi S(T - T_a)}{V} \qquad (4.21)$$

where c_p is the specific heat capacity, σ the density, T_0 the inflow temperature, χ the surface heat transfer coefficient, S the surface area, V the reactor volume, and T_a the ambient temperature in which the reactor is sitting.

Farr and Aris (1986) have shown that even the stationary-state behaviour of this system is remarkable rich, with up to seven possible branches of solutions: earlier similar studies are due to Hlavacek *et al.* (1972), Balakotaiah and Luss (1982, 1984), and Jorgensen *et al.* (1984) and some of these authors allowed one of the reactions to be endothermic ($Q_i < 0$).

This system also supports Hopf bifurcation (Hlaveck *et al.* 1972; Halbe and Poore 1981; Pismen 1980, 1984*a,b*) and the emerging periodic solutions can be followed by various techniques (see e.g. Doedel and Heinemann 1983). Jorgensen and Aris (1983) considered a specific case with the following simplifications: (i) no inflow of B or C, so $b_0 = 0$; (ii) both reactions have the same activation energy, $E_1 = E_2$; (iii) both reactions have the same exothermicity, $Q_1 = Q_2 > 0$; (iv) the temperature dependence of the reaction rate constants can be sufficiently well represented by the same exponential approximation as invoked above. Additionally we may assume that the inflow and ambient temperatures are equal, although unequal but fixed values for these two parameters can be incorporated into the following analysis without extra complexity.

Jorgensen and Aris chose to vary the heat transfer conditions, reflected in $\chi S/V$ or the Newtonian cooling time, as their bifurcation parameter. With this in mind, the residence time t_{res} can then be used as the reference timescale for the non-dimensionalization of the equation. An appropriate form is then

$$\frac{\mathrm{d}\alpha}{\mathrm{d}\tau} = 1 - \alpha - \frac{\alpha\,\mathrm{e}^{\theta}}{\tau_{\text{ch}}} \qquad (4.22)$$

$$\frac{\mathrm{d}\beta}{\mathrm{d}\tau} = \frac{\alpha\,\mathrm{e}^{\theta}}{\tau_{\text{ch}}} - \frac{\rho\beta\,\mathrm{e}^{\theta}}{\tau_{\text{ch}}} - \beta \qquad (4.23)$$

$$\frac{\mathrm{d}\theta}{\mathrm{d}\tau} = \frac{\theta_{\text{ad}}\alpha\,\mathrm{e}^{\theta}}{\tau_{\text{ch}}} + \frac{\theta_{\text{ad}}\rho\beta\,\mathrm{e}^{\theta}}{\tau_{\text{ch}}} - (1 + \tau_N^{-1}\theta). \qquad (4.24)$$

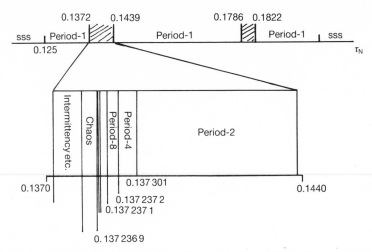

FIG. 4.11. The bifurcation sequences observed in a model of two consecutive exothermic reactions in a non-isothermal CSTR as the Newtonian cooling time τ_N is varied. (Reprinted with permission from Jorgensen and Aris (1983), © Pergamon Press.)

The dimensionless concentrations and temperature excess are given by

$$\alpha = a/a_0 \qquad b = b/a_0 \qquad \theta = E_1(T - T_a)/RT_a^2 \qquad (4.25)$$

and we have the following parameters

chemical time, $\qquad \tau_{ch} = 1/k_1(T_a)t_{res} \qquad (4.26)$

rate constant ratio, $\qquad \rho = A_2/A_1 \qquad (4.27)$

adiabatic temperature rise, $\qquad \theta_{ad} = Q_1 a_0 E_1/\sigma c_p RT_a^2 \qquad (4.28)$

cooling time, $\qquad \tau_N = t_N/t_{res} = c_p \sigma V/\chi St_{res}. \qquad (4.29)$

Jorgensen and Aris took $\tau_{ch} = 1.818$, $\theta_{ad} = 17.5$, and $\rho = 0.01$, and varied τ_N across the range

$$0.125 \leqslant \tau_N \leqslant 0.2. \qquad (4.30)$$

The stationary-state solution is unique and unstable across the parameter range of interest, with the upper and lower limits in eqn (4.30) corresponding to points of supercritical Hopf bifurcation. Near to the limits, the stationary state is surrounded by a stable period-1 limit cycle. The loss of stability of this oscillatory solution and the higher-order periodicities which emerge as τ_N is varied are summarized in Fig. 4.11.

The simple pattern of unstable stationary state with a stable period-1 limit cycle is found over almost the whole range of instability. There are just two windows in the parameter range for which more complex responses arise: these are $0.1372 \leqslant \tau_N \leqslant 0.1439$ and $0.1786 \leqslant \tau_N \leqslant 0.1822$. In the lower of these regions there is a subharmonic cascade on decreasing τ_N, beginning with the first period doubling at $\tau_N = 0.143\,897\,7$. This is accompanied by the critical Floquet multiplier (CFM) for the period-1 limit cycle passing through -1. The new period-2 limit cycle is born with a CFM equal to $+1$. This decreases rapidly as we reduce τ_N further. Eventually, the CFM approaches -1 again: a second period doubling occurs at $\tau_N = 0.137\,307\,1$ to give a period-4 solution. Further period doublings have been located at $\tau_N = 0.137\,237\,2$ (to period-8) and $0.137\,237\,1$ (to period-16). The ranges over which each of these higher-order periodicities holds sway reduce successively, as predicted by the Feigenbaum scenario. There appears to be a convergence of the cascade as τ_N is reduced to the value of $0.137\,236\,9$. Beyond this point the solution is chaotic. Aperiodicity does not survive across the whole range of Newtonian cooling time between its Feigenbaum limit and the return of simple period-1 (at $\tau_N = 0.137\,167$). Rather the character of the solution varies intermittently between chaotic and periodic. A period-doubling cascade based on an initial period-6 oscillation appears at $\tau_N = 0.137\,236\,7$, and converges back to chaos at $\tau_N = 0.137\,210\,5$: subsequent cascades based on period-5 and then period-4 also occur, with $0.137\,210\,5 < \tau_N < 0.137\,210\,4$ and $0.137\,210\,4 < \tau_N < 0.137\,196\,9$ respectively. Finally there is a very complex region, even by these standards, in which high-order periodicities and chaos seem to be mixed together.

The above sequence of exotica occurs, in reverse, in the upper window. The ranges of parameter values over which much of the 'interesting' patterns occur may be very small, so it is a tribute to the power of the Floquet analysis (and even more so to the patience and skill of Jorgensen and Aris) that such fine detail can be resolved.

Kahlert *et al.* (1981) also found a period-doubling cascade to chaos with two consecutive reactions. In their study, the second step was taken to be endothermic. Similar sequences also arise for a CSTR in which two non-competitive exothermic reactions ($A \rightarrow B$, $C \rightarrow D$) proceed, coupled only through the temperature rise and its effects on the two rate constants. The study by Lynch *et al.* (1982) again addressed a set of parameter values for which the stationary-state solution is unique and loses stability via a Hopf bifurcation.

Planeaux and Jensen (1986) considered an interesting variation on the non-isothermal reaction theme, and one that really does indicate that chaos may well be a most natural feature of the real world. Their CSTR has only one exothermic reaction, like the classical FONI system, but they imagine that the reactor is agitated by a stirrer with a finite thermal capacity. The exchange of heat with this otherwise inert component of the system provides a third interacting variable and this is again sufficient to allow for a complete range of periodicity and aperiodicity.

4.7.3. Willamowski and Rössler model

In a response to criticisms that the earliest model schemes, such as the Lorenz
and the Rössler models, contained features incompatible with chemical
systems, Willamowski and Rössler (1980; see also Aguda and Clarke 1988)
briefly investigated the following scheme:

$$A_1 + X \rightleftharpoons 2X$$

$$X + Y \rightleftharpoons 2Y$$

$$A_5 + Y \rightleftharpoons A_2$$

$$X + Z \rightleftharpoons A_3$$

$$A_4 + Z \rightleftharpoons 2Z.$$

The three variables here are the concentrations of the species X, Y, and Z:
A_1–A_5 are taken to be pool chemicals with constant concentrations that can
be adjusted as bifurcation parameters. This rather complex example satisfied
all chemical requirements (the concentrations do not become negative and
mass is conserved).

With suitable scalings, the rate equations for this scheme can be written as

$$dx/dt = (a_1 - k_{-1}x - z - y)x + k_{-2}y^2 + a_3 \qquad (4.31)$$

$$dy/dt = (x - k_{-2}y - a_5)y + a_2 \qquad (4.32)$$

$$dz/dt = (a_4 - x - k_{-5}z)z + a_3. \qquad (4.33)$$

Smith *et al.* (1983) considered the specific parameter choices $k_{-1} = 0.25$,
$k_{-2} = 10^{-3}$, $k_{-5} = 0.5$, $a_1 = 30$, $a_2 = 0.1$, $a_4 = 16.5$, and took a_5 as the
bifurcation parameter. There is a Hopf bifurcation as a_5 is reduced through
12.16, followed by successive period doublings (at $a_5 = 10.57$, 10.18, 10.136,
10.126, etc.) to chaos which is established for $a_5 = 10$.

4.7.4. Hudson and Rössler schemes

Hudson and Rössler (1984; Killory *et al.* 1987; Hudson *et al.* 1988) have
presented a family of schemes loosely based on autocatalytic feedback and
chemical inhibition (saturation) steps. These include three- and four-variable
models capable of supporting 'ordinary' chaos of the type described above
and also 'hyperchaos'. The distinction between the two forms of chaos is
linked to the 'dimensionality' of the underlying strange attractors in the two
cases or, equivalently, the number of positive Lyapounov experiments
characterizing the chaotic evolution.

The two three-variable schemes are

$$
\begin{array}{ll}
\text{P} \rightarrow \text{A} & \text{rate} = k_1 p \\
\text{Q} \rightarrow \text{B} & \text{rate} = k_2 q \\
\text{A} + \text{B} \rightarrow 2\text{B} & \text{rate} = k_3 ab \\
\text{B} \rightarrow \text{D} & \text{rate} = k_4 b/(K + b) \\
\text{A} \rightleftharpoons \text{C} & \text{rate} = k_5 a - k_{-5} c
\end{array}
$$

and a scheme based on a proposition by Turing (1952)

$$
\begin{array}{ll}
\text{P} \rightarrow \text{C} & \text{rate} = k_1 p \\
\text{C} \rightarrow \text{A} & \text{rate} = k_2 c \\
\text{C} + \text{A} \rightarrow 2\text{A} & \text{rate} = k^3 ac \\
\text{A} \rightarrow \text{B} & \text{rate} = k_4 a \\
\text{A} + \text{B} \rightarrow \text{D} + \text{B} & \text{rate} = k_5 ab/(K + a) \\
\text{B} \rightarrow \text{D} & \text{rate} = k_6 b.
\end{array}
$$

In each case, the corresponding two-variable system obtained if the conentration of C is held constant (or treated as a parameter) can show Hopf bifurcation and simple limit cycle oscillation. Recoupling C as a variable and hence allowing its time evolution to feed back on to the oscillatory response of A and B can convert the system to a 'designed' chaos.

Both of these three-variable models, like those discussed above, will have strange attractors of dimension larger than two (as argued earlier) but less than three (because the attractor must be embedded in the three-dimensional phase space), and will have a single positive Lyapounov exponent (plus one zero and one negative: there are three Lyapounov exponents for a three-variable system).

The algorithm of allowing a parameter to become a variable just applied to give chaos from a limit cycle can be applied a second time. Hudson and Rössler (1984) adapted the second of the above schemes by introducing a fourth variable E through which B is converted to A (B → E → A) and allowing a direct formation of B from a pool chemical (Q → B) and a further decay of C (C → F). This four-variable system must evolve in a four-dimensional phase space and has four Lyapounov exponents. Under suitable parameter values, two of the Lyapounov exponents may become positive (with then a third equal to zero and the fourth negative). Then the attractor will have dimension greater than three.

4.7.5. Rössler's model

Perhaps the simplest system leading to chaos is that invented by Rössler (1976a,b) with the 'rate equations'

$$\frac{\mathrm{d}x}{\mathrm{d}t} = -(y + z) \tag{4.34}$$

$$\frac{\mathrm{d}y}{\mathrm{d}t} = x + ay \tag{4.35}$$

$$\frac{\mathrm{d}z}{\mathrm{d}t} = b + xz - cz \tag{4.36}$$

with three parameters a, b, and c. This has only a single quadratic non-linearity. These equations cannot be easily represented as a mechanism of 'chemical' steps as, for instance, neither term in $\mathrm{d}x/\mathrm{d}t$ involves species X even though both are removal processes. In fact x, y, and z can become negative.

Many investigations of this scheme choose $a = b = 0.2$ and then vary c. Rössler demonstrated that for $c = 5.7$ the behaviour is chaotic: Crutchfield et al. (1980) showed that the route to chaos from the original period-1 limit cycle is a classic period-doubling cascade as c is increased through the range $2 \leqslant c \lesssim 4.2$, and that various periodic windows exist for higher c. The next-return map has a simple smooth hump and hence the subharmonic cascade reproduces the classic Feigenbaum scaling and the periodic windows follow the universal sequence described in Chapter 3. An example of this flow is shown in Fig. 4.12.

Celarier and Kapral (1987) have examined the 'basins of attraction' for various periodic attractors in this scheme for conditions such that there are coexisting limit cycles (birhythmicity). These can have extremely complex three-dimensional geometries.

The Rössler scheme could easily be overlooked in the present context because of the particular 'unchemical' aspects discussed above, but later we will examine a method due to Samardzija et al. (1989) for constructing 'equivalent chemical systems' appropriate to the same governing equations.

4.7.6. Pseudo 'two-variable' schemes

Having stated categorically that complex oscillations (i.e. anything more varied than period-1) and chaos are strictly ruled out for two-variable models, it is interesting to consider briefly a class of apparent exceptions to this rule. For this we must consider not the ordinary differential equations appropriate to well-stirred systems but the partial differential forms that

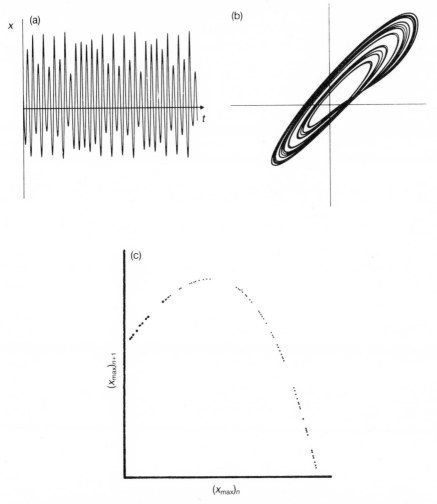

FIG. 4.12. The Rössler model showing aperiodicity: (a) the $x(t)$ time series; (b) a projection of the strange attractor; (c) the next-maximum map showing a simple quadratic maximum.

apply for reactions accompanied by diffusional processes. Three different examples have been considered in the literature: the Brusselator, the autocatalator model of Chapter 3, and a self-heating reaction that might occur in a porous catalytic pellet or in smouldering combustion.

The most straightforward case is that of the non-isothermal pellet (McGarry and Scott 1990). Here, we envisage a reaction zone in which a diffusing species A reacts exothermically to give a product species B and heat:

$$A \rightarrow B + \text{heat} \qquad k = k(T).$$

The heat released contributes to warming the catalytic zone and hence to increasing the local value of the reaction rate constant k. The reaction zone is embedded in a reservoir at constant temperature T_a and in which A has a constant concentration a_{ex}: no reaction occurs in the reservoir.

The competing effects of reaction and diffusion and of heat release and thermal conduction lead naturally to the development of concentration and temperature gradients. If the reaction zone is idealized to that of a one-dimensional slab (finite width but infinite length and breadth), the equations governing these evolutions can be written in the relatively simple form

$$\frac{\partial a}{\partial t} = D_A \frac{\partial^2 a}{\partial z^2} - k(T)a \tag{4.37}$$

$$c_p \frac{\partial T}{\partial t} = \kappa \frac{\partial^2 T}{\partial z^2} + Qk(T)a \tag{4.38}$$

where D_A is the diffusion coefficient, κ the thermal conductivity, Q the reaction exothermicity, and z the spatial coordinate. If the reaction zone extends over $-r_0 \leqslant z \leqslant +r_0$ we will consider the following boundary conditions:

$$\partial a/\partial z = \partial T/\partial z = 0 \qquad \text{at } z = 0$$

and

$$D_A(\partial a/\partial z) - k_A(a_{ex} - a) = 0 \qquad \kappa(\partial T/\partial z) + \chi(T - T_a) = 0 \qquad \text{at } z = \pm r_0.$$

The first of these simply uses the (assumed) symmetry of the concentration and temperature profiles about the mid-point $z = 0$. The second has the so-called Robin form and allows for resistances to mass and heat transfer across the boundary between the reaction zone and the reservoir.

Equations (4.37) and (4.38) can be written in the dimensionless form

$$\frac{\partial \alpha}{\partial \tau} = \frac{\partial^2 \alpha}{\partial \rho^2} - \gamma \lambda \alpha \exp[\theta/(1 - \varepsilon\theta)] \tag{4.39}$$

$$(Le)\frac{\partial \theta}{\partial t} = \frac{\partial^2 \theta}{\partial \rho^2} + \lambda \alpha \exp[\theta/(1 + \varepsilon\theta)] \tag{4.40}$$

with $\alpha = a/a_{ex}$, $\theta = (T - T_a)E/RT_a^2$. The parameters in these equations are: $\lambda = EQr_0^2 a_{ex}k(T_a)/\kappa RT_a^2$, a measure of the reaction zone size and the reservoir parameters; $\gamma = \kappa RT_a^2/EQa_{ex}D_A$, the inverse of the reaction exothermicity; $\varepsilon = RT_a/E$ and the Lewis number $(Le) = D_A c_p/\kappa$; $\tau = D_A t/r_0^2$ is dimensionless time; and $\rho = z/r_0$ is dimensionless distance.

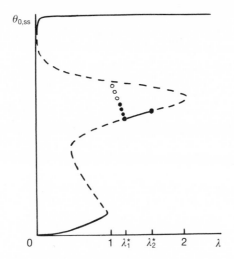

FIG. 4.13. Stationary-state bifurcation diagram for the model of an exothermic reation in a catalyst pellet, showing the central temperature excess $\theta_{0,ss}$ as a function of the parameter λ, with $\gamma = 0.16$, $\varepsilon = 0.02$, $\nu = 125$, $\mu = 15$, and $(Le) = \frac{2}{9}$: ——, stable stationary state; –––, unstable stationary state. There are two Hopf bifurcation points on the middle branch, the lower at λ_1^* sheds a stable limit cycle that undergoes further period doublings.

The boundary conditions become

$$\partial\alpha/\partial\rho = \partial\theta/\partial\rho = 0 \qquad \text{at } \rho = 0$$

and

$$(\partial\alpha/\partial\rho) - \nu(1 - \alpha) = 0 \qquad (\partial\theta/\partial\rho) + \mu\theta = 0 \qquad \text{at } \rho = \pm 1$$

containing two extra parameters (Biot numbers), $\nu = k_c r_0/D_A$ and $\mu = \chi r_0/\kappa$, that measure the relative importance of the transport across the surface and through the bulk of the reaction zone.

Equations (4.39) and (4.40) can have up to five simultaneous stationary-state solutions (Burnell *et al.* 1983; Brindley *et al.* 1990). Figure 4.13 shows the stationary-state temperature excess at the centre of the reaction zone θ_{ss} ($z = 0$) as a function of the parameter λ for the particular conditions $\gamma = 0.16$, $\varepsilon = 0.02$, $\nu = 125$, and $\mu = 15$. The Lewis number (Le) does not affect the stationary-state solutions but does influence their stability—Fig. 4.13 also shows the location of Hopf bifurcation points (two, both on the third branch) and the development of a period-1 limit cycle from one of these appropriate to $(Le) = \frac{2}{9}$.

The lower Hopf bifurcation is supercritical, giving rise to a stable period-1 limit cycle that grows as λ is decreased. Eventually, at some lower λ this periodic solution is destroyed via a homoclinic orbit bifurcation with the

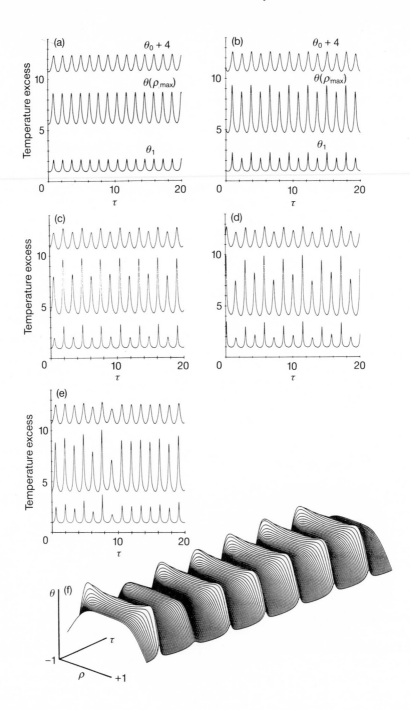

saddle point solution on the fourth branch, as shown in the figure. Before this, however, the period-1 limit cycle becomes unstable. This loss of stability marks the beginning of a period-doubling cascade that leads to the establishment of period-2, period-4, etc., and ultimately to a chaotic response. Figure 4.14 shows typical time series for various periodicities and for chaos.

The deterministic nature of the chaos in this system is revealed by a series of tests: a Poincaré section has been constructed by plotting the temperature excess at the edge $\theta(z = 1)$ against that at a particular point in the interior, $\theta(z = 0.773)$, every time the central temperature excess has a maximum (Fig. 4.15(a)). The successive maxima in $\theta(z = 0)$ have also been used to construct a next-amplitude map (Fig. 4.15(b)). Both of these are smooth curves indicative of slices through an almost two-dimensional attractor. The underlying strange attractor itself can be revealed (Fig. 4.15(c)), using data only from the central temperature excess time series. The latter consists of a series of points, determined numerically, at a corresponding set of times. We can reconstruct the attractor by plotting the value of $\theta(z = 0)$ at each time step τ against the value at some later time $\tau + \Delta\tau$. The choice of the delay time $\Delta\tau$ is somewhat arbitrary, and will be discussed further in Chapter 7: here we have taken $\Delta\tau = 0.35$, approximately one-quarter of the mean period between successive maxima. Finally, the Fourier power spectrum for the chaotic trace is compared with that for a simple periodic solution in Fig. 4.15(d) and (e).

It appears to be important that the boundary conditions have the above Robin form, rather than the simpler Dirichlet conditions $\theta = 0$, $\alpha = 1$, in order to obtain complex and chaotic responses and, more specifically, we also require $\mu < \nu$ so that resistance at the surface is more significant for mass transfer than heat transfer. It is certainly necessary that the Lewis number should be less than unity (Luss and Lee 1970), i.e. that mass diffusion should proceed on a shorter timescale than thermal conduction within the bulk of the catalyst.

The reaction–diffusion Brusselator has been examined in the present context by Kuramoto (1978) and by Nandapurkar *et al.* (1984, 1986) following earlier work by Herschkowitz-Kaufman and Nicolis (1972). The chemistry given in the model of Section 3.12.1 is again considered to take place with an infinite slab with the following assumptions. The concentrations

---

FIG. 4.14. Time series showing the temperature at three points across the catalyst reaction zone for different values of λ: in each case the temperature excess at the edge θ_1 and at the centre θ_0 (the latter offset for clarity) are shown along with that at the spatial location for which the system achieves its highest degree of self-heating $\theta(\rho_{max})$: (a) $\lambda = 1.15$, $\rho_{max} = 0.627$, period-1; (b) $\lambda = 1.13$, $\rho_{max} = 0.714$, period-2; (c) $\lambda = 1.126$, $\rho_{max} = 0.748$, period-4; (d) $\lambda = 1.125$, $\rho_{max} = 0.773$, period-8; (e) $\lambda = 1.124$, $\rho_{max} = 0.773$, aperiodic response; (f) a three-dimensional representation of the full spatiotemporal development of chaos in this model for the parameter values in (e).

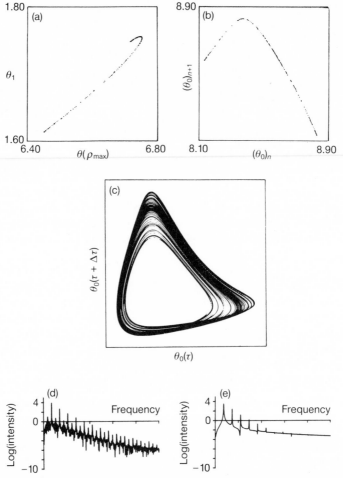

FIG. 4.15. Diagnostic tests for chaos for the exothermic catalyst pellet response in Fig. 1.14(e): (a) Poincaré section constructed from the values of θ_1 and $\theta(\rho_{max})$ as θ_0 attains successive maxima (i.e. we choose $d\theta_0/d\tau = 0$ to define the Poincaré plane); (b) next maximum map for θ_0; (c) reconstructed two-dimensional projection of the attractor using delay method with $\Delta\tau = 0.35$; (d) Fourier power spectrum; (e) Fourier power spectrum for the period-4 solution with $\lambda = 1.126$ for comparison.

of the pool chemical species A and B are maintained at a constant value throughout the reactions zone, e.g. by imagining that these diffuse quickly or react very slowly. The reaction zone is surrounded at each end by an infinite reservoir in which the reactive species X and Y have constant concentrations x_{ex} and y_{ex} respectively. Allowing for diffusion as well as reaction in the slab, the appropriate governing equations can be written in the form

$$\frac{\partial x}{\partial t} = D_x \frac{\partial^2 x}{\partial z^2} + A + yx^2 - Bx - x \qquad (4.41)$$

$$\frac{\partial y}{\partial t} = D_y \frac{\partial^2 y}{\partial z^2} + Bx - yx^2 \qquad (4.42)$$

where D_x and D_y are the diffusion coefficients. Again using symmetry at the centre of the slab ($z = 0$) and the Robin boundary condition at the ends ($z = \pm L$):

$$\partial x/\partial z = \partial y/\partial z = 0 \qquad \text{at } z = 0$$

$$D_x(\partial x/\partial z) + K_x(x_{ex} - x) = 0 \qquad D_y(\partial y/\partial z) + K_y(y_{ex} - y) = 0 \qquad \text{at } z = L$$

where K_x and K_y are mass exchange coefficients (Biot numbers) for the reaction zone–reservoir interface. Nandapurkar *et al.* found that suitable parameter variations could induce a change from periodic oscillations to chaos. In this case, the critical Floquet multipliers do not pass through -1 at each bifurcation, but two leave the unit circle simultaneously as a complex pair. This is characteristic of a bifurcation to a torus, which we will discuss in more detail below.

Kuramoto (1978) has introduced the ideas of phase chaos (or phase turbulence) and amplitude chaos appropriate to the above behaviour. When the simple limit cycle first loses stability for this model, the concentration profiles are such that the concentration gradients are relatively small. The onset of chaos here is phase chaos: at each location across the reaction zone there is an almost regular oscillation, but each coordinate point is typically out of phase with its neighbours to create an overall aperiodicity. If we vary an appropriate parameter to move further into the region of chaos, the aperiodicity becomes in some sense stronger. Larger concentration gradients can become established and the differences in phase between neighbouring locations can become augmented by individual locations exhibiting large variations in amplitude, i.e. we get both phase and amplitude turbulence.

Returning to systems with equal diffusion coefficients and without pool chemicals, Brindley *et al.* (1988) have studied the reaction–diffusion analogue of the autocatalytic model from Section 3.7:

$$A \rightarrow B$$

$$A + 2B \rightarrow 3B$$

$$B \rightarrow C$$

again in a slab with edges at $z = \pm r_0$. The dimensionless equations for this can be written as

$$\frac{\partial \alpha}{\partial t} = D \frac{\partial 2\alpha}{\partial \rho^2} - \alpha\beta^2 - \kappa_u\alpha \tag{4.43}$$

$$\frac{\partial \beta}{\partial \tau} = D \frac{\partial^2 \beta}{\partial \rho^2} + \alpha\beta^2 + \kappa_u\alpha - \kappa_2\beta. \tag{4.44}$$

It is convenient to express the boundary conditions on each side of the slab ($z = -r_0$ or $\rho = -1$ and $z = +r_0$ or $\rho = +1$ respectively) separately for what follows. We will assume that the concentration of A in the reservoir on each side of the slab is the same, so that $\alpha = \alpha_{ex} = 1$ (by definition) at $\rho = \pm 1$. For the autocatalyst, however, we will specify

$$\beta = \beta_L \quad \text{at } \rho = -1 \quad \text{and} \quad \beta = \beta_R \quad \text{at } \rho = +1.$$

In this way, we are ready to allow the reaction zone to experience different boundary conditions on the two faces.

With symmetric boundary conditions, i.e. $\beta_L = \beta_R$, the above scheme can support basic two-variable behaviour such as multistability and period-1 oscillations. Figure 4.16 shows a typical oscillatory response, for $\beta_L = \beta_R = 0.044$, $D = 0.001$, $\kappa_u = 0.0025$, and $\kappa_2 = 0.04$. The oscillatory amplitude and, particularly, the frequency depend on the various parameters, including the external concentration of the autocatalyst. Furthermore, if the dimensionless diffusion coefficient D is small, the most significant variations in concentration occur close to the edges of the reaction zone, as evidenced by the three-dimensional projection in Fig. 4.16. The central portion of the zone can become almost dormant, providing only a weak coupling between the two 'boundary layers'. If we change the concentration of the autocatalyst in the reservoir on one side of the reaction zone (only), the two boundary layers may still both wish to oscillate, but now will wish to have different frequencies. Under some conditions, these two basic frequencies may interact through the weak coupling to give a complicated spatio-temporal pattern, such as that shown in Fig. 4.17, which has $\beta_R = 0.41$ but the other parameters (including β_L) as given earlier.

Such biperiodic behaviour can be located systematically by looking for so-called 'Hopf–Hopf interactions' (Guckenheimer and Holmes 1983). Normally, the onset of periodic solutions occurs at a non-degenerate Hopf bifurcation point, where a principal pair of eigenvalues (those with real part of smallest magnitude) is complex with real part passing through zero. Reaction–diffusion systems, unlike their well-stirred (ordinary differential equation) counterparts, have an infinite number of eigenvalues because of the continuous spatial cordinate and so more than one complex pair may exist. If parameter values can be found such that two pairs have real parts passing through zero at the same time (or very close together) we expect—and find here—more complex responses.

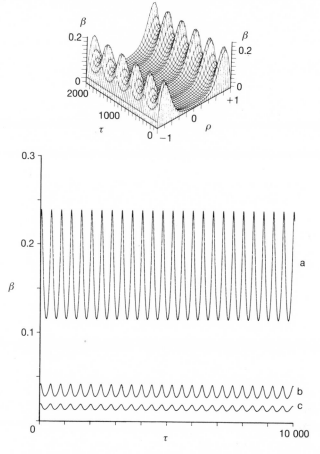

FIG. 4.16. Spatio-temporal oscillations in the two-variable autocatalator with symmetric boundary conditions, showing simple period-1 character and spatial exact symmetry. The largest-amplitude excursions are found near the edges with only small oscillations in the central region: a, $\rho = \pm 0.8$; b, $\rho = \pm 0.4$; c, $\rho = 0$.

4.8. Mixed-mode oscillations

If equations as simple as those of Rössler (Section 4.7.5) can give rise to period doubling, it is not too surprising that the subharmonic cascade route to chaos is the most common shown by small chemical models (three or four variables). Period-doubling scenarios have also been observed in some real chemical systems, but this is in fact not the most frequent manifestation of chaos in experiments. The simple model of Sections 4.1–4.5 can be used to introduce a more typical sequence of behaviour—that of mixed-mode oscillations (Tomlin 1990; Scott and Tomlin 1990a). There still exists no little

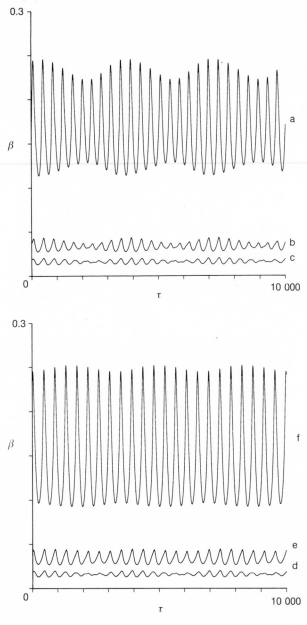

FIG. 4.17. Spatio-temporal oscillations in the two-variable autocatalator with asymmetric boundary conditions, showing 'biperiodic' character. The largest-amplitude excursions are again found near the edges, but differ on each side of the reaction zone. There is a weak 'coupling' across the central region: a, $\rho = -0.8$; b, $\rho = -0.4$; c, d, $\rho = 0$; e, $\rho = +0.4$; f, $\rho = +0.8$.

debate over the theoretical interpretation of the alternating periodic and chaotic responses that arise as a parameter is varied in, say, the Belousov–Zhabotinskii reaction (see Chapter 8) and although some firm conclusions can be drawn via the non-isothermal autocatalytic scheme, the full issue remains unsolved.

4.8.1. *Examples of mixed-mode responses*

The mixed-mode waveform typically has a repeating unit consisting of a number of both large- and small-amplitude excursions. Different periodicities are formed by combining different numbers of small and large peaks, and an intricate structure of bifurcations between these forms exists. Aperiodicity is also possible.

We can take the system with $\kappa_u = 10^{-3}$, $\gamma = 0.005$, and $\delta = 0.025$, corresponding to relatively low reaction exothermicity but also slow heat transfer. This has the same stationary-state behaviour as that seen earlier: two solutions exist for reactant concentratons $\mu \leqslant \gamma/\delta\, e = 2e^{-1} \approx 0.736$, merging in a saddle–node point at the upper end of this range. The upper branch consists of saddle points. Along the lower branch, the stationary state is stable, except for the range between the two primary Hopf bifurcation points $\mu_2^* \leqslant \mu \leqslant \mu_1^*$.

Over much of the range between μ_2^* and μ_1^* the stable solution is a simple period-1 limit cycle oscillation. However, more complex behaviour emerges for $0.5 \lesssim \mu \lesssim 0.56$. As the reactant concentration is increased above μ_2, the amplitude of the emerging (period-1) oscillation grows relatively quickly. A typical time series is shown in Fig. 4.18(a) for $\mu = 0.498$. For higher reactant concentrations, the period-1 limit cycle appears to become unstable and a mixed-mode form emerges. Figure 4.18(b) shows the waveform form $\mu = 0.51$ consisting of one large and one small peak: a 1^1 response (we will use the notation L^s where L and s are the number of large and small peaks per repeating unit respectively). The change from 1^0 to 1^1 is not a period doubling: the critical Floquet multiplier does not leave the unit circle through -1. In fact, it is still not yet clear whether there really is any bifurcation at all here, never mind what character it has. We will examine these apparent changes in waveform in more detail shortly.

As the parameter μ is increased further, more small peaks emerge per repeating unit. Figure 4.18(c), (d), (e), and (f) shows the 1^2, 1^3, 1^5, and 1^8 patterns respectively for $\mu = 0.53$, 0.535, 0.55, and 0.558: the 'missing' responses 1^4, 1^6, and 1^7 also occur for this system in the appropriate order in this sequence but have not been shown here. The basic trend, then, is fairly simple: as the reactant concentration increases so extra small excursions appear, one at a time, in between each large peak. For the moment we will leave the further changes in behaviour at higher μ.

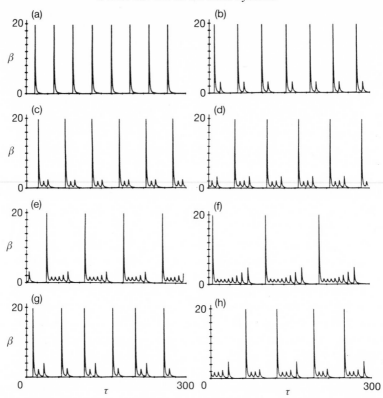

FIG. .4.18 A sequence of 'mixed-mode' oscillations for the three-variable non-isothermal autocatalator model with $\gamma = 0.05$, $\delta = 0.025$, and $\kappa_u = 10^{-3}$: (a) $\mu = 0.498$, 1^0; (b) $\mu = 0.51$. 1^1; (c) $\mu = 0.53$, 1^2; (d) $\mu = 0.535$, 1^3; (e) $\mu = 0.55$, 1^5; (f) $\mu = 0.558$, 1^8; concatenations of the basic states are found for intermediate parameter values: (g) $\mu = 0.52$, $1^1 1^2$; (h) $\mu = 0.54$, $1^3 1^4$.

4.8.2. *Concatenations*

Returning to the 1^1 and 1^2 patterns in Fig. 4.18(b) and (c), we can examine the behaviour for values of the reactant concentration between $\mu = 0.51$ and 0.53. The response for $\mu = 0.52$ (Fig. 4.18(g)) is a 'mixed' state with a repeating unit consisting of $1^1 1^2$. Such a mixing of the parent 1^1 and 1^2 states is known as a 'concatenation', yielding a daughter state. There are similar concatenated daughters between each of the pairs of parents shown in Fig. 4.18(b)–(f), e.g. the $1^3 1^4$ state (Fig. 4.18(h)) for $\mu = 0.54$ between the 1^3 and 1^4 (not shown).

Further stages of concatenation are not only possible but almost guaranteed as daughters mix with parents. Thus between the $1^1 1^2$ and 1^2 states there will be various mixed-mode responses of the general form $1^1(1^2)^n$ with n increasing with μ and becoming infinite before $\mu = 0.53$. There may be the mirror-image sequence $(1^1)^n 1^2$ with n increasing as μ decreases, but both are not always seen (this may be a problem of resolution even with numerical studies).

4.8.3. Firing number and the Devil's staircase

Typically, the range of the reactant concentration over which a given mixed-mode form exists decreases with the increasing level of concatenation. This latter point can be exemplified by means of the 'Devil's staircase'. The staircase merely plots some representation of the observed pattern against the parameter being varies—μ in the above sequence. The appropriate measure is the 'firing number' F, defined as the fraction of the peaks within the complete repeating unit that are small. Thus with the L^s notation, $F = s/(L + s)$: for the simple, large period-1 oscillation with $\mu = 0.498$ we have 1^0 so $F = 0$; F increases with μ giving $F = \frac{1}{2}$ for the 1^1 pattern up to $F = \frac{8}{9}$ for the 1^8. When F is plotted against μ, we see a series of steps (Fig. 4.19). Any given tread represents the range of the reactant concentration across which a particular concatenation, with a given firing number, is found. When the waveform changes, the firing number also changes, and does so abruptly.

Between each of the main treads, corresponding to the most easily observed (most widespread) patterns, are small substaircases corresponding to daughter-state concatenations. The concatenation *ad infinitum* described above means that there will be an infinite number of steps on the staircase— apparently the reason for it being attributed to the Devil.

The Devil's staircase can be 'complete', 'incomplete', or 'multi-valued' ('non-invertible'). In the first of these cases, we could imagine that at higher values of μ than those discussed above, we see patterns with more small

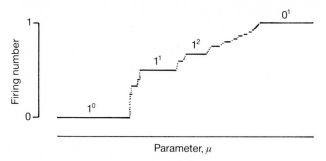

FIG. 4.19. The Devil's staircase showing how the firing number F typically varies with some parameter and the system proceeds through a sequence of concatenated mixed-mode states such as that in Fig. 4.18.

peaks, 1^n with n increasing without limit. As n tends to infinity, so the firing number will tend to unity, corresponding to period-1, small-amplitude oscillations 0^1. If, then, for all values of μ between the 1^0 and 0^1 states there exists some concatenated mixed-mode waveform, the treads of the staircase will fill up the whole range. If, furthermore, the treads do not overlap, then the staircase is said to be complete. Incompleteness can arise if the behaviour changes from the mixed-mode form to, say, quasiperiodicity at some values of μ, where the firing number would become irrational. We will discuss quasiperiodicity in the next section. If the treads begin to overlap, more than one mixed-mode pattern can be found for some ranges of the reactant concentration.

4.8.4. Farey sequences

There is a formal numerical relationship between parents and daughters in these processes, encapsulated in the idea of 'Farey arithmetic'. Farey addition of two rational numbers p_1/q_1 and p_2/q_2 follows the rule

$$(p_1/q_1) \oplus (p_2/q_2) = (p_1 + p_2)/(q_1 + q_2).$$

The firing numbers of concatenating states then always satisfies this rule: for example, the 1^1 and 1^2 states with $F = \frac{1}{2}$ and $\frac{2}{3}$ respectively yield a daughter $1^1 1^2$ with $F = (1 + 2)/(2 + 3) = \frac{3}{5}$. We will see Farey arithmetic and Devil's staircases in evidence when we discuss the Belousov–Zhabotinskii reaction in Chapter 8.

Another reason for breakdown of completeness in the Devil's staircase, frequently associated with a tendency for multi-valued behaviour, is the onset of aperiodicity. The regular mixing described above can be interrupted at some levels of the concatenation process by waveforms that apparently consist of aperiodic combinations of the parent—thus 1^1 and 1^2 units may appear in a 'random' or (more properly) a non-repeating sequence for some values of μ. Here there is some debate as to whether such chaotic responses are simply (long-lived) transients or caused by minute numerical perturbations (or real perturbations in experimental systems) or true aperiodic solutions— again this will be an important point of discussion in Chapter 8.

4.8.5. Other transitions to chaos

In fact, for the present parameter values, our model scheme does not continue its Farey sequence to higher reactant concentration. The 1^8 response may add more small peaks, but soon this periodic behaviour gives way to a chaotic mixed-mode state. The bifurcation causing this transition is not understood, but a typical selection of time series is shown in Fig. 4.20(a)–(c). The development of the phase-plane attractor through this sequence is given in Fig. 4.21(a)–(c). For the periodic 1^8 solution, the limit cycle has a tightly

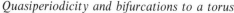

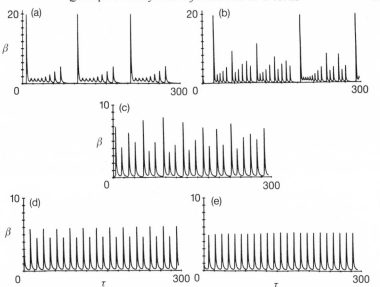

FIG. 4.20. The transition to chaos and a subsequent period-halving cascade for the model from Fig. 4.18 at high precursor reactant concentrations: (a) $\mu = 0.558$, 1^8; (b) $\mu = 0.5586$, large-amplitude chaos; (c) $\mu = 0.5587$, small-amplitude chaos; (d) $\mu = 0.5595$, period-2; (e) $\mu = 0.56$, period-1.

wound section corresponding to the small peaks and then a large excursion which moves away underneath the spiral but is reinjected back into the top. In a sense this appears like the slow winding through the central hole of a torus followed by a large excursion around the external surface. As μ is increased, the trajectory appears to stay longer at the lower end of the spiral, and for $\mu = 0.5586$ the behaviour here has become as complicated as that of a strange attractor. Interestingly, at larger μ, the big excursion disappears leaving a small, fractal chaotic attractor.

Figure 4.20(d) and (e) also indicates how the chaos is resolved back to cosmos for yet larger μ, passing down a period-halving cascade to relatively small-amplitude period-1 oscillations.

4.9. Quasiperiodicity and bifurcations to a torus

The discussion of the form of the attractor at the end of the last section mentioned the idea of the system moving across the surface of a torus in the phase plane. Here we will examine the role of tori in more detail, partly because they have an important place in the historical development of this subject.

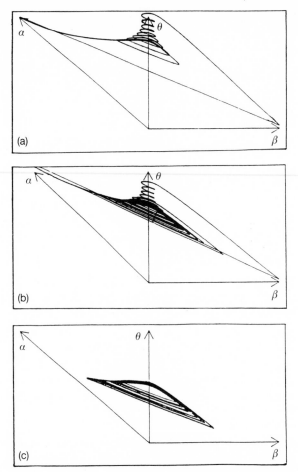

FIG. 4.21. The changing structure of the attractor for the transition from mixed-mode oscillation through large-amplitude chaos to small-amplitude chaos as evidenced in Fig. 4.20(a–c): (a) 1^8 limit cycle with an unwinding 'core' followed by a single large excursion; (b) the flow at the exit of the unwinding 'core' has become very complex leading to strange attractor behaviour with a single large excursion then reinjecting the flow to the top of the core; (c) the large-amplitude excursion has vanished leaving only the complex structure of the strange attractor.

4.9.1. Couette–Taylor turbulence

Some of the earliest studies of chaos arose in the area of fluid flow, and with turbulent flow in particular (turbulence and chaos are still occasionally used synonymously, e.g. chemical turbulence). A transition from laminar to turbulent flow can be neatly demonstrated in the Couette–Taylor system (Taylor 1923) which has a viscous fluid contained within two concentric

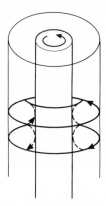

FIG. 4.22. Schematic representation of the Couette–Taylor experiment showing two concentric cylinders, the inner one being rotated, and the establishment of rotating tori of fluid.

cylinders. If the inner cylinder is rotated slowly, the fluid also rotates in some relatively smooth flow pattern—with an angular velocity that falls to zero at the outer cylinder. On increasing the rotation rate of the inner cylinder, the flow pattern becomes more complex—pairs of vortices are born and the system develops the appearance of a stack of doughnuts, as sketched in Fig. 4.22. The fluid contained within a given 'cell' moves with the rotation of the inner cylinder, but also rotates in a direction perpendicular to that motion, so as to wind around the doughnut. The sense of this second rotation, clockwise or anticlockwise, alternates between neighbouring cells. (These 'doughnuts' have a toroidal geometry, but this is not the same as the phase-plane object mentioned above—this is an unfortunate clash, and in fact the bifurcation just described in the Couette–Taylor system is really a primary Hopf bifurcation.)

Further increases in the rotation velocity can cause a second bifurcation. Now the doughnuts distort with an additional superimposed motion, becoming wavy. Beyond this, the motion becomes turbulent

4.9.2. *Rayleigh–Bénard turbulence*

A second classical example from fluid dynamics is that of the convective turbulence in a Rayleigh–Bénard cell (Bénard 1900; Rayleigh 1916; Libchaber and Maurer 1982; Behringer 1985). Here a viscous fluid is confined to a fixed volume between two plates perpendicular to a gravitational field—see Fig. 4.23. The upper and lower plates can be heated (or cooled) and a temperature difference δT is established between them, the lower plate being at a higher temperature than the upper one. The local heating in the vicinity of the lower plate causes an expansion of the fluid, with a consequent decrease in density.

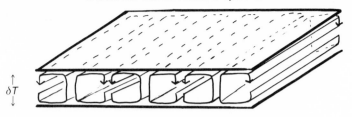

Fig. 4.23. Schematic representation of the Rayleigh–Bénard experiment showing convection cells established between two plates for which the lower has a slightly high temperature.

Fluid elements near to the bottom plate thus experience an upward force, whilst fluid moving to the upper plate cools and increases in density, then tending to feel a downard force. The motion between the plates that these forces might induce is resisted, in part, by frictional or viscous forces. Thus small temperature differences may not be sufficient to cause fluid motion; heat transfer remains a conductive process.

The characteristic (dimensionless) measure of the temperature difference is the Rayleigh number $(Ra) = (\sigma g \alpha d^3 / \eta D_T) \, \delta T$, where σ is the mean density, g the gravitational constant, α the coefficient of thermal expansion, d the distance between the plates, η the dynamic viscosity, and D_T the thermal diffusivity. For low (Ra) there is no convective motion as explained above. As the temperature difference is increased beyond a critical threshold value $(Ra)_{cr}$, corresponding to a Hopf bifurcation, convection appears in the system as the friction no longer counters the gravitational forces. The system develops counter-rotating pairs of cylindrical convecting cells. With $(Ra) \gg (Ra)_{cr}$, the ordered cylindrical motion gives way to turbulence (Gollub 1983). Two other important physical parameters are the aspect ratio of the system L/d and the Prandtl number for the fluid $(Pr) = v/D_T$ where v is the kinematic viscosity.

4.9.3. *Theoretical interpretations of the transition to turbulence*

The original interpretation of the transition to turbulence is credited to Landau (Landau and Lifshitz 1959). He suggested that the process involves successive Hopf bifurcations. The first introduces a frequency v_1. At the second Hopf bifurcation, the system develops a second characteristic frequency v. If v_1 and v_2 are incommensurate (i.e. if v_1/v_2 is irrational) the combined effect will be to cause the system to move on a torus in phase space. This can be envisaged in the following way. The first Hopf bifurcation gives rise to motion on a limit cycle in some appropriate phase space. At the second Hopf point, the limit cycle itself begins to move on a closed curve. These

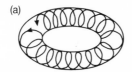 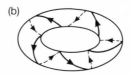

FIG. 4.24. Construction of a torus from two incommensurate oscillatory motions: (a) if the frequencies are irrationally related the trajectory will eventually fill the complete two-dimensional surface of the torus; (b) if the frequencies become rationally related, the trajectory closes on itself and does not explore the full surface (in the example $v_1/v_2 = 1/5$).

simultaneous rotations will describe a torus as shown in Fig. 4.24. (Notice that if the frequencies are rationally related, the resulting motion leads to a trajectory that eventually closes up and does not cover the complete surface of the torus—this is phase-locked motion, which we will discuss later.)

Motion on the surface of a torus with incommensurate frequencies corresponds to quasiperiodicity. This is in some respect more complex than simple period-1 oscillations. Thus we might feel that the next stage in complexity is the addition of a third incommensurate frequency v_3 at a third Hopf bifurcation. With then more and more Hopf bifurcations, the evolution obtains more underlying frequencies, and Landau's suggestion was that chaos results from an infinite cascade much like the geometrical convergence of the subharmonic cascade in the period-doubling route.

Such a 'simple' scenario does not accord with more recent experimental measurements. Gollub and Swinney (1975; Swinney and Gollub 1978; Fenstermacher *et al.* 1979) measured local velocities through laser Doppler anemometry and analysed the resulting time series with some of the methods discussed in Chapter 7. They clearly observed the appropriate changes for the primary Hopf bifurcation, with a relatively simple power spectrum based on a single fundamental and its overtones emerging. The emergence of a second incommensurate frequency corresponding to the development of a (two-dimensional or T^2) torus was also detected. However, no further simple Hopf bifurcations were found. Instead the power spectrum changed from one with sharp peaks corresponding to two fundamental frequencies to one with a broad-banded structure. The implication here, then, is for the sequence period-1 → quasiperiodicity → chaos, or in terms of the attractor in the phase plane, limit cycle → torus → strange attractor.

4.9.2. *Ruelle–Takens–Newhouse (RTN) scenario*

Current interpretations are based on the ideas of Ruelle and Takens (1971; Ruelle 1980) as embodied in the RTN scenario (Newhouse *et al.* 1978). In rather abstract terms, they suggested (and, in fact, proved) that only a small number of bifurcations are needed to bring about chaos. Essentially, the T^3 torus created after the third Hopf bifurcation is not a particularly stable object and this can quite readily be 'perturbed' to produce a strange attractor. The (fractal) strange attractor is then stable to 'perturbations'. Gollub and Benson (1979, 1980) were able to detect each stage of the RTN scenario during Rayleigh–Bénard experiment with water: with an aspect ratio of 3.5 and a Prandtl number $(Pr) = 5$ they observed bifurcations at $(Ra) = 30$, 39.5, and 41.5. For Rayleigh numbers slightly larger than this latter value, they obtained a clear three-fundamental-frequency power spectrum although in general it does not appear to be easy to find this section of the transition sequence as the T^3 torus seems to break up rather quickly.

Curry and Yorke (1977) later showed that a slightly different scenario is also possible (at least in the abstract mappings they studied) whereby no three-frequency quasiperiodicity need occur. Instead, they found that the third Hopf bifurcation can lead to a phse-locked state, yielding a periodic solution which then may become chaotic.

4.9.5. Mode locking, Poincaré sections, and circle maps

The concept of phase locking versus quasiperiodicity is of fundamental significance in the present context. To examine this point further, we should dissect a typical torus. Cutting once opens the attractor out to a cylinder, whilst a second cut yields a 'rectangle' comprising the surface of the original structure, as depicted in Fig. 4.25, whose two sides have lengths related to

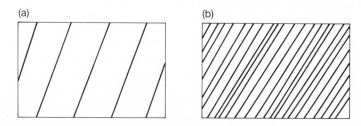

(a) (b)

FIG. 4.25. The torus can be opened with two cuts to form a topological rectangle. Paths cannot cross on this surface (unless at singular points) so must form parallel trajectories: (a) for a frequency-locked state with a rational relationship between v_1 and v_2 the paths close up (shown here is a 1:4 locking); (b) for irrationally related component frequencies the trajectory fully covers the rectangular surface.

the two fundamental frequencies of the system. Trajectories are now represented by lines on the rectangle. Just as trajectories cannot in general cross in the phase plane, so they must also not cross on this surface (if they do it must be at a singular point). Thus motions on the torus must give parallel lines in Fig. 4.25.

If the two fundamental frequencies have some rational relationship, i.e. if $v_1/v_2 = p/q$ for some integer numbers p and q, then the trajectory will 'join up' as shown in Fig. 4.25(a). The system does not explore the whole surface area of the torus (or rectangle) but just describes a one-dimensional line corresponding to a periodic or 'mode-locked' response. On the other hand, if v_1/v_2 is irrational, the trajectory will cover the whole surface without self-intersection in infinite time—corresponding to quasiperiodicity (Fig. 4.25(b))

A slightly gentler treatment of our torus is to construct a Poincaré section (similar to making our first cut above). Placing a plane transversely through the torus gives a 'circle' of intersection. We then note where the system cuts the plane each time it completes a circuit of the attractor. Let us go one stage back in the bifurcation sequence, so that the system is following a period-1 limit cycle. This will give a point on the Poincaré plane (Fig. 4.26(a)), a so-called invariant point, because any trajectory starting there

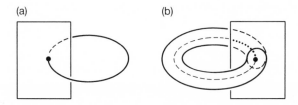

FIG. 4.26. The construction of a Poincaré section for (a) a limit cycle, giving an invariant point, and (b) a torus, giving an invariant circle and an unstable invariant point corresponding to the unstable period-1 limit cycle surrounded by the torus.

returns to exactly the same point. If small disturbances decay, the invariant point is stable (and so is the corresponding limit cycle): the decay of the disturbance will give rise to a transient motion (or more accurately a mapping) on the Poincaré plane reminiscent of the approach to a stable node or focal stationary state in the phase plane.

When the period-1 solution bifurcates to yield a torus, the limit cycle remains, but becomes stable. We thus have an unstable invariant point on the Poincaré plane. This will be surrounded by the image of the new torus. If the motion on the torus is truly quasiperiodic, so the whole of the surface is covered, the repeated intersections will never coincide with earlier ones and a full 'invariant circle' (or at least some closed loop) will be mapped out. The Poincaré plane now has the same appearance (Fig. 4.26(b)) as a phase plane with a focal singularity surrounded by a limit cycle. It is partly this resemblance that leads to the nomenclature 'secondary Hopf bifurcation' for the birth of a torus in this way, although the term Naimark–Sacker torus bifurcation is also used (Langford 1983).

The characteristic signature of the bifurcation from a period-1 solution to a torus can be seen by following the Floquet multipliers discussed in Section 3.6. For a three-variable system there will be three Floquet multipliers, but one of these always has the value $+1$ as perturbations 'along' the periodic solution (i.e. in the phase) do not decay or grow on average over the period. The remaining two multipliers are either both real or form a complex pair. While the period-1 solution is stable, the magnitude of these two multipliers is such that they remain within the unit circle. If the multipliers pass out of the unit circle, the period-1 solution loses stability. We have seen that a single Floquet multiplier leaving the unit circle through -1 characterizes a period-doubling bifurcation. If the multipliers are a complex pair, they must leave simultaneously at $\alpha \pm i\beta$ say. This exiting of the unit circle as a complex pair characterizes the bifurcation to a torus. The second fundamental frequency of the new solution will be related to the quotient α/β: the mode-locked state may emerge directly, for instance if the Floquet multiplier happens to coincide with an nth root of unity.

Should mode locking occur at any stage, the form of the Poincaré section will change. As the trajectory in the phase plane follows a fixed one-dimensional path on the torus, the intersections with the Poincaré plane eventually repeat themselves. Only a finite number of discrete points appear in the plane, If v_1 is the frequency of rotation transverse to the Poincaré plane and v_2 is that in the plane, and if $v_1/v_2 = p/q$, there will be p invariant points in the Poincaré section (assuming p and q have no common factor). For every stable mode-locked trajectory on the surface of the torus there must be a corresponding unstable (saddle-like) solution, i.e. that emerges from saddle–node bifurcations in the Poincaré plane.

If we have a mode-locked trajectory, its evolution in time has a corresponding sequence or mapping of the intersections described above in the Poincaré plane. There are two more useful characteristic quantities that can be associated with this motion: the return angle θ_n and the winding number W. By selecting some arbitrary zero angle on the Poincaré plane, the angle θ_n made by successive returns can be specified with respect to this. If we then plot the $(n + 1)$th angle against the nth we obtain the 'circle map' $\theta_{n+1}(\theta_n)$. This has a similar interpretation and significance to the next-amplitude maps mentioned earlier, and to the mappings of Chapter 2 in general.

A typical circle map is shown in Fig. 4.27. The ordinate and abscissa run from $-\pi$ to $+\pi$ and there is an apparent discontinuity in the mapping curve that arises merely from our arbitrary choice of reference angle. Again, intersections of the mapping locus with the identity line $\theta_n = \theta_{n+1}$ are equivalent to invariant solutions that may be stable or unstable depending on the gradient of the locus through such points. The exact position of the mapping locus with respect to the identity line will depend on the various experimental or model parameter values.

Another typical and interesting situation arises when the mapping locus

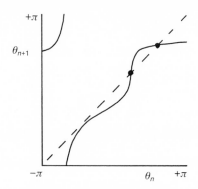

Fig. 4.27. A circle map formed by plotting angles (with respect to some arbitrary reference direction) described by successive returns to the Poincaré section: intersections with the identity line indicate invariant points (frequency-lock responses) that are stable or unstable if the magnitude of the gradient of the mapping curve is less than or greater than unity respectively. Pairs of such points arise through 'tangent bifurcations' such as that almost formed in the lower corner; before such a bifurcation there is a narrow channel in the mapping through which the system evolves very slowly.

approaches the identity line, preparing for a tangential touching, as shown in the figure. For parameter values just above those at which the intersection occurs, there will be a narrow 'channel' in the circle map. Iterating the map will give rise to a 'slow motion' through this channel as indicated. The closer the two curves comes to tangency, the more iterative steps are needed to negotiate this channel. At tangency a pair of invariant points will be born: one stable (with |gradient| < 1) and one unstable (with |gradient| > 1). This is known as a tangent bifurcation and we will see its importance to mode locking in later chapters.

The circle map also leads to the winding number W. If we follow the movement of the return angle θ around the invariant circle in the Poincaré section, it should be possible to obtain a mean rate of rotation per iteration. (We can only measure return angles modulo 2π, so θ, $2\pi + \theta$, $4\pi + \theta$, and indeed $2n\pi + \theta$ are the same for any n.) This 'angular velocity' is then related to the winding number (formally we must slightly modify the circle map so that the apparent discontinuity is removed (e.g. by adding 2π to the mapping function for those iterates $\theta_n > \theta_0$ in Fig. 4.27) and then we find the limit $W = \lim(f^n/n)/2\pi$ as $n \to \infty$, where f^n is the nth iterate of the (new) map).

Any mode-locked (synchronized) state will have a rational winding number: quasiperiodic states have irrational values for W. We can show how the qualitative behaviour of a given system changes with a given parameter by following the change in winding number. This may again give rise to the Devil's staircase: each tread on this staircase corresponds to a range of the parameter for which a particular mode-locked state holds sway, and steps

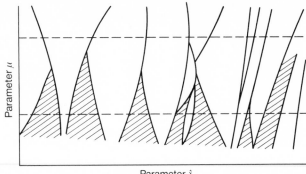

FIG. 4.28. Schematic representation of an 'excitation diagram' showing regions of the λ–μ parameter plane corresponding to different forms of frequency locking etc. and (shaded) areas of quasiperiodicity. Two horizontal cuts are marked corresponding to variations of the parameter λ for two different values of μ: for the upper traverse the system is always in an Arnol'd tongue and so moves directly from one frequency-locked state to another; for the lower path, the system emerges from the frequency locking states into quasiperiodicity at various stages.

up and down correspond to bifurcations at which the mode locking changes to a different synchronization. The significance of such a response is that particular mode lockings have a 'structural stability'. Thus an evolution that sees the trajectory make one circuit of the torus in one plane and exactly three in the perpendicular plane (so $v_1/v_2 = W = \frac{1}{3}$) occurs for a range of parameter values not just for one specific value. Without this structural stability, and given finitely imperfect control over experimental parameters in real systems, mode locking would not be observable.

The structural stability of mode locking can be represented with respect to a second parameter simultaneously using a parameter plane or excitation diagram. Now the treads of mode-locked states from the Devil's staircase become areas or regions in the two-parameter plane, as shown in Fig. 4.28. Each region is known as a 'resonance horn' or Arnol'd tongue. For any parameter combination within a given horn, the same degree of mode locking is obtained. Different horns may intersect or even overlap, but in general they lie separated in a background corresponding to irrational winding numbers and quasiperiodicity. The Devil's staircase can now be seen to correspond to a particular traverse across this excitation diagram. If the path remains completely within resonance horns, the staircase will be complete: if the parameter variation takes us through a region or regions of quasiperiodicity, it will have incompleteness.

4.10. Homoclinicity and heteroclinicity

In order to study qualitative features of the transitions to turbulence discussed in the previous section, Lorenz (1963, 1964) produced a severely truncated set of ordinary differential equations from the (partial differential) Navier–Stokes equations of motion appropriate to fluids. (The particular interest, in fact, was to obtain a simplified model appropriate to the atmosphere and meteorology). The story of Lorenz's discovery of chaotic behaviour has been reported elsewhere (Gleick 1988; Stewart 1989) and the Lorenz system has been widely and comprehensively studied (Sparrow 1982). Various aspects, such as the derivation of the equations, can be found clearly expostulated (see, for instance, Bergé *et al.* 1984, pp. 301–12; Thompson and Stewart 1986, pp. 212–35) and the fascinating structure underlying the evolution is now well understood.

4.10.1. *The Lorenz equations*

The Lorenz model is a three-variable system whose equations can be written in the form

$$ds/dd = (Pr)(y - x) \tag{4.45}$$

$$dy/dt = -xz + rx - y \tag{4.46}$$

$$dz/dt = xy - bz. \tag{4.47}$$

The variables x, y, and z are related to the velocity field and the temperature excess, (Pr) is the Prandtl number, r is related to the Rayleigh number, and b is a parameter that is determined by the size of the Rayleigh–Bénard cells. It is sometimes stressed that equations for the form of (4.45)–(4.47) do not represent 'acceptable chemistry' because of terms such as $-xz$ in dy/dt, which suggest a removal process not involving the species being removed (here y): this, in turn, allows the variables to become negative. We return to this point later.

Much of the analysis of the Lorenz equations has taken the particular parameter values $(Pr) = 10$ and $b = \frac{8}{3}$, varying r and the bifurcation parameter. For $0 \leqslant r \leqslant 1$, there is only one stationary state and that has $x = y = z = 0$, the origin in the phase space. This state is stable. As r passes through 1, there is a pitchfork bifurcation as shown in Fig. 4.29. Two non-zero states emerge with $x_{ss} = y_{ss} = \pm[(r - 1)b]^{1/2}$, $z = r - 1$, and correspond to convecting states with opposite senses of rotation. These are symmetrically placed in the phase space.

The zero state is now unstable: it has two negative eigenvalues and one positive one. The eigenvectors associated with the eigenvalues then create a situation whereby there is an attracting two-dimensional plane approaching

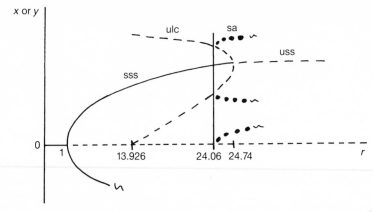

FIG. 4.29. Representation of the bifurcation diagram for the Lorenz model. The base state $x_{ss} = y_{ss} = 0$ is unique and stable for $0 < r < 1$, but a pitchfork bifurcation produces two non-zero states that frow for $r > 1$ for which the zero state becomes unstable: the subsequent behaviour is symmetrical about the $x, y = 0$ axis. At $r = 13.926$ a homoclinic bifurcation forms (two) unstable limit cycles from the zero (saddle) branch: each of these surrounds one of the stable non-zero solutions. At $r = 24.06$ a second global bifurcation occurs between the saddle solution and the unstable limit cycles: this sheds a chaotic attractor which grows as r increases and coexists with the two stable stationary states. On the chaotic attractor, the trajectory constantly crosses the $x, y = 0$ plane. At $r = 24.74$, the two non-zero stationary states undergo subcritical Hopf bifurcations with the unstable limit cycle. For $r > 24.74$ there are no stable states other than the strange attractor, although this undergoes a complex sequence of further bifurcations as r increases to large values.

the origin (an attracting 'centre manifold') and a one-dimensional outset corresponding to the positive eigenvalue. The origin thus has a saddle point character and the attracting plane or inset separates the three-dimensional phase space into two separate 'basins of attraction'—one for each of the two, stable, non-zero stationary states.

Local stability analyses can take us one stage further. The non-zero states undergo a Hopf bifurcation (simultaneously) at $r = 24.74$. These are subcritical bifurcations and an unstable period-1 limit cycle emerges and grows as r decreases. For $r > 24.74$, the system has three locally unstable stationary states. To understand the remaining bifurcations we need to examine non-local features of the phase space (Kaplan and Yorke 1979; Afraimovich *et al.* 1977).

The first important global bifurcation occurs at $r = 13.926$. Homoclinic orbits for two-variable systems have been discussed in Section 3.9, but these structures also occur with higher dimensions. As r is increased through 13.926, so the inset and outset of the saddle point at the origin close up to form a homoclinic loop, which then detaches itself to become a limit cycle.

In fact, because of the symmetry of the Lorentz model, a pair of loops and hence of limit cycles are formed simultaneously—one on each side of the inset plane. The limit cycles born at this homoclinic bifurcation are both unstable: and these ultimately are the cycles that undergo the Hopf bifurcations at $r = 24.74$. For two-variable systems, the unstable limit cycles describe the basin of attraction of the stationary state they surround. In a three-variable system, however, the flow is not confined to a plane and so trajectories can go round these unstable cycles and still approach the respective stable non-zero states. In fact, for $r > 13.926$, the (three-dimensional) basins of attraction of the two stable points become very complex. The separating inset of the saddle at the origin is no longer a simple sheet, but now one curiously interleaved in itself (Abraham and Shaw 1985): two trajectories with very close starting points can be attracted to different final states and the transient behaviour in each case shows much winding about both points. Once a trajectory has come sufficiently close to one of the non-zero states it will be 'captured' and spiral in to the appropriate stable focus.

For $13.926 < r < 24.06$, the system can show chaotic transients as described above. A further non-local bifurcation occurs at the upper end of this range. Now the outset from the saddle point in each directon can form a 'heteroclinic orbit' with each of the unstable limit cycles: in other words, the outset no longer is asymptotic to the stable non-zero stationary states, but for this particular parameter value falls on to the limit cycle. The breaking up (perturbation) of this heteroclinic structure probably gives rise to a strange attractor and motion of this attractor is chaotic. For $24.06 < r < 24.74$, the system has stable chaos and two stable stationary states coexisting. After the Hopf bifurcation at $r = 24.74$, the stationary states lose stability and the only attractor is the chaotic one.

There is yet more versatility in this scheme—even if we stay restricted to the particular (Pr) and b—at larger values of r. Here, all the stationary states remain unstable, but the strange attractor undergoes various bifurcations. Many of these involve inverse period doubling or subharmonic cascades which yield a periodic window: windows of various periodicities have been observed (Sparrow 1982). Ultimately a stable period-1 limit cycle persists for highest r $(r > 313)$.

4.10.2. Homoclinicity in the Shilnikov scenario

A second and general example of a route to chaos involving homoclinic structures is that due to Shilnikov (1965, 1967, 1970, 1976) and which has been discussed in a more chemical context by Gaspard and Nicolis (1983; Gaspard *et al.* 1984).

Here we will be interested in a given stationary-state solution, perhaps unique, in a three-dimensional (or higher) system. This solution should be of a saddle-focus character. Such a state may occur, for instance, after a

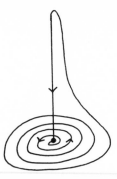

Fig. 4.30. Representation of a Shilnikov homoclinicity to a saddle focus stationary state. In this example the unstable focal plane forms a slow manifold on which the system unwinds: as the trajectory lifts off this plane it is ultimately reinjected along the strongly attracting stable manifold to the stationary point. When this structure is perturbed, a strange attractor may form with the initial point for each slow unwinding depending sensitively on the rapid reinjection process. This example corresponds to the eigenvalues being such that the (negative) real eigenvalue has much greater magnitude than the (positive) real parts of the complex pair: reversing the arrows gives the homoclinic structure appropriate to a weakly repelling (positive) real eigenvalue and strongly attracting complex focal pair with large (negative) real parts..

stable solution has undergone a Hopf bifurcation. In such a case the three eigenvalues would be arranged thus: the outset from the stationary state corresponding to the unstable focal character typical after a Hopf bifurcation will be represented by a complex pair of eigenvalues with positive real part $\lambda_\pm = \mathrm{Re}_\pm \pm \mathrm{Im}_\pm \mathrm{i}$, whilst the inset will have a real, negative eigenvalue λ_3. For the Shilnikov condition we are interested in situations for which the sttraction is relatively strong, so the following inequalities amongst the real parts hold:

$$-\mathrm{Re}_3 \gg \mathrm{Re}_\pm > 0. \qquad (4.48)$$

This suggests a rapid injection of the flow into the plane of the outset and a subsequent relatively slow winding out in that plane.

Figure 4.30 shows the special case of homoclinicity for a saddle focus satisfying the Shilnikov condition (4.48). Now the outset unwinds locally but away from the stationary state is reinjected along the inset. A similar structure with the arrows corresponding to the direction of motion reversed is also of interest: now the inset lies in the focal plane and we must have a rapid winding-in with the ejection along the slow, one-dimensional outset (and hence we require $-\mathrm{Re}_\pm \gg \mathrm{Re}_3 > 0$).

The simplest form for a set of equations that can satisfy this scenario is the three-variable system

$$dx/dt = \rho x - \omega y + P(x, y, z) \tag{4.49}$$

$$dy/dt = \omega x + \rho y + Q(x, y, z) \tag{4.50}$$

$$dz/dt = \lambda z + R(x, y, z) \tag{4.51}$$

where P, Q, and R are arbitrary functions of the variables that must each be zero and have zero derivative for the stationary state $(x, y, z) = (0, 0, 0)$, i.e. all terms must be quadratic or higher. The inequalities in (4.48) are equivalent to $\lambda > -\rho > 0$ (or $-\lambda > \rho > 0$). Gaspard and Nicolis have shown that the Rössler model, with suitable parameter combinations, can also show this behaviour and they also present a number of 'mass-action' schemes.

The homoclinic orbit structure is unstable, in that it will clearly break up under arbitrary small variations in the parameter values. Shilnikov was able to show that, for some parameter values close to those for the homoclinicity, there exist chaotic trajectories which are obtained by perturbing the homoclinic structure.

Gaspard and Wang (1987) have widened the idea of homoclinicity beyond that for a stationary-state solution. Figure 4.31 shows possible homoclinic

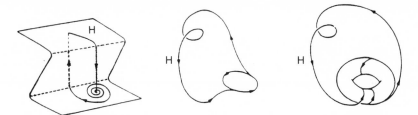

FIG. 4.31. Representations of various homoclinic structures: (a) saddle focus loop formed from a singular point; (b) homoclinicity to a saddle limit cycle; (c) homoclinicity to a saddle torus. (Reprinted with permission from Gaspard and Wang (1987), © Plenum Publishing Corporation.)

structures associated with a saddle focus (as above), a limit cycle, and a torus. In a three-dimensional (or more) system, a limit cycle can develop saddle character, losing stability. With a stable limit cycle all trajectories approach the attractor, arriving along the insets or stable manifolds: there will be $n - 1$ of these for an n-variable system. If one of these manifolds loses stability it becomes an outset, and the limit cycle has saddle character, being approached along all other manifolds as time increases and along the outset as time runs backwards. Figure 4.32 sketches the situation for a saddle limit cycle in three dimensions. The stable and unstable manifolds can intersect away from the limit cycle, forming a homoclinic loop. This is quite a general phenomenon

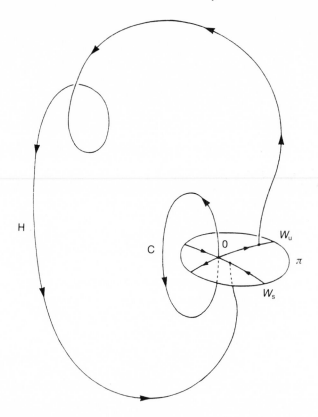

FIG. 4.32. Details of formation of a homoclinic loop H to a saddle limit cycle C involving the connection between the outset (unstable manifold) W_u and the inset (stable manifold) W_s of the limit cycle. The inset and outset lie on the tangent plane π that rotates with the motion along the limit cycle. (Reprinted with permission from Gaspard and Wang (1987), © Plenum Publishing Corporation.)

if the two manifolds cross each other transversely. If, however, they meet tangentially we have a homoclinic tangency—the condition for an important bifurcation in the phase space. In general phase space structures with homoclinic tangencies are unstable objects and break up under 'perturbation'. They may then yield a periodic or aperiodic solution for parameter values 'close to' that for the homoclinic tangency.

We will discuss homoclinic tangency with regard to a particular model for a real chemical system, the oxidation of acetaldehyde, in Chapter 9. It is likely, however, that some homoclinic tangency of the manifolds of a non-stationary state solution underlines the mixed-mode oscillations in the model examined earlier in this chapter (Section 4.8) and also those we will describe later for the Belousov–Zhabotinskii reaction.

4.11. Intermittency

To conclude the list of routes to chaos (of those that are presently understood) we should consider the class categorized as intermittency. A full discussion is given in the book by Bergé *et al.* (1984), where three distinct types are recognized and distinguished according to the changes in the Floquet multipliers at the bifurcation point. Type I intermittency arises when a Floquet multiplier leaves the unit circle along the real axis through $+1$. Type II intermittency involves complex multipliers leaving the unit circle, whilst type III occurs if the multiplier leaves through -1. These latter conditions are also those for bifurcation to a torus (secondary Hopf) and period doubling respectively. Intermittency arises if these are subcritical rather than supercritical events.

With each of the intermittent bifurcations, the previous periodic solution actually disappears. For type I, this has the same form as a saddle–node bifurcation. Close to the bifurcation the system appears to retain some 'memory' of the previous periodic state. The temporal evolution now appears as long sequences of almost periodic behaviour (the laminar phase) inter-rupted by 'turbulent bursts'. Pomeau *et al.* (1981) have reported this form of response for the Belousov–Zhabotinskii reaction. Types II and III differ in slight detail from type I, but all seem more common in fluid dynamical situations rather than those of most direct chemical interest. Type II, in particular, requires rather more variables and hence high-dimensional phase spaces than those we have been following so far.

4.12. Chemical mechanisms from classical models

Many of the most widely known (and generally simplest) model schemes that have been studied in a chaotic context are derived from problems in mechanics, electronics, and other 'non-chemical' contexts. The Lorenz and Rössler schemes discussed above are examples. Often the very features that bring about the simplicity of the equations appear to be those that disqualify them as chemistry—removal terms that do not involve the variable being removed etc. However, Samardzija *et al.* (1989) have devised an ingenious approach through which a chemical mechanism can be found that has the same basic dynamics as a particular set of governing equations. Thus they have shown that the same dynamics that drives a simple harmonic oscillator will also determine the primary behaviour of the following reaction set:

$$A + X + Y \rightarrow 2X + Y$$

$$X \rightarrow P$$

$$X + Y \rightarrow X + P$$

$$B + Y \rightarrow 2Y.$$

The Lorenz and Rössler schemes require slightly more extensive mechanisms, with 13 and nine steps respectively, but in each of these cases, the rate constants for the various steps can also be identified in terms of the 'original' parameters, so all the collected wisdom from the earlier 'non-chemical' studies carried through directly. A suitable 'Lorenz chemical' model is

$$X + Y + Z \rightarrow X + 2Z$$
$$A + X + Y \rightarrow 2X + Y$$
$$B + X + Y \rightarrow X + 2Y$$
$$X + Z \rightarrow X + P$$
$$Y + Z \rightarrow 2Y$$
$$2X \rightarrow P$$
$$2Y \rightarrow P$$
$$2Z \rightarrow P$$
$$C + X \rightarrow 2X$$
$$X \rightarrow Q$$
$$Y \rightarrow Q$$
$$D + Y \rightarrow 2Y$$
$$E + Z \rightarrow 2Y.$$

4.13. Conclusions

Whilst the foregoing gives a relatively detailed introduction to complex oscillations, quasiperiodicity, and chaos relevant to the present context, it is certainly incomplete. There are clearly transition sequences that are still not understood, exotica such as hyperchaos (two positive Lyapounov exponents) or worse are possible, but even before that models can happily combine more than one of the different scenarios given above simultaneously just to make life difficult (but interesting).

References

Abraham, R. H. and Shaw, C. D. (1985). *Dynamics: the geometry of behavior*, Parts 1–3. Aerial Press, Santa Cruz, CA.

Afraimovich, V. S., Bykov, V. V., and Shilnikov, L. P. (1977). On the origin and structure of the Lotenz attractor. *Sov. Phys. Dokl.*, **22**, 253–5.

Aguda, B. D. and Clarke, B. L. (1988). Dynamic elements of chaos in the Willamowski–Rössler network. *J. Chem. Phys.*, **89**, 7428–34.

Balakotaiah, V. and Luss, D. (1982). Exact steady-state multiplicity criteria for two consecutive or parallel reactions in lumped-parameter-systems. *Chem. Eng. Sci.*, **37**, 433–45.

Balakotaiah, V. and Luss, D. (1984). Global analysis of the multiplicity features of multi-reaction lumped-parameter chemically reacting systems. *Chem. Eng. Sci.*, **39**, 865–81.

Behringer, R. P. (1985). Rayleigh–Bénard convention and turbulence in liquid helium. *Rev. Mod. Phys.*, **57**, 657–87.

Bénard, H. (1900). *Rev. Génerale des Sciences Pures et Appliquées*, **12**, 1261, 1309.

Bergé, P., Pomeau, Y., and Vidal, C. (1984). *Order within chaos*. Wiley, New York.

Brindley, J., Kaas-Petersen, C., Merkin, J. H., and Scott, S. K. (1988). Bi-periodic behaviour in the diffusive autocatalator. *Phys. Lett.*, A **128**, 260–5.

Brindley, J., Jivraj, N. A., Merkin, J. H., and Scott, S. K. (1990). *Prc. R. Soc.*, A **430**, 459–77.

Burnell, J. G., Lacey, A. A. and Wake, G. C. (1983). Steady-states of the reaction–diffusion equations, part 1: questions of existence and continuity of solution branches. *J. Aust. Math. Soc.*, **B24**, 374–91.

Celarier, E. A. and Kapral, R. (1987). Bistable limit cycle oscillations in chemical systems. I. Basins of attraction. *J. Chem. Phys.*, **86**, 3357–65.

Crutchfield, J., Farmer, D., Packard, N., Shaw, R., Jones, G., and Donnelly, R. J. (1980). Power spectral analsis of a dynamical system. *Phys. Lett.*, **76A**, 1–4.

Curry, J. M. and Yorke, J. A. (1977). A transition from Hopf bifurcation to chaos: computer experiments with maps in $\mathbf{R}^5$. In *The structure of attractor in dynamical systems*, p. 48. Springer, Berlin.

Doedel, E. J. and Heinemann, R. F. (1983). Numerical computation of periodic solution branches and oscillatory dynamics of the stirred tank reactor with $A \rightarrow B \rightarrow C$ reactions. *Chem. Eng. Sci.*, **38**, 1493–9.

Farr, W. W. and Aris, R. (1986). 'Yet who would have thought the old man to have so much blood in him?'—reflections on the multiplicity of steady states of the stirred tank reactor. *Chem. Eng. Sci.*, **41**, 1384–402.

Fenstermacher, P. R., Swinney, H. L., and Gollub, J. P. (1979). Dynamical instabilities and the transition to chaotic Taylor vortex flow. *J. Fluid Mech.*, **94**, 103–28.

Gaspard, P. and Nicolis, G. (1983). What we can learn from homoclinic orbits in chaotic dynamics? *J. Stat. Phys.*, **31**, 399–518.

Gaspard, P. and Wang, X.-J. (1987). Homoclinic orbits and mixed-mode oscillations in far-from-equilibrium systems. *J. Stat. Phys.*, **48**, 151–99.

Gaspard, P., Kapral, R., and Nicolis, G. (1984). Bifurcation phenomena near homoclinic systems: a two parameter analysis. *J. Stat. Phys.*, **35**, 697–727.

Gleick, J. (1988), *Chaos: making a new science*. Heinemann, London.

Gollub, J. P. (1983). Recent experiments on the transition to turbulent convection. In *Nonlinear dynamics and turbulence*, (ed. G. I. Barenblatt, G. Iooss, and D. Joseph), p. 156–71. Pitman, Boston.

Gollub, J. P. and Benson, S. V. (1979). Phase-locking in the oscillations leading to turbulence. In *Pattern formation*, (ed. H. Haken), p. 74. Springer, Verlag.

Gollub, J. P. and Benson, S. V. (1980). Many routes to turbulent convection. *J. Fluid Mech.*, **100**, 499.

Gollub, J. P. and Swinney, H. L. (1975). Onset of turbulence in a rotating fluid. *Phys. Rev. Lett.*, **35**, 927–30.

Guckenheimer, J. and Holmes, P. (1983). *Nonlinear oscillations and bifurcations of vector fields.* Springer, New York.

Halbe, D. C. and Poore, A. B. (1981). Dynamics of the continuous stirred tank reactor with reactions A → B → C. *Chem. Eng. J.*, **21**, 241–53.

Herschkowitz-Kaufman, M. and Nicolis, G. (1972). Localized spatial structures and non-linear chemical waves in dissipative systems. *J. Chem. Phys.*, **56**, 1890.

Hlavacek, V., Kubicek, M., and Visnak, K. (1972). Modelling of chemical reactors—XXVI. *Chem. Eng. Sci.*, **27**, 719–42.

Hudson, J. L. and Rössler, O. E. (1984). Chaos in simple three- and four-variable chemical systems. In *Modelling of patterns in space and time*, (ed. W. Jäger and J. D. Murray). Springer, Berlin.

Hudson, J. L., Rössler, O. E., and Killory, H. C. (1988). Chaos in a four variable piecewise-linear system of differential equations. *IEEE Trans. Circ. Syst.*, **35**, 902–8.

Jorgensen, D. V. and Aris, R. (1983). On the dynamics of a stirred tank with consecutive reactions. *Chem. Eng. Sci.*, **38**, 45–53.

Jorgensen, D. V., Farr, W. W., and Aris, R. (1984). More on the dynamics of the stirred tank with consecutive reactions. *Chem. Eng. Sci.*, **39**, 1741–52.

Kahlert, C., Rössler, O. E., and Varma, A. (1981). Chaos in a continuous stirred tank reactor with two consecutive first-order reactions, one exo-, one endothermic. In *Modelling of chemical reaction systems*, (ed. K. H. Ebert, P. Deuflhard, and W. Jäger). Springer, New York.

Kaplan, J. L.and Yorke, J. A. (1979). Preturbulence, a regime observed in a fluid flow model of Lorenz. *Commun. Math. Phys.*, **67**, 93–108.

Killory, H. C., Rössler, O. E., and Hudson, J. L. (1987). Higher chaos in a four-variable chemical reaction model. *Phys. Lett.*, A **122**, 341–5.

Kuramoto, Y. (1978). Diffusion-induced chaos in reaction systems. *Suppl. Prog. Theor. Phys.*, **64**, 346–67.

Landau, L. D. and Lifshitz, E. M. (1959). *Fluid mechanics.* Pergamon, Oxford.

Langford, W. F. (1983). A review of interactions of Hopf and steady-state bifurcations. In *Nonlinear dynamics and turbulence*, (ed. G. I. Barenblatt, G. Iooss, and D. Joseph), pp. 215–37. Pitman, Boston.

Libchaber, A. and Maurer, J. (1982). A Rayleigh–Bernard experiment: Helium in a small box. In *Nonlinear phenomena at phase transitions and instabilities*, (ed. T. Riste), pp. 259–96. Plenum, New York.

Lorenz, E. N. (1973). Deterministic nonperiodic flow. *J. Atmos. Sci.*, **20**, 130–41.

Lorenz, E. N. (1964). The problem of deducing the climate from the governing equations. *Tellus*, **16**, 1–11.

Luss, D. and Lee, J. C. M. (1970). The effects of Lewis number on the stability of a catalytic reaction. *AIChE J.*, **4**, 620–5.

Lynch, D. T., Rogers, T. D., and Wanke, S. E. (1982). Chaos in a continuous stirred tank reactor. *Math. Modelling*, **3**, 103–16.

McGarry, J. K. and Scott, S. K. (1990). Period-doubling and chaos in a catalyst pellet. *Phys. Lett.*, **A143**, 23–6.

Nandapurkar, P. J., Hlavacek, V., and van Rompay, P. (1984). Complex oscillations over an isothermal catalyst pellet. *Chem. Eng. Sci.*, **39**, 1151–6.

Nandapurkar, P. J., Hlavacek, V., and van Rompay, P. (1986). Chaotic behaviour in a diffusion-reaction system. *Chem. Eng. Sci.*, **41**, 2747–60.

Newhouse, S., Ruelle, D., and Takens, F. (1987). Occurrence of strange axiom A attractors near quasiperiodic flows on T^m, $m \geqslant 3$. *Commun. Math. Phys.*, **64**, 35–40.

Peng, B., Scott, S. K., and Showalter, K. (1990). Period doubling and chaos in a three-variable autocatalator. *J. Phys. Chem.*, **94**, 5243–6.

Pismen, L. M. (1980). Kinetic instabilities in man-made and natural reactors. *Chem. Eng. Sci.*, **35**, 1950–78.

Pismen, L. M. (1984a). Dynamics of lumped chemically reacting systems near singular bifurcation points. *Chem. Eng. Sci.*, **39**, 1063–77.

Pismen, L. M. (1984b). Dynamics of lumped chemically reacting systems near singular bifurcation points—II. Almost Hamiltonian dynamics. *Chem. Eng. Sci.*, **40**, 905–16.

Planeaux, J. B. and Jensen, K. F. (1986). Bifurcation phenomena in CSTR dynamics: a system with extraneous thermal capacitance. *Chem. Eng. Sci.*, **41**, 1497–523.

Pomeau, Y., Roux, Rossi, A., Bachelart, S., and Vidal, C. (1981). Intermittent behaviour in the Belousov–Zhabotinsky reaction. *J. Phys. Lett.*, **42**, L271–3.

Rayleigh, Lord (1916). On convective currents in a horizontal layer of fluid when the higher temperature is on the under side. *Philos. Mag.*, **32**, 52–66.

Rössler, O. E. (1976a). An equation for continuous chaos. *Phys. Lett.*, **57A**, 397–8.

Rössler, O. E. (1976b). Chaotic behavior in simple reaction systems. *Z. Naturforsch.*, **31a**, 259–64.

Ruelle, D. (1980). Strange attractors. *Math. Intell.*, **2**, 126–37.

Ruelle, D. and Takens, F. (1971). On the nature of turbulence. *Commun. Math. Phys.*, **20**, 167–92.

Samardzija, N., Greller, L. D., and Wasserman, E. (1989). Nonlinear chemical kinetic schemes derived from mechanical and electrical dynamical systems. *J. Chem. Phys.*, **90**, 2296–304.

Scott, S. K. and Tomlin, A. S. (1990a). Period doubling and other complex bifurcations in non-isothermal chemical systems. *Phil. Trans. R. Soc.*, **A332**, 51–68.

Scott, S. K. and Tomlin, A. S. (1990b). Mixed-mode oscillations in the non-isothermal autocatalator. In (ed. R. Aris, H. L. Swinney). *Patterns in chemical reactors*. Institute for Mathematics and its Applications, University of Minnesota, Minneapolis, USA.

Shilnikov, L. P. (1965). A case of the existence of a denumerable set of periodic motions. *Sov. Math. Dokl.*, **6**, 163–6.

Shilnikov, L. P. (1967). A case of the existence of a denumerable set of periodic motions in four-dimensional space in the extended neighborhood of a saddle-focus. *Sov. Math. Dokl.*, **8**, 54–8.

Shilnikov, L. P. (1970). A contribution to the problem of the structure of an extended neighborhood of a rough equilibrium state of the saddle-focus type. *Math. USSR Sb.*, **10**, 91–102.

Shilnikov, L. P. (1976). Theory of the bifurcation of dynamical systems and dangerous boundaries. *Sov. Phys. Dokl.*, **20**, 674–6.

Smith, C. B., Kuszta, B., Lyberatos, G., and Bailey, J. E. (1983). Period doubling and complex dynamics in an isothermal chemical reaction system. *Chem. Eng. Sci.*, **38**, 425–30.

Sparrow, C. (1982). *The Lorenz equations: bifurcations, chaos and strange attractors.* Springer, New York.

Stewart, I. (1989). *Does God play dice? The mathematics of chaos.* Blackwell, Oxford.

Swinney, H. L. and Gollub, J. P. (1978). The transition to turbulence. *Phys. Today,* **31**, 41–9.

Taylor, G. I. (1923). Stability of a viscous liquid contained between two rotating cylinders. *Philos. Trans. R. Soc.,* **A223**, 289–343.

Thompson, J. M. T. and Stewart, H. B. (1986). *Nonlinear dynamics and chaos.* Wiley, Chichester.

Tomlin, A. S. (1990). Bifurcation analysis for non-linear chemical kinetics. *Ph.D. Thesis,* University of Leeds.

Turing, A. M. (1952). The chemical basis for morphogenesis. *Phil. Trans. R. Soc.,* **B237**, 37–72.

Willamowski, K. D. and Rössler, O. E. (1980). Irregular oscillations in a realistic quadratic mass action system. *Z. Naturforsch.,* **35a**, 317–18.

5

Forced systems

In the previous chapters we have considered models that have distinct variables, such as concentrations or local temperature, and parameters. The parameters are under our control and stay fixed when we want them to—only varying should we wish so. Thus the inflow concentrations or ambient temperature are fixed in time. In this chapter we investigate the types of behaviour possible when one or more of the parameters is varied in time (or does so of its own volition in spite of our wishes). Specifically we will be interested in the case where this time variation is periodic and in situations where the system without this additional periodicity can also oscillate or show multistability. Again, this chapter will deal mainly with models and theory: practical aspects and real experimental systems come where appropriate in Chapters 8 to 12.

Some preliminary comments can be made. The idea of forcing a chemical system is not just of academic interest. There has been much interest in the possibility of improving on stationary-state process operation in industrial settings. Substantial improvements in overall yield have been achieved in the laboratory or pilot scale and more have been predicted in modelling studies; others have examined the possibility of increased selectivity amongst a range of possible reaction products and pathways. Reviews have been given by Bailey (1973, 1977), Schneider (1985), Kevrekidis *et al.* (1986*b*), Aris (1990). Forcing has become a popular 'game' amongst workers with heterogeneous catalysts in particular: regular perturbations can be imposed on several of the experimental constraints—this may provide additional information and thus help to discriminate between different possible mechanistic interpretations.

With low-amplitude forcing, the original system feels only a minor perturbing influence. If the period of the forcing is close to the natural frequency of the system itself, or to some simple multiple or fraction, we may expect some form of synchronization. Otherwise, the two frequencies can be expected to combine to produce quasiperiodic behaviour rather naturally (Kuramoto 1984). At the other extreme, large-amplitude forcing will eventually dominate sufficiently to bring about full entrainment, relegating the natural non-linear behaviour to the level of finer detail.

For the intermediate, and most interesting, cases the interactions between the natural and imposed time dependences will differ quantitatively from model to model (or real system to real system) but many 'universal' qualitative features can also be identified.

5.1. Governing equations for forced systems: the Brusselator

We are primarily interested in the interaction of an applied forcing with natural non-linearities. Such features can be readily investigated in a simple way through the various model schemes discussed in earlier chapters. Chapter 3 presented a number of two-variable schemes that display autonomous oscillations or multistability, and many of these have also been studied in a forced scenario. Practically, time-dependent periodic forcing can arise in many was, whether deliberate or inevitable. There has been much discussion with regard to solution-phase oscillatory reactions (see Chapter 8) concerning the use of peristaltic pumps that by their very mode of action impose a rhythmic perturbation to the inflow. Also in the vein of experimental control, many temperature controllers operate in such a way that they execute continuous (but generally small) overshoots and undershoots of the target temperature.

A particularly clean form of forcing is the deliberate addition of a sinusoidal modulation of inflow pumping speed or concentration that can be made under computer control. Experimentally, we may then vary parameters such as the amplitude and forcing frequency.

One of the earliest studies followed the behaviour of the Brusselator model (Section 3.12.1). Tomita and Kai (1979; Kai and Tomita 1979; Tomita 1982) modified the autonomous model equations (3.54)–(3.55), imagining that the concentration of the pool chemical A executed a time-dependent form

$$A = \langle A \rangle [1 + \alpha \cos(\omega_f t)] \tag{5.1}$$

where $\langle A \rangle$ is the mean concentration, α the forcing amplitude, and ω_f the forcing frequency. The forced non-autonomous equations for this version of the scheme are thus

$$\frac{dx}{dt} = \langle A \rangle [1 + \alpha \cos(\omega_f t)] - Bx + x^2 y - x \tag{5.2}$$

$$\frac{dy}{dt} = Bx - x^2 y. \tag{5.3}$$

Tomita and Kai considered in some detail the particular parameter values $\langle A \rangle = 0.4$, $B = 1.2$ for which the unforced system has a stable limit cycle (see condition (3.59)) with natural frequency $\omega_0 = 0.3776$ (period = 16.64) and amplitude $A_x = 0.242$, $A_y = 0.658$. They produced an 'excitation diagram' (Fig. 5.1) that summarizes the range of different responses shown by this model as a function of the forcing amplitude and frequency. Note that

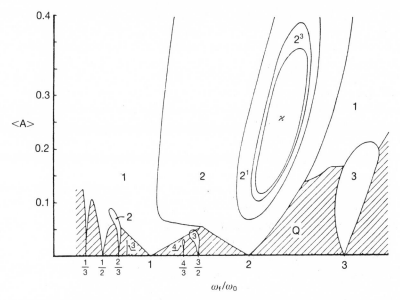

F_IG. 5.1. The excitation diagram for the Brusselator model with parameter values $A = 0.4$ and $B = 1.2$ showing regions of quasiperiodicity (shaded), entrainment, and chaos (denoted by χ). (Reprinted with permission from Tomita (1986), © Manchester University Press.)

the abscissa has the forcing frequency scaled by the natural frequency and the amplitude is naturally scaled by the parameter $\langle A \rangle$.

The shaded region in Fig. 5.1 corresponds to combinations of the forcing parameters such that the system displays quasiperiodic motion. Amongst this sea of quasiperiodicity, however, are embedded a number of islands of periodic evolution. In fact there are strictly an infinite number of these 'Arnol'd tongues' or 'resonance horns', one emerging from each rational number along the abscissa. When the forcing and natural frequencies are rationally related, the system will become phase locked. Only the most important tongues are shown in the figure and these are those corresponding to some of the simplest (or 'nicest') rational numbers. Thus, there is a closed tongue emerging from $\omega_f/\omega_0 = 3$ and within which there is a 3:1 phase-locked oscillation. (The full repeating unit requires three forcing periods.) Importantly, the tongues open out as the forcing amplitude is increased, so phase locking of a given type can occur across a range of forcing frequency not just for the appropriate exact ratio. We will return to this later.

The other most significant feature in Fig. 5.1 is the set of nested closed loops approximately above the 2:1 resonance horn. These correspond to solutions of 2^n:1 phase locking, with n increasing with the degree of nesting. If the forcing amplitude or frequency is varied so as to traverse this region, the system undergoes a period-doubling sequence. This has all the universal features described earlier for maps (Chapter 2) and autonomous flows (Chapter 4), including a geometric convergence reproducing the Feigenbaum constant. At the centre of the nesting is a region of chaotic evolution, interrupted by windows of periodicity.

Additional work on this scheme is given by Aronson *et al.* (1986) who computed the various boundaries in the excitation diagram directly and by Schreiber *et al.* (1988) who used a different form of forcing.

5.2. Other forced model schemes

The qualitative form of the excitation diagram in Fig. 5.1 is reproduced by many other models for which the unforced autonomous system has a unique unstable stationary state surrounded by a stable limit cycle. Hence, the cubic autocatalator scheme of Section 3.1 can be forced in a similar way to that described above through the parameter μ, again corresponding to a periodic modulation of the concentration of a pool chemical reactant (Daido and Tomita 1979*a, b*; Tomita and Daido 1980) or, perhaps, of the rate constant k_0 through a temperature variation. Temperature and concentration forcing can also be envisaged for the Salnikov scheme of Section 3.5. In both cases the excitation diagrams have the same basic features as described for the Brusselator.

Slightly more sophisticated is the forcing of the flow reactor models (cubic autocatalysis, Section 3.7, and the exothermic FONI system, Section 3.1.22) and the Takoudis–Schmidt–Aris model (Section 3.12.4). These exhibit multiple stationary states as well as sustained oscillations. In the case of the flow systems, a convenient form of the forcing operates on the flow rate, whilst the non-isothermal studies have frequently considered forcing the coolant temperature in a reactor.

As an example of this latter case, a series of papers beginning with Sincic and Bailey (1977) and then developed by Mankin and Hudson (1984) and Kevrekedis *et al.* (1984, 1986*a, b, c*) considers a single exothermic reaction in a CSTR. The appropriate equations are slightly different from those given in Section 3.12.2 in that they allow for the inflow temperature and the cooling temperature to be unequal. Using the same variables and parameters as in Section 3.12.2 (but different from those of the above authors) the autonomous equations for the unforced system are

$$\frac{d\alpha}{d\tau} = \frac{(1 - \alpha)}{\tau_{res}} - \alpha \exp[\theta/(1 + \varepsilon\theta)] \tag{5.4}$$

$$\frac{d\theta}{d\tau} = B\alpha \exp[\theta/(1 + \varepsilon\theta)] - \frac{\theta}{\tau_{res}} - \frac{(\theta - \theta_c)}{\tau_N}. \tag{5.5}$$

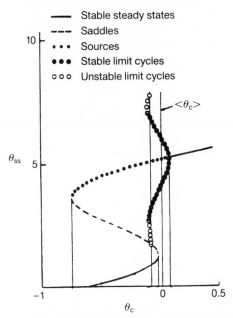

FIG. 5.2. The autonomous bifurcation diagram for an exothermic CSTR showing stationary-state temperature excess as a function of the cooling temperature. (Reprinted with permission from Kevrekidis *et al.* (1986*a*), © Pergamon Journals.)

Here, θ_c represents the dimensionless temperature difference between the cooling temperature T_c and the inflow temperature T_a: $\theta_c = (T_c - T_a)(E/RT_a^2)$. (Previously we have assumed $T_c = T_a$.) Much of the forcing work has further employed the exponential approximation to the Arrhenius temperature dependence, letting $\varepsilon = 0$, although Mankin and Hudson have also investigated the effects of the full form.

Sincic and Bailey employed so-called 'big bang' forcing, with a square-wave variation of the cooling temperature parameter θ_c. Similar results are obtained through sinusoidal forcing on which we will concentrate here. Thus both of the later groups substituted the non-autonomous form $\theta_c = \langle \theta_c \rangle + q \cos(\omega_f \tau)$ into eqn (5.5)—notice the slightly different form from that given above, allowing forcing of the cooling temperature excess about a mean value of zero. Mankin and Hudson considered three sets of parameters, one of which has subsequently been exploited in some detail by Kevrekidis *et al.*: typically $\tau_{res} = 0.085$, $B = 22$ (or 24.4 in one case), $\varepsilon = 0$ (0.02 in one case), and $\tau_N = 0.0283$ (0.0567 in one case). If the cooling temperature θ_c is treated as a bifurcation parameter in the unforced system, the response is as shown in Fig. 5.2. The system displays a hysteresis loop and hence multistability over the range $-0.749\,635 \leqslant \theta_c \leqslant -0.010\,088$ (note that negative values for this temperature difference are entirely acceptable physically). Along the upper branch is a supercritical Hopf bifurcation at $\theta_c = 0.063\,036$, leading to the creation and growth of a stable limit cycle as θ_c is decreased. The stable limit cycle disappears by coalescence with an unstable cycle born via a homoclinic orbit bifurcation at $\theta_c = -0.051\,245\,46$ (Kevrekidis *et al.* 1986*a*).

For the main case of interest the various authors above took $\langle\theta_c\rangle = 0$. Thus the autonomous system has a unique and unstable stationary state surrounded by a stable limit cycle and exhibits sustained oscillations (Fig. 5.3(a)) with period $= 0.093\,075$ ($\omega_0 = 67.5$). Typical non-autonomous behaviour for such systems is found with a forcing frequency slightly higher than this natural value, e.g. $\omega_f = 1.1\omega_0 = 74.25$ (Mankin and Hudson 1984, case I).

With the lowest, non-zero forcing amplitudes ($0 < q < 0.0355$), the autonomous and forcing periodicities interact only to yield a quasiperiodic solution. An example for $q = 0.025$ is shown in Fig. 5.3(b). The waveform consists of an oscillation with basic period close to that of the autonomous system but whose amplitude is modulated with an incommensurate period. The Fourier power spectrum of such a trace shows a classic form based on two fundamental frequencies.

With increasing amplitude q, the evolution becomes entrained with (or phase locked to) the forcing. Figure 5.3(c) shows an example for $q = 0.15$. After an initial quasiperiodic transient, the system settles to a period-1 response with a frequency now equal to the forcing frequency. Such behaviour occurs across the parameter range $0.0355 < q < 0.176$.

With higher forcing amplitudes, the system changes from simple entrainment and undergoes a period-doubling sequence: period-2 and period-4 responses for $q = 0.185$ and 0.21 respectively are shown in Fig. 5.3(d) and (e). By $q = 0.212$, the cascade has converged to yield a chaotic solution that survives over a range of forcing amplitude—Fig. 5.3(f) for $q = 0.215$.

In addition to the behaviour just described, this system and set of parameter values displays yet more versatility. For the upper range of the forcing amplitude discussed, there is a stable period-10 solution (Fig. 5.3(g)) coexisting with the period-doubling state. The coexisting attractors for $q = 0.215$ (for which the original solution has a strange attractor) are shown in Fig. 5.4. The period-10 solution has a much larger amplitude, with θ attaining a maximum value of about 17 during the oscillation. This exemplifies how a chemical system can act as a non-linear amplifier: here

FIG. 5.3. Temperature evolution for the forced non-isothermal CSTR: (a) unforced, autonomous oscillation; (b) low-amplitude forcing $q = 0.025$ leading to quasiperiodicity; (c) $q = 0.15$ 1:1 phase locking between response and forcing period; (d) $q = 0.185$, period-doubled solution; (e) $q = 0.21$, period-4; (f) $q = 0.215$, chaos; (g) $q = 0.215$, period-10 oscillation coexisting with chaotic response.

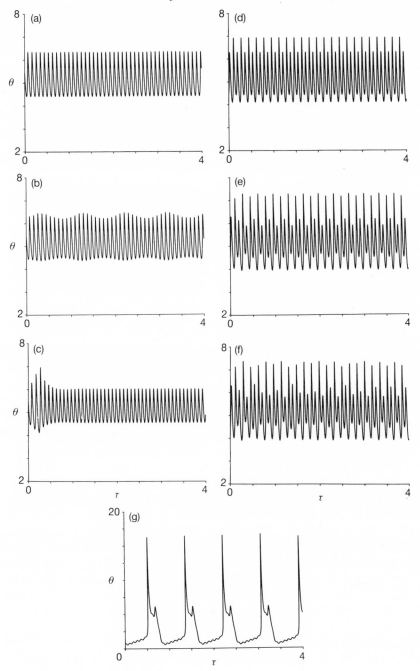

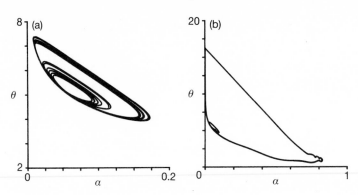

FIG. 5.4. Coexisting attractors for $q = 0.215$: (a) chaotic attractor; (b) period-10 limit cycle. Note change of scale between phase portraits: the strange attractor lies close to the small windings of the period-10 limit cycle.

the response is larger by a factor of about 80 than the imposed variation in cooling temperature. Mankin and Hudson quote data for the oscillatory decomposition of H_2O_2 for which $q = 0.215$ might correspond to a variation of $1.26\,°C$ in T_c leading to an excursion of about $100\,°C$ in the temperature within the reactor.

As q is increased further, the period-10 solution also undergoes a period-doubling sequence, yielding a large-amplitude chaotic attractor. For sufficiently large amplitudes, however, 1:1 entrainment with the forcing period must again emerge.

The chaotic nature of the attractor shown in Fig. 5.3(f) and the sub-harmonic cascade leading to it have been confirmed by a variety of the techniques discussed in the previous section. A special case of the Poincaré section, known as a 'stroboscopic map', is especially useful for forced systems—and will be discussed in detail below. This map shows the values of the two variables—here concentration α and temperature excess θ—at the beginning of each forcing cycle. For a quasiperiodic solution the system evolves on a torus in the (three-dimensional) phase space and (eventually) yields a closed curve in the stroboscopic plane (Fig. 5.5(a)). An entrained of phase-locked solution leads only to a finite number of discrete points on the stroboscopic map: a solution that repeats every n forcing periods gives rise to n such intersections (Fig. 5.5(b)). For a chaotic solution, the stroboscopic map has the form shown in Fig. 5.5(c) corresponding to a slice through a strange attractor and appearing much as a normal Poincaré section for a three (or more) variable system with autonomous chaos.

A return map can then be derived from the stroboscopic map, by plotting the value of θ, say, at the mth period against that at the $(m + 1)$th period. Figure 5.6 shows a typical, single extremum curve (here a minimum rather than a maximum, but having the same qualitative features) as seen in earlier

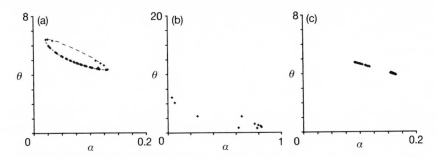

FIG. 5.5. Stroboscopic plots for (a) quasiperiodic solution $q = 0.025$; (b) period-10 entrainment $q = 0.215$; (c) chaotic response $q = 0.215$.

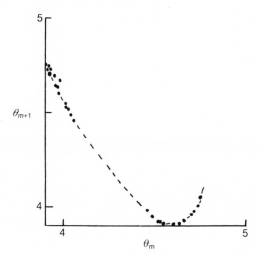

FIG. 5.6. Next-return map to stroboscopic plane constructed from chaotic trace for $q = 0.215$.

sections. Mankin and Hudson also calculated maximum Lyapounov exponents, showing that this becomes positive for the solution in Fig. 5.3(f) and the variation of the critical Floquet multiplier with q: the CFM passes through -1 on each period doubling, as with an autonomous cascade.

Kevrekidis *et al.* (1986a, b, c) extended this investigation to cover additional forcing frequencies and produced an excitation diagram as shown in Fig. 5.7. Such an extensive study was possible only because of the relative ease of application to direct path following of the bifurcations in forced systems as described in these papers. These also yield a detailed mathematical insight to the way in which qualitative changes in the behaviour occur and some of this will be described in the next section.

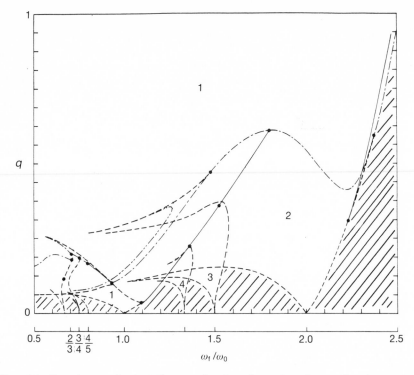

Fig. 5.7. Complete excitation diagram for forced exothermic reaction in a non-isothermal CSTR. (Reprinted with permission from Kevrekidis *et al.* (1986*c*), © Elsevier Science Publishers.)

The same group have also investigated the behaviour of other model schemes that have similar gross autonomous behaviour, i.e. an autonomous oscillation close to a Hopf bifurcation point. Summaries of this work are given in McKarnin *et al.* (1988*a*) and Aris (1990). In addition to the FONI model, these authors examined the Brusselator (Section 5.1) and the Takoudis–Schmidt–Aris scheme (Section 3.12.4). The latter has Hopf bifurcation and multiple stationary states, but McKarnin *et al.* (1988*b*) concentrated on parameter values such that the autonomous system has a single stationary state. In terms of the dimensionless parameters introduced earlier, they took $p = 0.019$, $\kappa_1 = 0.001$, $\kappa_2 = 0.002$, and forced r about a mean value of 0.028. This latter lies almost half-way across a relatively narrow region of stationary-state instability ($0.025\,903 \leqslant r \leqslant 0.030\,728$) and for which the autonomous system has a stable limit cycle. The forcing is of the form $r = \langle r \rangle + \rho \cos(\omega_f \tau)$. The excitation diagram for this scheme is shown in Fig. 5.8 and has clear qualitative resemblances to those for the earlier two models. These grossly similar features have allowed 'universal' behaviour to be postulated.

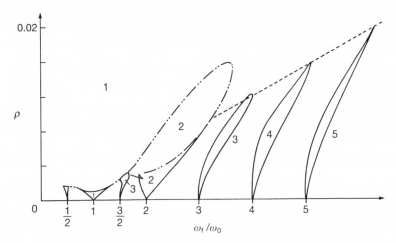

FIG. 5.8. Excitation diagram for forced Takoudis–Schmidt–Aris model. (Reprinted with permission from Aris (1990), © Manchester University Press.)

5.3. The stroboscopic map and bifurcations

In order to understand the mathematical basis for the various changes in behaviour observed as a parameter is varied it is particularly convenient for forced systems to work in terms of the phase space and the stroboscopic plane. Although we have typically two dynamical variables—concentrations or temperatures—the phase space for these systems is three dimensional. The third dimension arises from the time dependence of one of the 'parameters'— that which is being forced. Thus the cooling temperature or the concentration of a pool chemical provides a third variable. There is, however, a significant difference between a forced system and an unforced, autonomous three-variable scheme as in the present case we know exactly how the extra variable (the parameter) behaves: in particular we know the period in which it repeats itself. Thus at the end of every (known) forcing period the forced parameter has the same value.

In the previous chapter we worked in terms of a Poincaré section, defined as a surface in the parameter space on which a chosen variable attained a particular value. In that case the return to the Poincaré surface served to determine the period. In the present situation, we can choose the parameter as the 'variable' whose fixed value defines the plane of section. Now we know the time of return; this is the stroboscopic plane. We observe the state of the real variables at the beginning of each forcing period.

Returning to the idea of the phase space, the unforced autonomous system is confined to a simple plane in this space, corresponding to the fixed value of the parameter: there may be a simple limit cycle in this plane as shown

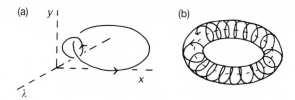

FIG. 5.9. Schematic representation of origin of a torus in a forced autonomous oscillator.

in Fig. 5.9(a). If we now begin to force the system with a low amplitude, the effect is to add a cycling motion to the parameter, i.e. to the position of the plane in which the original limit cycle is being executed. The total motion combining these two cycles is equivalent to motion on a torus (Fig. 5.9(b)). If the two frequencies of periods are irrationally related, the system will eventually cover the complete surface of the torus: if they happen to be rationally related, the path along the surface will close up on itself, so giving a line wrapped on the torus.

The stroboscopic section in this construction corresponds to a plane cutting the torus transversally at the chosen value of the forced parameter. A quasiperiodic motion covering the full surface of the torus will have a complete circle of intersection with the stroboscopic plane. This is known as an invariant circle (Fig. 5.10(a)). If the motion becomes phase locked on

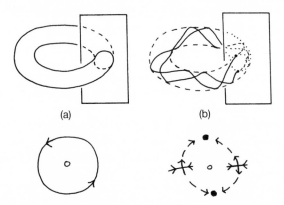

(a)

(b)

FIG. 5.10. Poincaré section or stroboscopic mapping for quasiperiodic and phase-locked motions on a torus.

the torus, the circle 'collapses' (in a sense) to a number of discrete points—one for each forcing period before the trajectory repeats. In fact, the torus and hence the circle of intersection remain even for a phase-lock state (Fig. 5.10(b)). The change is that the circle now has developed a number of fixed points around its circumference, corresponding to the iterated intersections of the entrained solution. Additionally, there are an equal number

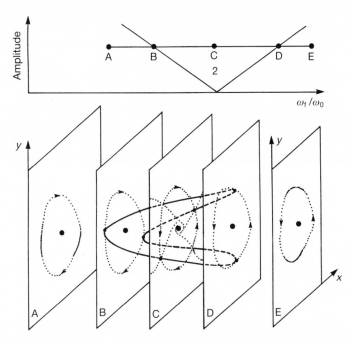

FIG. 5.11. Evolution of invariant points and invariant circles in stroboscopic plane during a typical traverse through an Arnol'd tongue (here the 2:1 entrainment region). (Reprinted with permission from Aris (1990), © Manchester University Press.)

of unstable fixed points on the circle. The remainder of the circumference now corresponds to transient motion as the system moves to the stable fixed points.

There will also typically be a second feature of the stroboscopic plane: an unstable point within the circle corresponding to an unstable state of complete 1:1 entrainment of the system with the forcing frequency. This gives the stroboscopic plane (map) for a quasiperiodic system a similar appearance to that of a phase plane for a two-variable system with an unstable stationary state surrounded by a stable limit cycle. This similarity has important consequences as we discuss below.

The various boundaries of the Arnol'd tongues etc. in the excitation diagram can now be categorized. Let us imagine starting with a system corresponding to a point in the quasiperiodic 'sea' between the resonance horns and then consider the variation in, say, forcing frequency such that we pass through such a horn (Fig. 5.11). Initially, the system has a stable motion on a torus with a full invariant circle. As the system evolves, so it gradually winds around the circle, the direction of rotation being fixed. As we vary the frequency to pass through the boundary of the Arnol'd tongue, so the structure of the circle is altered. If the emerging phase-locked state corresponds to entrainment over n forcing periods, then n saddle–node points appear on the circle. These separate to give n nodes, with n saddle points, one between each pair of nodes. The entrained solution corresponds to the system sequentially visiting the nodes: the saddles separate initial conditions that visit different nodes first. Figure 5.11 shows the appearance of two saddle–node pairs for a period-2 entrainment. As we move further across the tongue, so the original saddle–node pairs separate more. As we reach the other side, so each node reforms with a saddle point, but a different one from that with which it was born. The system then develops a full invariant circle as it moves back into the quasiperiodic sea, but now the direction of rotation around the circle is of opposite sense to that found on the other side of the horn.

The sides of the entrainment horns described by the saddle–node bifurcations above can be parametrized mathematically, by considering the critical Floquet multiplier: in this case the CFM passes out of the unit circle at $+1$. This parametrization allows the boundary to be computed directly using path-following techniques (Kevrekidis *et al.* 1986a, b, c).

Another way of leaving the quasiperiodic sea is to increase the forcing amplitude. Eventually, the system will become fully entrained to the forcing period. The entrained state corresponds to the fixed point lying inside the invariant circle in the stroboscopic map, as described above. As the forcing amplitude is increased, so the invariant circle shrinks in size. In the limit, the circle shrinks on to the invariant point, which then becomes stable. This sounds like the reverse of a supercritical Hopf bifurcation, and indeed is termed a Hopf transition. The reverse process sees the birth of an invariant circle and can also be categorized in terms of the CFM of the entrained state. Now, there is a complex pair of such multipliers, and these leave the unit circle away from the real axis. For a fully quasiperiodic state to develop, the points of exit must correspond to an angle $2\pi/\bar{\omega}$ where $\bar{\omega}$ is irrational. If the angle of exit is $2\pi p/q$, where p and q are (mutually prime) integers, a period-p phase-locked state is formed with p saddle–node pairs.

This latter point becomes important when we consider how the system leaves an Arnol'd tongue with increasing amplitude. For the higher periodicities, $p \geqslant 4$, th appropriate tongue appears simply to close up to a point lying on the line of Hopf bifurcations just described, corresponding to parameter values at which the CFM passes out of the unit circle with an appropriate angle (the appropriate, complex pth root of unity).

The 1:1 entrainment horn has a relatively extensive top. Passing out of this we see a saddle–node bifurcation in which the saddle on the invariant circle merges with the unstable node corresponding to an unstable 1:1 entrained state lying inside the circle—no change is seen in the behaviour across this boundary. We may also note that the cusp points at which the top and sides of this horn meet correspond to 'Bogdanov' points, where two CFMs become $+1$ together—these play a similar organizing role for the stroboscopic plane as that of double zero eigenvalues for the phase plane (Section 3.10).

The 2:1 Arnol'd tongue is crowned with a set of nested period doublings. Across these various boundaries, the CFM for the bifurcating state leaves the unit circle through -1. Aris has conjectured the existence of such period-doubling nests above each $2:q$ resonance horn (Aris 1990; McKarnin *et al.* 1988*a*).

The Hopf point for which the CFM pair leaves through the complex third root of unity, which should give rise to the 3:1 entrainment (the angle is $\pm 2\pi/3$), actually lies within the Arnol'd tongue. Now there is a separation of the saddle–node pairs from the invariant circle, so period-3 and quasi-periodic states coexist for some parameter ranges (Kevrekidis *et al.* 1986*b*).

Additional, and yet more sophisticated, features have also been identified and understood by these groups, but perhaps the most important message is that this type of scenario is essentially independent of both the model and the form in which the forcing is applied.

5.4. Forced bistable (non-oscillatory) systems

Cordonier *et al.* (1990) investigated a form of the cubic autocatalysis scheme (Section 3.7) for parameter values such that the autonomous system has multiple stationary states in the form of a hysteresis loop, but no limit cycle. They considered four cases, forcing the flow rate (the inverse of the residence time). In the first, the mean flow rate was chosen to lie exactly half-way between the two saddle–node points (Fig. 5.12): the other cases involved forcing about points at one-quarter and three-quarters of the way across the range of bistability and finally at one of the saddle–node points. Although, the autonomous system has no Hopf bifurcation, it does have (stable) focal states and these appear to be crucial to the ability to display complex behaviour under forcing.

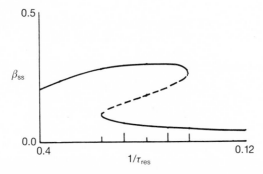

FIG. 5.12. Autonomous bifurcation diagram for cubic autocatalator model showing multistability but no Hopf bifurcation. (Reprinted with permission from Cordonier *et al.* (1990), © Pergamon Press.)

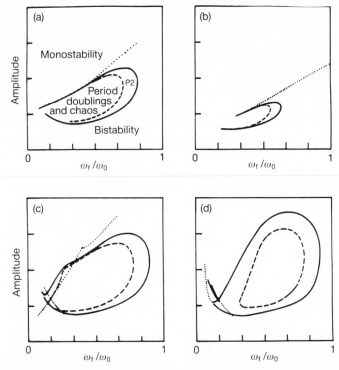

FIG. 5.13. Excitation diagrams for a forced bistable system—cubic autocatalator:
(a) forcing at mid-point of bistable region; (b) and (c) forcing at 25 per cent and 75
per cent points of bistable region; (d) forcing at a saddle–node parameter value at
end of region of bistability. (Reprinted with permission from Cordonier *et al.* (1990),
© Pergamon Press.)

Figure 5.13 shows the four excitation diagrams for the different mean flow
rates: in each of the diagrams the forcing amplitude is scaled by the distance
(in terms of flow rate) from the mid-point to the saddle–node turning points.
Figure 5.13(a) corresponds to mid-point forcing. The dashed line emerges
for unity at zero forcing amplitude. This corresponds to a locus of saddle–
node bifurcations in the stroboscopic plane: above the line, the forcing
amplitude is sufficiently high for the system to pass beyond the region of
autonomous bistability and just a single forced oscillatory solution exists—
fully entrained to the forcing period. At lower amplitudes, below the line,
the forcing about each of the autonomous stationary states is not sufficient
for the system to escape, so we see two possible entrainments: one about the
upper state and one about the lower. (There is also an unstable solution
corresponding to a forced oscillation about the unstable autonomous saddle
point.) As we move back towards the dashed line, it appears that the upper

state (which has focal character in the autonomous system) disappears with the saddle point.

The second feature of Fig. 5.13(a) is a nested set of period-doubling loops. These correspond to bifurcations of the forced oscillation about the upper (focal) state. Full cascades to chaos have been confirmed, even satisfying the Feigenbaum number, along with periodic windowing. These loops must all lie below the saddle–node locus in the figure, as the upper state disappears at this latter line.

In Fig. 5.13(b), the system is forced about a mean much closer to the upper saddle–node bifurcation in the autonomous system. The upper state, that which period doubled in the previous case, has a narrower region of separate existence (multiplicity disappears for a forcing amplitude greater than one-half with these scalings). The regions of multistability and period doubling are smaller than for mid-point forcing.

With forcing closer to the other saddle–node point of the autonomous scheme (Fig. 5.13(c)), a cusp develops in the dashed locus corresponding to loss of multistability on the excitation diagram. Above the cusp (at higher forcing periods) on this locus, the upper state disappears at the saddle–node point as before, but for low forcing periods, it is the lower and relatively uninteresting state that departs with the saddle point. For this reason, the period-doubling region now overlaps into the region of monostability in the excitation diagram—as it does in the case of forcing about the mean flow rate for the lower saddle–node point in Fig. 5.12 whose excitation diagram is shown in Fig. 5.13(d).

Another example of forcing non-oscillatory systems is that of Dolnik *et al.* (1989), who considered an autonomously excitable model (Section 3.12.3). Sufficiently small perturbations will not move the system away from the narrow basin of attraction of the stationary state, so the concentrations simply 'dance' around this state. If, however, the forcing is of sufficient amplitude (and period) to move the system across the excitability threshold, the trajectories will escape from the stationary state. An excitation diagram showing regions of entrainment that do not emerge from zero amplitude is thus obtained. Some additional mathematical points for such systems have been derived by Othmer (1990).

5.5. Analytical studies

The power and detail of the numerical studies discussed above notwithstanding, various analytical results for forced systems can also be obtained for special circumstances. Much of the latter has been concerned with forcing a two-variable scheme with parameter values such that the autonomous system is at or close to a double zero eigenvector (see Section 3.10). The importance of such a degeneracy is that 'close by' the existence of a homoclinic orbit formed by the inset and outset of the two-dimensional saddle point can be guaranteed.

Typical of such analyses is that of Baesens and Nicolis (1983). These authors examined a 'normal form' pair of equations

$$dx_1/dt = x_2 \tag{5.6}$$

$$dx_2/dt = \gamma_1 + \gamma_2 x_1 + x_1^2 - \mu^{1/2} x_1 x_2. \tag{5.7}$$

For the special case $\mu = 0$, this is a 'conservative' or 'Hamiltonian' system, having a constant (energy) of motion. The corresponding phase portrait for the conservative system has the structure shown in Fig. 5.14: there are two

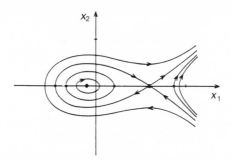

FIG. 5.14. Phase portrait for Hamiltonian reference system showing centre, nested closed curves and homoclinic orbit connecting the inset and outset of the saddle point. (Reprinted with permission from Baesens and Nicolis (1983), © Springer-Verlag.)

singular points, a saddle and a 'centre'; the centre is surrounded by an infinite number of nested loops, each of which corresponds to a periodic orbit or oscillatory evolution. One special loop links the inset and outset of the saddle—the homoclinic orbit. For non-zero μ, we recover our more familiar case of a dissipative system. The infinite nesting of loops collapses, to yield a single limit cycle. Typically this limit cycle will not touch the saddle, i.e. the homoclinic orbit is destroyed. However, for the special set of cases such that $\gamma_1 = -0.2\gamma_2^2$ the homoclinic orbit is retained (with no other limit cycle).

Baesens and Nicolis then considered the effect of forcing the parameter μ in the following form:

$$dx_1/dt = x_2 \tag{5.8}$$

$$dx_2/dt = \gamma_1 + \gamma_2 x_1 + x_1^2 - \mu^{1/2}[x_1 x_2 - b \sin(\omega_f t)]. \tag{5.9}$$

They were particularly interested in the case of asymptotically small forcing ($\mu \ll 1$) so the system could then be continuously compared with the well-characterized unperturbed reference equations with $\mu = 0$—the Hamiltonian case. Similar analyses based on Melnikov's theorem (Melnikov 1963) have been performed by Keener (1981), Chow and Hale (1982), Chow *et al.* (1980), and Greenspan and Holmes (1983), but we continue to follow Baesens and Nicolis.

The approach is to assume that for sufficiently small μ the actual solution $(x_1(t), x_2(t))$ can be written as 'small' departures from the Hamiltonian system $(x_{1,0}(t), x_{2,0}(t))$ as

$$x_1(t) = x_{1,0}(t) + u \qquad x_2(t) = x_{2,0}(t) + v$$

where u and v are functions of time that can be written as an expansion in $\mu^{1/2}$, i.e.

$$\begin{pmatrix} u \\ v \end{pmatrix} = \mu^{1/2} \begin{pmatrix} u_1 \\ v_1 \end{pmatrix} + \mu \begin{pmatrix} u_2 \\ v_2 \end{pmatrix} + \cdots. \tag{5.10}$$

To leading order, the governing equations for (u_1, v_1) are linear with non-constant coefficients, whose solution therefore comprises a particular integral and a general integral. These provide a 'solvability' criterion that the forcing parameters must satisfy if the homoclinic orbit structure is to be preserved.

If the forcing amplitude b is less than a critical value b_{cr}, the solvability condition cannot be satisfied. For $b = b_{cr}$, some initial conditions do satisfy the requirement and the stable and unstable manifolds of the saddle form a homoclinic tangency. 'Close by' such a tangency, there will exist an infinite number of periodic orbits, some of which may be stable. With $b > b_{cr}$, the saddle inset and outset cross transversally an infinite number of times, to give a 'homoclinic tangle'—Fig. 5.15. This structure allows for the appearance of non-periodic (chaotic) attractors.

As well as working with this very simple normal form, Baesens and Nicolis showed that a more chemical scheme, namely

$$A + 2X \rightarrow 3X$$

$$X + Y \rightarrow Y + D$$

$$B + X \rightarrow Y + X$$

$$C + 2Y \rightarrow 3Y$$

$$Y \rightarrow F$$

where only X and Y are variables (the rest are pool chemicals), will have the same underlying dynamics in the vicinity of its double zero eigenvalue.

5.6. Quasiperiodic forcing

The idea of imposing a quasiperiodic parameter variation on an autonomous oscillator or bistable system may seem a rather esoteric extension of the

Forced systems

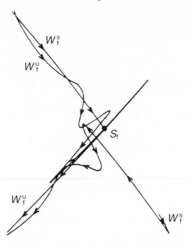

FIG. 5.15. Homoclinic tangle formed from inset and outset of saddle point for forced system. (Reprinted with permission from Baesens and Nicolis (1983), © Springer-Verlag.)

previous sections, but in fact it has immediate physical relevance. Rather than interpret such a cause as a single parameter being varied, the same situation arises if two different parameters undergo 'normal' periodic forcing with incommensurate periods. Thus a reactor in which both the inflow and the ambient temperature are subject to independent perturbations would be equivalent to a quasiperiodically forced system.

The equivalent of an entrained or, more generally, a phase-locked response in this case would be a quasiperiodic motion on a T^2 torus. If the forcing and forced frequencies remain incommensurate then we may expect three-frequency (T^3) quasiperiodicity. Two important quantities used to categorize the flow are the Lyapounov exponent spectrum and the winding number W (mean angular frequency of the circle map). Romeiras *et al.* (1987) examined a particularly simple, quasiperiodically forced equation system of the form

$$\frac{d\phi}{dt} = \cos\phi + \varepsilon\cos 2\phi + K + V(\cos\omega_1 t + \cos\omega_2 t) \qquad (5.11)$$

with irrational forcing frequencies $\omega_1 = \frac{1}{2}(\sqrt{5} - 1)$ and $\omega_2 = 1$. In the special case $\varepsilon = 0$, the resulting equation can be transformed to a form of the Schrödinger equation with a quasiperiodic potential, but otherwise this system is not specifically chosen for its chemical relevance.

Three qualitatively different forms of response could be distinguished in terms of the Lyapounov spectra and winding number for different combinations of K and V. The winding number typically shows the Devil's staircase dependence of the parameters—K in Fig. 5.16. This consists of phase-locked

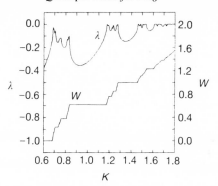

FIG. 5.16. Evidence for non-chaotic strange attractors in quasiperiodically forced oscillator: the winding number becomes irrational between the phase-locked treads of the Devil's staircase, but the Lyapounov exponent does not become positive. (Reprinted with permission from Romeiras *et al.* (1987), © Elsevier Science Publishers.

intervals for which the winding number can be written in terms of the two forcing frequencies

$$W = (l/n)\omega_1 + (m/n)\omega_2$$

where l, m, and n are integers. The system then shows two-frequency quasiperiodicity and the Lyapounov spectrum consists of two zero values and a negative exponent.

In between the steps of the staircase are the typical regions where the winding number increases continuously with K. The winding number does not have a relationship to the forcing frequencies as given above. All three frequencies of the motion are incommensurate. If the third Lyapounov exponent is also zero for these states, we find a T^3 quasiperiodicity. For some values of K, however, we may have such 'irrational' winding numbers but still have a negative Lyapounov exponent. The corresponding evolution shows a volume contraction in the full three-dimensional phase space as the sum of the exponents is negative, but does not collapse to a two-dimensional T^2 torus. Instead the attractor is a fractal object: a strange attractor. However, motion on this attractor is not chaotic, as there is no positive Lyapounov exponent. This new form of behaviour typifies a 'non-chaotic strange attractor'—one for which neighbouring trajectories do not show the exponential divergence characteristic of chaos.

Non-chaotic strange attractors occur as special (degenerate) cases in simple mappings and period-doubling systems of the form seen earlier (exactly at the accumulation points at the end of any given period-doubling cascade) but are not then 'realizable' or structurally stable responses. With quasiperiodic systems, such states do appear to be of 'non-zero' measure and hence realizable features of physical situations.

References

Aris, R. A. (1990). Forced oscillations of chemical reactors. In *Spatial inhomogeneities and transient behaviour in chemical kinetics*, (ed. P. Gray, G. Nicolis, F. Baras, P. Borckmans, and S. K. Scott), Ch. 2, pp. 11–28. Manchester University Press.

Aronson, D., McGehee, R., Kevrekidis, I., and Aris, R. (1986). Entrainment regions for periodically forced oscillators. *Phys. Rev.*, **A33**, 2190–2.

Baesens, C. and Nicolis, G. (1983). Complex bifurcations in a periodically forced normal form. *Z. Phys.*, B **52**, 345–54.

Bailey, J. E. (1973). Periodic operation of chemical reactors: a review. *Chem. Eng. Commun.*, **1**, 111–24.

Bailey, J. E. (1977). Periodic phenomena in chemical reactor theory. In *Chemical reactor theory*, (ed. L. Lapidus and N. R. Amundson). Prentice-Hall, Englewood Cliffs, NJ.

Chow, S. N. and Hale, J. K. (1982). *Methods of bifurcation theory*. Springer, New York.

Chow, S. N., Hale, J. K., and Mallet-Paret, J. (1980). An example of bifurcation to homoclinic orbits. *J. Diff. Equations*, **37**, 351–73.

Cordonier, G. A., Schmidt, L. D., and Aris, R. (1990). Forced oscillations of chemical reactors with multiple steady states. *Chem. Eng. Sci.*, **45**, 1659–76.

Daido, H. and Tomita, K. (1979a). Thermal fluctuation of a self-oscillating reaction system entrained by a periodic external force. *Prog. Theor. Phys.*, **61**, 825–41.

Daido, H. and Tomita, K. (1979b). Thermal fluctuation of a self-oscillating reaction system entrained by a periodic external force, II. *Prog. Theor. Phys.*, **62**, 1519–32.

Dolnik, M., Finkeova, J., Schreiber, I., and Marek, M. (1989). Dynamics of forced excitable and oscillatory chemical reaction systems. *J. Phys. Chem.*, **93**, 2764–74.

Greenspan, B. D. and Holmes, P. J. (1983). Homoclinic orbits, subharmonics and global bifurcations in forced oscillations. In *Nonlinear dynamics and turbulence*, (ed. G. I. Barenblatt, G. Iooss, and D. D. Joseph), Ch. 10, pp. 172–214. Pitman, Boston.

Kai, T. and Tomita, K. (1979). Stroboscopic phase portrait of a forced nonlinear oscillator. *Prog. Theor. Phys.*, **61**, 54–73.

Keener, J. P. (1981). Infinite period bifurcation and global bifurcation branches. *SIAM J. Appl. Math.*, **41**, 127–44.

Keener, J. P. (1982). Chaotic behavior in slowly varying systems of nonlinear ordinary differential equations. *Stud. Appl. Math.*, **67**, 25–44.

Kevrekidis, I. G., Schmidt, L. D., and Aris, R. (1984). On the dynamics of periodically forced chemical reactors. *Chem. Eng. Commun.*, **30**, 323–30.

Kevrekidis, I. G., Schmidt, L. D., and Aris, R. (1986a). The stirred tank forced. *Chem. Eng. Sci.*, **41**, 1549–60.

Kevrekidis, I. G., Schmidt, L. D., and Aris, R. (1986b). Some common features of periodically forced reacting systems. *Chem. Eng. Sci.*, **41**, 1263–76.

Kevrekidis, I. G., Schmidt, L. D., and Aris, R. (1986c). Forcing an entire bifurcation diagram: case studies in chemical oscillators. *Physica*, **23D**, 391–5.

Kuramoto, Y. (1984). *Chemical oscillations, waves and turbulence*. Springer, Berlin.

McKarnin, M. A., Schmidt, L. D., and Aris, R. (1988a). Response of nonlinear oscillators to forced oscillations: three chemical reaction case studies. *Chem. Eng. Sci.*, **43**, 2833.

McKarnin, M. A., Schmidt, L. D., and Aris, R. (1988*b*). Forced oscillations of a self-oscillating bimolecular surface reaction model. *Proc. R. Soc.*, **A417**, 363–88.

Mankin, J. C. and Hudson, J. L. (1984). Oscillatory and chaotic behaviour of a forced exothermic chemical reaction. *Chem. Eng. Sci.*, **39**, 1807–14.

Melnikov, V. K. (1963). On the stability of the center for periodic perturbations. *Trans. Moscow Math. Soc.*, **12**, 1–57.

Othmer, H. G. (1990). The dynamics of forced excitable systems. In *Nonlinear wave processes in excitable media*, (ed. A. V. Holden, M. Markus, and H. G. Othmer). Plenum, London.

Romeiras, F. J., Bondeson, A., Ott, E., Antonsen, T. M., and Grebogi, G. (1987). Quasiperiodically forced dynamical systems with strange nonchaotic attractors. *Physica*, **26D**, 277–94.

Schneider, F. W. (1985). Periodic perturbation of chemical oscillators: experiments. *Annu. Rev. Phys. Chem.*, **36**, 347–78.

Schreiber, I., Dolnik, M., Choc, P., and Marek, M. (1988). Resonance behaviour in two-parameter families of periodically forced oscillators. *Phys. Lett.*, A **128**, 66–70.

Sincic, D. and Bailey, J. E. (1977). Pathological dynamic behavior of forced periodic chemical processes. *Chem. Eng. Sci.*, **32**, 281–6.

Tomita, K. (1982). Chaotic response of a nonlinear oscillator. *Phys. Rep.*, **86**, 114–67.

Tomita, K. (1986). Periodically forced nonlinear oscillators. In *Chaos*, (ed. A. V. Holden), pp. 211–36. Manchester University Press.

Tomita, K. and Daido, H. (1980). Possibility of chaotic behavior and multibasins in forced glycolytic oscillations. *Phys. Lett.*, A **79**, 133–7.

Tomita, K. and Kai, T. (1979). Chaotic response of a nonlinear oscillator. *J. Stat. Phys.*, **21**, 65–86.

6

Coupled systems

In the previous chapter we looked at the range of behaviour exhibited by chemical systems exposed to some periodic external input. A related experimental arrangement is that in which one or more reactors are coupled together in some way. Then oscillations in one or more of the individual units might provide a periodic stimulus to the others. In this case, however, there is an extra subtlety as the other units can 'fight back' in the sense that their response will also influence the first.

Coupling may occur in a variety of forms (Fig. 6.1). With chemical coupling one may imagine a direct connection between the individual units through which there may be exchange of all or perhaps just some of the components. With thermal coupling there would be heat transfer across common surfaces (Mankin and Hudson 1986). Systems of electrodes have been used to couple reactions electrically (Crowley and Field 1986). When more than two units are involved, there will be different geometrical arrangements with different consequences to consider.

Some of the responses typical of coupled systems that we will find here are familiar from earlier chapters—such features as quasiperiodicity and period-doubling cascades. There are new, interesting, and important questions: for instance, if identical oscillators are coupled together will the behaviour in each necessarily remain the same or can there be a spontaneous transition to inhomogeneity so the units do different things?

6.1. Coupled autocatalators

As before, we begin with particularly simple models and consider two reactors in each of which a cubic autocatalator mechanism is operating. If we denote the two reactors as cells 1 and 2 we can write the governing reaction rate equations for the two concentrations in each while they are still uncoupled in dimensionless form:

for cell 1

$$\frac{d\alpha_1}{d\tau} = \mu_1 - \alpha_1\beta_1^2 - \kappa_1\alpha_1 \tag{6.1}$$

$$\frac{d\beta_1}{d\tau} = \alpha_1\beta_1^2 + \kappa_1\alpha_1 - \beta_1 \tag{6.2}$$

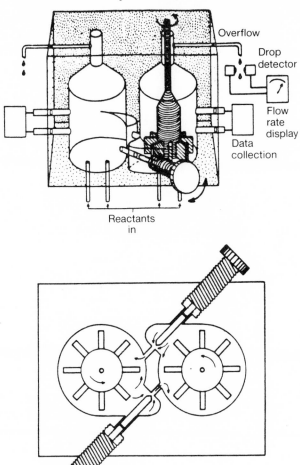

FIG. 6.1. Typical experimental arrangement for a coupled solution-phase reaction showing two reaction cells connected by channels and needle valves. (Reprinted with permission from Crowley and Epstein 1989, © American Chemical Society.)

and for cell 2

$$\frac{d\alpha_2}{d\tau} = \mu_2 - \alpha_2\beta_2^2 - \kappa_2\alpha_2 \tag{6.3}$$

$$\frac{d\beta_2}{d\tau} = \alpha_2\beta_2^2 + \kappa_2\alpha_2 - \beta_2. \tag{6.4}$$

Differences in the experimental conditions, such as in the pool chemical concentration, within the two reactors would be represented by different values for the parameters μ_1 and μ_2 (or κ_1 and κ_2 if appropriate).

Initially, however, we will be interested in coupled identical cells, so we can take

$$\mu_1 = \mu_2 = \mu \quad \text{and} \quad \kappa_1 = \kappa_2 = \kappa \qquad (6.5)$$

dropping the subscripts on the parameters (but not on the variables).

Again for simplicity, we will imagine there is some semi-permeable membrane at out disposal that allows us to couple the two cells in such a way that only one species is exchanged. First we consider the case in which coupling occurs through the autocatalyst B. We must introduce a coupling term into the rate equations: this will only enter into $d\beta_1/d\tau$ and $d\beta_2/d\tau$ and the rate of exchange will be proportional to the difference in concentration of the autocatalyst in the two cells. Thus we have the system of four equations

$$\frac{d\alpha_1}{d\tau} = \mu - \alpha_1 \beta_1^2 - \kappa\alpha_1 \qquad (6.6)$$

$$\frac{d\beta_1}{d\tau} = \alpha_1 \beta_1^2 + \kappa\alpha_1 - \beta_1 - \chi(\beta_1 - \beta_2) \qquad (6.7)$$

$$\frac{d\alpha_2}{d\tau} = \mu - \alpha_2 \beta_2^2 - \kappa\alpha_2 \qquad (6.8)$$

$$\frac{d\beta_2}{d\tau} = \alpha_2 \beta_2^2 + \kappa\alpha_2 - \beta_2 - \chi(\beta_2 - \beta_1) \qquad (6.9)$$

where χ is the dimensionless exchange coefficient. Thus, if the concentration of B in cell 1 exceeds that in cell 2 there will be e net 'diffusion' of autocatalyst from the first to the second (and vice versa for $\beta_2 > \beta_1$). If, however, the behaviour of the two cells becomes identical, so $\beta_2 = \beta_1$, there is no net exchange.

6.2. Coupling through autocatalyst: stationary states

At the most straightforward interpretation, we now have an overall four-variable model. We again expect to proceed by determining the stationary states and then assessing their local stability. If Hopf bifurcations arise, as we have every reason to expect given the results of Chapter 3, we can then follow time-dependent solutions corresponding to the emerging limit cycles. Furthermore, as we have more than two variables we may also expect generically that bifurcations from period-1 to waveforms with higher-order complexity and maybe even aperiodicity may feature.

At this stage we will reduce the model further by assuming that the uncatalyzed reaction converting A to B (step 1 of Chapter 3) has zero rate, so $\kappa = 0$. (With this simplification, the uncoupled reactor has the stationary-state solution $\alpha_{ss} = \mu^{-1}$, $\beta_{ss} = \mu$, and a single Hopf bifurcation point at $\mu = 1$: the stationary state is unstable and a stable limit cycle exists for $\mu < 1$.) The coupled system also has only one possible stationary-state solution, given by

$$\alpha_{1,ss} = \alpha_{2,ss} = \mu^{-1} \qquad \beta_{1,ss} = \beta_{2,ss} = \mu. \qquad (6.10)$$

Thus, the concentrations of A and of B in each reactor are equal: under stationary-state conditions the cells behave identically and in the same way as the uncoupled system.

Because we have a four-variable system, the local stability of a given stationary-state is determined by the four eigenvalues of the 4×4 Jacobian matrix. These $\lambda_1 - \lambda_4$ are given by the roots of a quartic equation. For the present model with the above stationary state, we find that the quartic can be factorized into two quadratic terms:

$$[\lambda^2 + (\mu^2 - 1)\lambda + \mu^2][\lambda^2 + (\mu^2 - 1 + 2\chi)\lambda + (1 + 2\chi)\mu^2] = 0. \quad (6.11)$$

Thus, we can write this as

$$\lambda^2 + (\mu^2 - 1)\lambda + \mu^2 = 0 \qquad (6.12)$$

and

$$\lambda^2 + (\mu^2 - 1 + 2\chi)\lambda + (1 + 2\chi)\mu^2 = 0. \qquad (6.13)$$

Taking eqn (6.12) first we see that the first part of the Hopf condition, i.e. that one pair of eigenvalues should become purely imaginary, is satisfied for

$$\mu = \mu_1^* = 1 \qquad (6.14)$$

i.e. the same value of the precursor reactant concentration as in the uncoupled system (this result is independent of χ). We must also ensure that the two remaining eigenvalues, given by the roots of eqn (6.13) have negative real parts for $\mu = \mu_1^*$. With $\mu = 1$, these two eigenvalues are given by

$$\lambda_{3,4} = -\chi \pm (\chi^2 - 2\chi - 1)^{1/2} \qquad (6.15)$$

and therefore satisfy this extra condition.

In the coupled system, we may also look for additional Hopf bifurcations from this second pair of eigenvalues. Equation (6.13) will have the desired imaginary roots when

$$\mu = (1 - 2\chi)^{1/2} \qquad (6.16)$$

(the negative root is clearly unacceptable physically). For this value of μ, however, the roots of eqn (6.12) are given by

$$\lambda_{1,2} = \chi \pm (\chi^2 + 2\chi - 1)^{1/2}. \qquad (6.17)$$

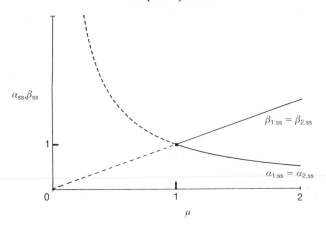

FIG. 6.2. The stationary-state and primary Hopf bifurcation characteristics for a coupled autocatalator with exchange only through the autocatalyst B.

These have positive real parts, are real and positive, or are real and have opposite sign (for $0 < \chi < \sqrt{2} - 1$, $\sqrt{2} - 1 < \chi < \frac{1}{2}$, and $\chi > \frac{1}{2}$ respectively), but in each of these cases the additional Hopf requirement is not met.

The stationary state and Hopf behaviour described above is shown in Fig. 6.2 and does not appear to differ from that for a single uncoupled cell. Furthermore, the Hopf bifurcation is always supercritical, leading to the creation and growth of a stable limit cycle for $\mu \leqslant \mu_1^*$.

The coupled system does, however, have some tricks left (Leach *et al.* 1990). As μ is decreased from the Hopf point, so the oscillations in each cell are identical. At some lower reactant concentration, the simple period-1 limit cycle loses stability. Two Floquet multipliers leave the unit circle as a complex conjugate pair and so there is bifurcation to a *torus*. The motion of the system on this torus gives rise to *quasiperiodic* time traces. Figure 6.3 shows some of these quasiperiodic solutions and Fig. 6.4 shows the region in the μ–χ plane for which they are found. This closed loop comes out of the rather degenerate point $\mu = 1$ at $\chi = 0$ (the uncoupled limit) and only survives for weak coupling, $\chi \lesssim 0.08$.

As the precursor concentration is decreased further, the torus loses stability again transferring it to the simple period-1 limit cycle. At yet lower μ there is the heteroclinic bifurcation mentioned in the previous chapter at which the conversion from A to B ceases and we have just the two cells producing A from B. This bifurcation occurs for $\mu = 0.900\,32$ for the uncoupled system and at the same value if the coupling is strong($\chi \gtrsim 0.09$). Weak coupling, on the other hand, seems to stabilize the system to some degree against this behaviour and the heteroclinic orbit locus in Fig. 6.4 shows a distinct dip with a minimum at $\chi \approx 0.05$.

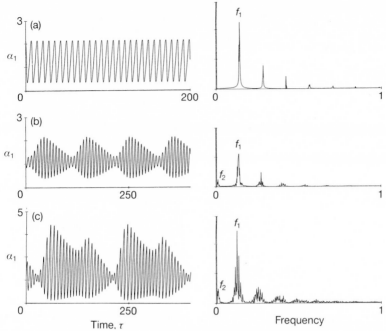

FIG. 6.3. Evolution of (a) periodic and (b),(c) quasiperiodic responses for two autocatalators coupled through B. Also shown are the corresponding power spectra on which the development of a second frequency can be seen.

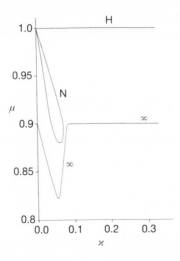

FIG. 6.4. Parameter plane showing loci of primary Hopf (H), secondary Hopf or Neimark (N), and heteroclinic orbit (∞) bifurcations separating ranges corresponding to period-1 and quasiperiodic evolution for two autocatalators coupled through B.

To summarize, with coupling through the autocatalyst the stationary-state and Hopf bifurcation phenomena correspond exactly to those of the uncoupled system. With weak coupling ($0 < \chi \lesssim 0.08$) a secondary Hopf bifurcation to quasiperiodic behaviour exists, but with stronger coupling the two cells act as one.

6.3. Coupling through A

The obvious 'other case' to study now is that in which coupling is achieved through exchange of the intermediate A. This provides a number of important new lessons. Still retaining identical cells, the governing equations are now

$$\frac{d\alpha_1}{d\tau} = \mu - \alpha_1\beta_1^2 - \kappa\alpha_1 - \chi(\alpha_1 - \alpha_2) \tag{6.18}$$

$$\frac{d\beta_1}{d\tau} = \alpha_1\beta_1^2 + \kappa\alpha_1 - \beta_1 \tag{6.19}$$

$$\frac{d\alpha_2}{d\tau} = \mu - \alpha_2\beta_2^2 - \kappa\alpha_2 - \chi(\alpha_2 - \alpha_1) \tag{6.20}$$

$$\frac{d\beta_2}{d\tau} = \alpha_2\beta_2^2 + \kappa\alpha_2 - \beta_2. \tag{6.21}$$

The uncoupled behaviour is, of course, the same as that discussed above.

For the coupled system, with $\chi > 0$, we can now have up to five stationary-state solutions—although these really constitute only three different cases. The sets of solutions are (with $\kappa = 0$):

(a) $\alpha_{1,ss} = \alpha_{2,ss} = \mu^{-1} \qquad \beta_{1,ss} = \beta_{2,ss} = \mu \tag{6.22}$

(b) $\begin{cases} \alpha_{1,ss} = \dfrac{1}{\mu + (\mu^2 - 2\chi)^{1/2}} \qquad \alpha_{2,ss} = \dfrac{1}{\mu - (\mu^2 - 2\chi)^{1/2}} \\ \beta_{1,ss} = \mu - (\mu^2 - 2\chi)^{1/2} \qquad \beta_{2,ss} = \mu + (\mu^2 - 2\chi)^{1/2} \end{cases} \tag{6.23}$

(c) $\begin{cases} \alpha_{1,ss} = \dfrac{1}{\mu - (\mu^2 - 2\chi)^{1/2}} \qquad \alpha_{2,ss} = \dfrac{1}{\mu + (\mu^2 - 2\chi)^{1/2}} \\ \beta_{1,ss} = \mu + (\mu^2 - 2\chi)^{1/2} \qquad \beta_{2,ss} = \mu - (\mu^2 - 2\chi)^{1/2} \end{cases} \tag{6.24}$

(d) $\begin{cases} \alpha_{1,ss} = (2\mu)^{-1} \qquad\qquad \beta_{1,ss} = 2\mu \\ \alpha_{2,ss} = \chi^{-1}\mu + (2\mu)^{-1} \qquad \beta_{2,ss} = 0 \end{cases} \tag{6.25}$

(e) $\begin{cases} \alpha_{1,ss} = \chi^{-1}\mu + (2\mu)^{-1} \qquad \beta_{1,ss} = 0 \\ \alpha_{2,ss} = (2\mu)^{-1} \qquad\qquad \beta_{2,ss} = 2\mu. \end{cases} \tag{6.26}$

The two pairs ((b) and (c)) and ((d) and (e)) can be interchanged simply by swapping over the subscripts 1 and 2. As we cannot really distinguish between the reactors, each pair only gives one new solution.

The first stationary-state solution is that apparently equivalent to the uncoupled system, with the same behaviour in each cell. When we consider the lower stability, the eigenvalues are again given by a quartic which can be factorized into two quadratics:

$$[\lambda^2 + (\mu^2 - 1)\lambda + \mu^2][\lambda^2 + (\mu^2 + 2\chi - 1)\lambda + \mu^2 - 2\chi] = 0. \quad (6.27)$$

Thus we require

$$\lambda^2 + (\mu^2 - 1)\lambda + \mu^2 = 0 \quad (6.28)$$

and

$$\lambda^2 + (\mu^2 + 2\chi - 1)\lambda + \mu^2 - 2\chi = 0. \quad (6.29)$$

Equation (6.28) can be recognized from the uncoupled system. If this pair of eigenvalues will support a Hopf bifurcation it will occur for

$$\mu = \mu_1^* = 1 \quad (6.30)$$

provided the corresponding roots of eqn (6.29) are negative or have negative real parts. The latter are given, for $\mu = \mu_1^*$, by

$$\lambda_{3,4} = -\chi \pm (\chi^2 + 2\chi - 1)^{1/2}. \quad (6.31)$$

These will only satisfy the additional requirement provided

$$0 \leqslant \chi < \tfrac{1}{2}. \quad (6.32)$$

The possibility of further Hopf bifurcation points arising from the eigenvalues from eqn (6.29)

$$\lambda_{3,4} = \tfrac{1}{2}\{1 - 2\chi - \mu^2 \pm [(\mu^2 - 1 + 2\chi)^2 - 4(\mu^2 - 2\chi)]^{1/2}\} \quad (6.33)$$

having zero real parts requires $\mu = (1 - 2\chi)^{1/2}$. With this condition, however, the other two roots from eqn (6.28) are given by

$$\lambda_{1,2} = \chi \pm (\chi^2 + 2\chi - 1)^{1/2} \quad (6.34)$$

and these always have positive real parts (with $\chi < \tfrac{1}{2}$). Thus the Hopf behaviour along this branch is as shown in Fig. 6.5.

Turning now to the pair of states (b) and (c) given by eqns (6.23) and (6.24), the eigenvalues for this state are given by the full quartic equation

$$\lambda^4 + [4\mu^2 - 2(\chi + 1)]\lambda^3 + [4\chi(\mu - 1) + 1]\lambda^2 + 2(4\chi^2 + 3\chi - 2\mu^2)\lambda$$
$$+ 4\chi(2\chi - \mu^2) = 0. \quad (6.35)$$

This pair of states is born at a 'pitchfork' bifurcation at $\mu = (2\chi)^{1/2}$. They only exist provided $\mu^2 > 2\chi$. Thus the final coefficient in characteristic

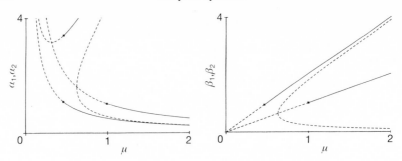

<small>F</small>IG. 6.5. Stationary-state and primary Hopf bifurcation structure for two auto-catalators coupled through reactant A, showing up to five branches.

equation (6.35) must be negative. This in turn means that there will be at least one positive eigenvalue: these are saddle point solutions.

An interesting situation arises for the special case $\chi = \frac{1}{2}$ as then the pitchfork bifurcation along the homogeneous branch (a) at which the pair of saddles (b) and (c) are born exactly coincides with the Hopf bifurcation point $\mu = 1$ (eqn (6.30)). This thus corresponds to a degenerate point where stationary-state and Hopf bifurcations can interact. Langford and Iooss (1980) have shown that quasiperiodic solutions can be expected 'close to' such parameter conditions as the degeneracy is 'unfolded'.

Thirdly we come to the solution pair (d) and (e) given by eqns (6.25) and (6.26). A particularly interesting feature is that the stationary-state concentration of the autocatalyst in one of the cells is zero. In fact we will see below that this concentration remains zero throughout the whole course of the reaction (it cannot, of course, become negative). Thus the reaction in the corresponding cell only goes 'half-way': the pool chemical reactant P can produce the first intermediate A, but this does not react further to B and ultimately C. Despite this last point, which means that A is continually being produced but not being consumed by reaction, we can notice that its concentration retains a finite stationary-state value. The production can be limited by the net exchange into the second reaction cell. Because one of the species is not ultimately involved in the overall system, we can expect a reduction to a three-variable system.

In fact the quartic equation for the eigenvalues of the 4×4 Jacobian in this case can be factorized into a cubic and a linear term:

$$(\lambda + 1)\{\lambda^3 + (2\chi + 4\mu^2 - 1)\lambda^2 + [4\mu^2(1 + \chi) - 2\chi]\lambda + 4\mu^2\chi\} = 0. \quad (6.36)$$

rapid relaxation from a four- to three-variable system. The condition for the remaining cubic to have complex roots with vanishing real parts then yields the following quadratic in μ^2:

$$8(1 + \chi)\mu^4 + 2(2\chi^2 - 2\chi - 1)\mu^2 + \chi(1 - 2\chi) = 0 \quad (6.37)$$

which has one positive root

$$\mu^2 = \frac{1}{8(1 + \chi)} \{1 + 2\chi - 2\chi^2 + [(1 + 2\chi - 2\chi^2)^2 + 8\chi(1 + \chi)(2\chi - 1)]^{1/2}\}.$$

(6.38)

An additional requirement is

$$\mu^2 > \tfrac{1}{4}(1 - 2\chi)$$ (6.39)

which clearly implies that we need $\chi < \tfrac{1}{2}$.

The Hopf bifurcation point described by eqn (6.38) for $\chi < \tfrac{1}{2}$ appears always to be supercritical, with a stable limit cycle emerging as the precursor reactant concentration is decreased below this value, again as shown for the uppermost branch in Fig. 6.5.

6.4. Period doublings and secondary Hopf bifurcations for coupling through A

The analysis in the previous section has revealed the existence of new, multiple stationary-state solutions in the coupled autocatalator system and of primary Hopf bifurcation points along two of the branches. We need now to follow the periodic solutions which emerge from these Hopf points and to determine whether any higher-order bifurcations arise from them.

We can begin with the stable limit cycle emerging from the symmetric stationary state (a). Figure 6.6 shows the corresponding μ–χ parameter plane (Leach *et al.* 1990). Below the line of Hopf points $\mu = 1$ ($\chi < \tfrac{1}{2}$) lies a curve labelled PD. This is a curve of period-doubling bifurcations at which a stable period-2 oscillation bifurcates from the period-1 as μ is decreased and the latter loses stability.

For small values of the coupling constant χ subsequent bifurcations are also of the period-doubling type (a Floquet multiplier leaving the unit circle through -1 in each case) and we find a typical cascade to aperiodic behaviour. Such a sequence for $\chi = 0.05$ is shown in Fig. 6.7. At sufficiently low values of the precursor concentration, the system leaves the vicinity of this attractor and moves to one of the (stable) asymmetric stationary states (d) or (e).

A slightly different scenario is observed with high coupling, e.g. for $\chi = 0.4$ (Fig. 6.8). Following the first period doubling there is a secondary Hopf bifurcation to quasiperiodic motion on a torus. A second incommensurate frequency arises in the power spectrum. For lower μ there can be a second of these bifurcations, with a complex pair of Floquet multipliers leaving the unit circle. However, the resulting T^3 torus (a four-dimensional object with a three-dimensional surface) is not structurally stable and so resolves into

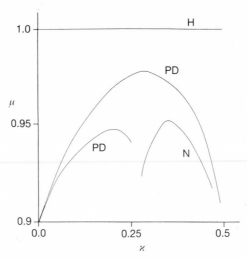

FIG. 6.6. Parameter plane showing loci of primary Hopf (H), period-doubling (PD), subharmonic cascade to chaos, and Neimark (N) bifurcation.

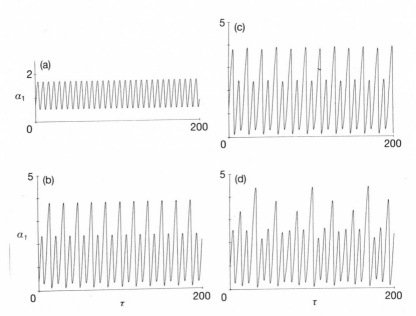

FIG. 6.7. Typical time series through a sequence of period doublings for weakly coupled autocatalators as the precursor concentration μ is reduced: (a) period-1; (b) period-2; (c) period-4; (d) period-8.

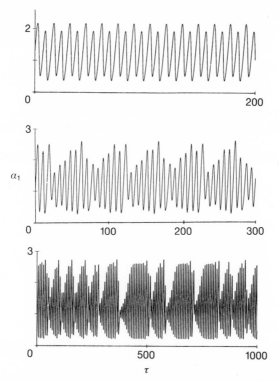

FIG. 6.8. Bifurcation sequence for strongly coupled autocatalators showing transition from period-2 to chaos via a torus as the precursor concentration is decreased.

a strange attractor with fractal dimension and leading to chaotic concentration histories. Again, the extinction of this attractor sees the system move to one of the pair of stable asymmetric states.

And so what of the periodic solution emerging from the asymmetric state? Just below the primary (supercritical) Hopf point there exists a stable period-1 limit cycle. As μ is decreased, so this undergoes a series of period doublings which might be expected to converge to chaos. The system here is now only a three-dimensional phase space and hence we can display the actual attractors as shown in Fig. 6.9. We have tracked the development of the limit cycles up to period-16. By this stage the parameter ranges for any given waveform are tiny. The convergence of the doubling sequence has also to compete with the formation of the heteroclinic orbit remnant from the uncoupled model. Thus if μ becomes too low, the concentration of autocatalyst tends to zero in both cells and the system settles into the steady production of A from (the inexhaustible supply of) P.

Entirely similar responses to those described in these three sections (destabilization of the uniform situation, onset of complex oscillations,

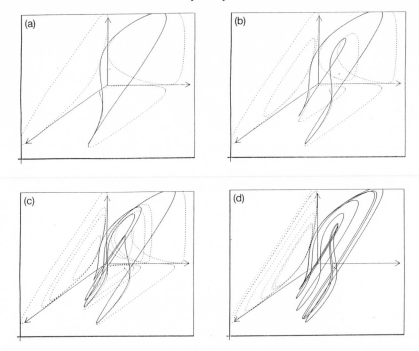

FIG. 6.9. Three-dimensional phase portraits for period-doubling sequence along the asymmetric branch of stationary states for coupled autocatalators. (Figures 6.2–6.9 from Leach *et al.* 1990.)

quasiperiodicity, period doubling, and chaos) has been seen in other models of coupled oscillators. For instance Waller and Kapral (1984) have reported such sequences for two coupled 'Rössler oscillators': the latter are three-variable, and rather unchemical, schemes coupled through one species z with the governing equations

$$dx_i/dt = -(y_i + z_i) \qquad dy_i/dt = x_i - ay_i$$
$$dz_i/dt = bx_i - cz_i + x_i z_i + \mu z_i(x_j - x_i)$$

with $i, j = 1$ or 2 and $i \neq j$.

A dramatic example of a quasiperiodic attractor for two non-isothermal CSTRs (FONI models of Section 3.12.2), which may be simply coupled by imagining that the reactors are in thermal contact but have no mass exchange (Mankin and Hudson 1986), is shown in Fig. 6.10.

6.5. Influence of uncatalysed reaction: cell mitosis

Tyson and Kauffman (1975) have also studied some aspects of coupled pool chemical autocatalator models—in fact Tyson can be said to have introduced

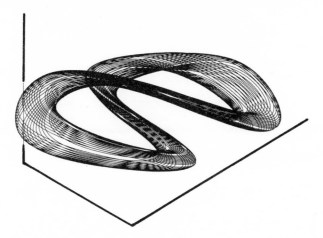

FIG. 6.10. Three-dimensional projection of quasiperiodic torus for thermally coupled non-isothermal CSTRs. (Reprinted with permission from Mankin and Hudson 1986, © Pergamon Journals.)

the form of the model with the uncatalysed step (Tyson 1973; Tyson and Light 1973) as a development of the Brusselator scheme and an earlier model of Sel'kov (1968) for cell division (mitosis). Their work also illustrates another general technique which can be applied to coupled systems. The experimental parameters μ and κ are again taken to be the same in each cell, but Tyson and Kauffman (1975) allowed for exchange of both species A and B and, furthermore, with different coupling coefficients χ_a and χ_b respectively. Similar studies have also been made by Ashkenazi and Othmer (1978).

The question of particular interest for this work was whether asymmetric or 'inhomogeneous' stationary states can exist for the model, such as those seen in the previous sections. Rather than working directly with the dimensionless concentration, we can write the governing rate equations in terms of the *sum* and *difference* of the particular concentrations in the two cells. Thus we define

$$S = \alpha_1 + \alpha_2 \qquad \Sigma = \beta_1 + \beta_2$$
$$D = \alpha_1 - \alpha_2 \qquad \Delta = \beta_1 - \beta_2. \tag{6.40}$$

The governing equations then become

$$\frac{dS}{d\tau} = 2\mu - \kappa S - \tfrac{1}{4}(S\Sigma^2 + 2D\Sigma\Delta + S\Delta^2) \tag{6.41}$$

$$\frac{dD}{d\tau} = -(\kappa + 2\chi_a)D - \tfrac{1}{4}(D\Sigma^2 + 2S\Sigma\Delta + D\Delta^2) \tag{6.42}$$

$$\frac{d\Sigma}{d\tau} = \kappa S - \Sigma + \tfrac{1}{4}(S\Sigma^2 + 2D\Sigma\Delta + S\Delta^2) \qquad (6.43)$$

$$\frac{d\Delta}{d\tau} = \kappa D - (1 + 2\chi_b)\Delta + \tfrac{1}{4}(D\Sigma^2 + 2S\Sigma\Delta + D\Delta^2). \qquad (6.44)$$

Clearly from the form of the new variables in eqn (6.40), a homogeneous stationary state will have $D_{ss} = \Delta_{ss} = 0$: these will be non-zero for an inhomogeneous state.

The homogeneous state, with $S_{ss} = 2\mu/(\mu^2 + \kappa)$ and $\Sigma_{ss} = 2\mu$, always exists. For inhomogeneous solutions we require the roots of a biquadratic:

$$\Delta^4 + 8(\kappa - \mu^2 + \chi_a)\Delta^2 + 16[(\mu^2 + \kappa)^2 + 2\chi_a(\mu^2 + \kappa) - 2\mu^2\xi] = 0 \quad (6.45)$$

where $\xi = 2\chi_a/(1 + 2\chi_b)$. Then

$$\Sigma = 2\mu \qquad D = -\Delta/\xi \qquad \text{and} \qquad S = 4\mu(2 + \xi^{-1}\Delta^2)/(4\mu^2 + 4\kappa + \Delta^2) \qquad (6.46)$$

(dropping the subscript ss for convenience). Equation (6.45) has one positive root, corresponding to a single pair of inhomogeneous stationary states, if

$$2\chi_b < 1 \qquad \chi_a > \tfrac{1}{2}\mu^2 \frac{1 + 2\chi_b}{1 - 2\chi_b} \qquad \text{and} \qquad \kappa < -(\mu^2 + \chi_a) + (\chi_a^2 + 2\mu^2\xi)^{1/2}. \qquad (6.47)$$

Two pairs of inhomogeneous solutions, for which eqn (6.45) must have two positive roots, exist provided

$$\chi_a > 8\mu^2 \frac{1 + \chi_b}{1 + 2\chi_b} \qquad \text{and} \qquad -(\mu^2 + \chi_a) + (\chi_a^2 + 2\mu^2\xi)^{1/2} < \kappa < \frac{\chi_a^2}{2\mu} - \chi_a + \tfrac{1}{2}\xi. \qquad (6.48)$$

Tyson and Kauffman also showed that limit cycle solutions arise from Hopf bifurcations. By numerically suppressing the coupling and starting each cell with identical initial conditions a homogeneous oscillation is obtained. If the coupling is then introduced by allowing $\chi_{a,b}$ to become non-zero, this splits into an inhomogeneous oscillation as shown in Fig. 6.11. 'Complicated' oscillations were also observed for isolated parameter values.

6.6. Coupled Brusselators

Schreiber and Marek (1982a) undertook a detailed investigation of coupled

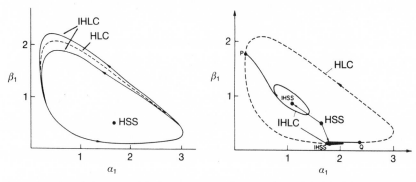

FIG. 6.11. Coexisting unstable homogeneous limit cycle and stable inhomogeneous oscillations for coupled cell mitosis model. In each case there are unstable stationary states (HSS, homogeneous stationary state; IHSS, inhomogeneous stationary state). The selection between the stable cycles depends on the initial conditions including any initial phase difference between the cells. (Reprinted with permission from Tyson and Kauffman 1975, © Springer-Verlag.)

versions of the Brusselator model (Section 3.12.1) in the form

$$\frac{dx_1}{dt} = A - (1 + B)x_1 + x_1^2 y_1 + D_x(x_2 - x_1) \tag{6.49}$$

$$\frac{dy_1}{dt} = Bx_1 - x_1^2 y_1 + D_y(y_2 - y_1) \tag{6.50}$$

$$\frac{dx_2}{dt} = A - (1 + B)x_2 + x_2^2 y_2 + D_x(x_1 - x_2) \tag{6.51}$$

$$\frac{dy_2}{dt} = Bx_2 - x_2^2 y_2 + D_y(y_1 - y_2). \tag{6.52}$$

They took $A = 2$, $B = 5.9$, fixed the ratio of exchange coefficients such that $D_x/D_y = 0.1$, and followed the response to changing D_x. Referring back to Section 3.12.1, the values of A and B are such that an uncoupled system will have a stable limit cycle solution surrounding the unstable stationary state.

The particular range of the coupling coefficient studied was $0.9 < D_x < 1.5$ and the behaviour observed is summarized in Figs 6.12 and 6.13. The stationary-state condition is equivalent to a quintic in, say, the concentration x_1. One solution is the homogeneous stationary-state $x_1 = x_2 = A$ and $y_1 = y_2 = B/A$: this exists for all D_x and is unstable as in the uncoupled system. The other solutions are inhomogeneous, with a pair of saddles and a second pair of states which are stable for low coupling coefficients. As D_x is increased, this latter pair undergo a Hopf bifurcation (occurring at the same condition for each of these two branches as they are simply mirror

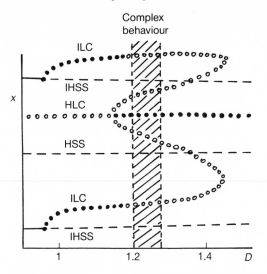

FIG. 6.12. The coupled Brusselator, showing development of homogeneous and inhomogeneous stationary states and limit cycles. In the shaded regions the inhomogeneous limit cycles have lost stability and chaotic attractors emerge. (Reprinted with permission from Schreiber and Marek 1982*a*, © North-Holland Publishing Co.)

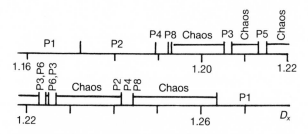

FIG. 6.13. Summary of period-doubling cascades to chaos and periodic windowing observed for inhomogeneous time-dependent solution with coupled Brusselators. (Reprinted with permission from Schreiber and Marek 1982*a*, © North-Holland Publishing Co.)

images) from which a stable period-1 solution emerges (at $D_x = 0.9542$).

The period-1 solution undergoes a period doubling at $D_x = 1.172$ and there is a full cascade culminating in chaotic behaviour at $D_x = 1.1933$. The chaotic region runs over the range $1.1933 \leqslant D_x \leqslant 1.2635$ but is interrupted by various periodic windows each of which have their own cascades back to aperiodicity. Thus the response shows the same universal behaviour as the simple one-dimensional mapping of Chapter 2. Indeed, the rate of convergence even reproduced the Feigenbaum constant 4.699 remarkably well. The chaotic attractors have one positive Lyapounov exponent and fractal dimensions slightly (but significantly) in excess of two.

As well as the responses just described, we have also to recognize that there can be multiplicity of attractors for some parameter values. For instance, with $D_x > 1.1219$, there is a stable period-1 limit cycle corresponding to a homogeneous oscillation coexisting with the various inhomogeneous periodicities and aperiodicities.

The same authors (Schreiber and Marek 1982b) also found a bifurcation from a torus to a strange attractor in this model, for slightly different parameter values. Taking $A = 2$, $B = 5.5$, and again keeping $D_x/D_y = 0.1$ as D_x varies, the following sequence occurs in $0.05 \leqslant D_x \leqslant 0.055$. For $D_x > 0.052\,95$, the system displays a homogeneous period-1 oscillation. As D_x is reduced the periodic solution changes to one in which the two cells oscillate out of phase. For $D_x = 0.052\,50$, this gives way to a two-frequency motion on a torus. The motion has two zero Floquet multipliers and hence corresponds to a two-dimensional attractor, as expected. Phase locking occurs at $D_x = 0.052\,46$, with a 3:1 frequency ratio emerging. The dimension of the attractor then reduces to one for this regular limit cycle. We return to a two torus at $D_x = 0.052\,44$, but this bifurcates to a fractal strange attractor at $D_x = 0.052\,38$. The highest Floquet multiplier becomes positive and grows in magnitude as the coupling coefficient is decreased further, and the dimension of the attractor increases above two. If D_x is reduced below $0.052\,06$, however, the system moves from this aperiodic behaviour and finds a symmetric stationary-state solution.

The onset of fractal character from the torus can be demonstrated from the Poincaré plane projection. The relatively simple and connected cross-section of the regular two torus becomes more and more 'buckled' as we approach the bifurcation (Fig. 6.14).

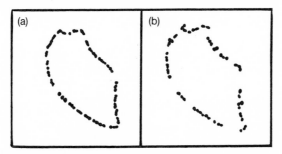

FIG. 6.14. Buckling of invariant circle in Poincaré plane as attractor becomes strange for coupled Brusselators. (Reprinted with permission from Schreiber and Marek 1982b, © North-Holland Publishing Co.)

In a remarkable paper, Wang and Nicolis (1987) have shown both analytically and numerically that birhythmicity and a stable three torus can also arise in coupled chemical oscillators generically and for this model in particular. Using the simplification of coupling only through the autocatalyst the system has only one stationary-state solution (see Section 6.2). Parameter conditions such that two different Hopf bifurcation points interact can be identified by requiring both pairs of eigenvalues to have vanishing real parts simultaneously. Previous theoretical studies (Takens 1974; Iooss 1981) had indicated that the above responses can be expected as the system is unfolded about such a codimension-2 degeneracy by varying the parameters slightly.

Wang and Nicolis discovered a suitable transformation of the original equations into the appropriate normal forms for this unfolding and could thus predict two different scenarios. In both of these, the first bifurcation is a primary Hopf taking the system from a stable stationary state to an unstable state with stable limit cycle. If the now unstable stationary state undergoes another bifurcation, the emerging solution will be unstable, at first, but may then bifurcate itself to reveal a second stable periodic state, with a frequency different from the first. There will then be two coexisting periodic solutions—a phenomenon known as birhythmicity.

Alternatively, the secondary bifurcation introducing the second frequency can occur for the periodic state. If the two frequencies are incommensurate, the resulting attractor will be a two torus as usual. There is then also the possibility of a third Hopf bifurcation in this (four-dimensional) system. If the parameters are varied sympathetically, the three torus can be stabilized so that it does not convert to a strange attractor following the usual Ruelle–Takens–Newhouse route (Ruelle and Takens 1971; Newhouse *et al.* 1978). The three torus here has sufficient 'structural stability' for examples to have been computed by Wang and Nicolis and a Fourier transform obtained (Fig. 6.15).

6.7. Coupled cross-shaped diagrams

Another model which has been used to interpret bistability and oscillations for reactions in CSTRs is that introduced by Boissonade and De Kepper (1980). This leads to a characteristic 'cross-shaped diagram' and has the differential form for a single or uncoupled cell

$$\frac{dx}{dt} = -x^3 + \mu x - \lambda - ky \qquad (6.53)$$

$$\frac{dy}{dt} = \frac{x - y}{\tau}. \qquad (6.54)$$

The first equation introduces a cubic nullcline in the concentration phase plane of the form discussed in Section 3.12.3 for the Oregonator scheme. The second equation can be tuned by varying the parameter τ to allow unstable intersections and relaxation oscillations. Figure 6.16 shows the curves of

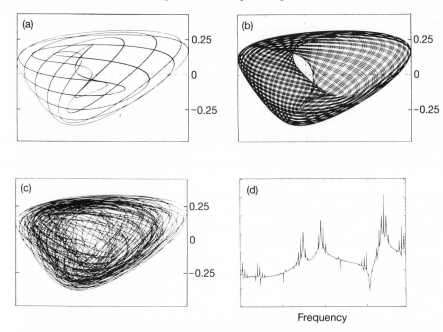

FIG. 6.15. Development of attractor from (a) phase locking through (b) T^2 torus to (c) T^3 torus; (d) shows the power spectrum for the T^3 torus (three-frequency quasiperiodicity). (Reprinted with permission from Wang and Nicolis 1987, © North-Holland Publishing Co.)

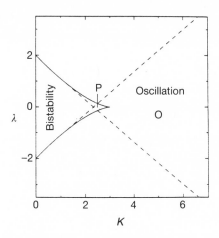

FIG. 6.16. Cross-shaped diagram showing loci of saddle–node bifurcation points (solid lines) and Hopf bifurcations (dashed lines) for the 'normal form' equations of Boissonade and De Kepper. (Reproduced with permission from Hocker and Epstein 1989, © American Institute of Physics.)

saddle–node and Hopf bifurcations in the λ–k parameter plane for this system with $\mu = 3$ and $\tau = 1$. The saddle–node curves form a cusp given by

$$\lambda = \pm \frac{2}{3\sqrt{3}} (\mu - k)^{3/2}.$$

Within this the system has three stationary-state solutions, of which two may be stable (leading to bistability). The two Hopf point loci have the form

$$\lambda = \pm (\tfrac{2}{3}\mu - k + 3\tau^{-1})[\tfrac{1}{3}(\mu - \tau^{-1})]^{1/2}.$$

These originate from points along one of the saddle–node curves (at double zero eigenvalue degeneracies with $\lambda = \pm (2/3\sqrt{3})(\mu - \tau^{-1})^{3/2}$). The Hopf curves cross the saddle–nodes close to the cusp point, although there is a small region P in which multiple stationary-state solutions and limit cycles coexist. In the region O, there is a stable limit cycle solution surrounding a unique stationary state. With the above equations, the two Hopf branches continue to separate for increasing k (Guckenheimer 1986): experimentally, the region of oscillations tends to join up into a closed loop.

The simple form of this scheme has aided the systematic design of many inorganic oscillators by modifying the wider class of bistable reactions (see e.g. De Kepper and Boissonade 1985). The equations can also be transformed through a redefinition of parameters and variables to the form of the Fitzhugh–Nagumo model for neural systems (Fitzhugh 1961; Nagumo *et al.* 1962).

Hocker and Epstein (1989) considered the coupling of two cells in each of which the above model operates. In most of their work they made the simplifying assumption that $\lambda = 0$ and took $\mu = 3$ and $\tau = 1$ as above. The resulting rate equations for two coupled cells are then

$$\frac{dx_1}{dt} = -x_1^3 + 3x_1 - k_1 x_1 - D(x_1 - x_2) \qquad (6.55)$$

$$\frac{dy_1}{dt} = x_1 - y_1 - D(y_1 - y_2) \qquad (6.56)$$

$$\frac{dx_2}{dt} = -x_2^3 + 3x_2 - k_2 x_2 - D(x_2 - x_1) \qquad (6.57)$$

$$\frac{dy_2}{dt} = x_2 - y_2 - D(y_2 - y_1). \qquad (6.58)$$

With $k_1 = 1$ and $k_2 = 6$, cell 1 lies in the region of bistability whilst cell 2 is in that for oscillations. For non-zero D, therefore, the model couples these two dissimilar responses. For small coupling coefficients, $D \lesssim 0.9$, period-1 oscillations occur in both cells. The oscillations in cell 1 differ in form and amplitude from those in cell 2 because of the underlying asymmetry in the parameters. Hocker and Epstein refer to this as birhythmicity but we will reserve that nomenclature for a different situation of coexisting stable limit cycles. For large D, greater than about 1.2, the cells become fully coupled and move on the same limit cycle. In between these states, however, the system shows a period-doubling sequence as D is increased which leads to chaos. The maximum Lyapounov exponent clearly becomes positive and the dimension of the strange attractor was calculated to lie in the range 2.03 to 2.08. Various periodic windows with complex oscillations were also observed.

Hocker and Epstein were able to explain this pattern of response in terms of homoclinicity. The simplification $\lambda = 0$ in the rate equations means that one stationary-state solution is just $x_1 = x_2 = y_1 = y_2 = 0$, the origin in the phase plane. There are two other stationary states which are non-zero. For low values of D, these non-zero states are unstable and surrounded by a stable limit cycle. As D is increased, these limit cycles pass ever closer to the origin. At $D = 1$ both limit cycles will simultaneously form a homoclinic loop with the saddle at the origin. Figure 6.17 sketches this double homo-

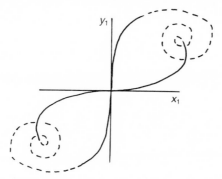

FIG. 6.17. Double homoclinic loop for coupled Boissonade–De Kepper model. (Adapted with permission from Hocker and Epstein 1989, © American Institute of Physics.)

clinic loop. For parameter values close to this condition, the system can exhibit characteristic 'homoclinic chaos' (see Section 4.10). The unstable manifold (or outset) from the saddle takes the trajectory away from the origin towards one of the non-zero states. This stationary state is unstable in an absolute sense, with two eigenvalues having positive real parts, but the other two may have negative real parts and allow for a lower-dimensional attracting submanifold: these solutions are saddle foci and so are approached in one plane but left in another. There is a *reinjection* of the trajectory back to the vicinity of the origin along its saddle inset. This mechanism for the creation of chaotic orbits is also found in the Lorenz model (Section 4.10.1).

In a second traverse through the parameter plane for this model, Hocker and Epstein considered the coupling of two oscillators. Again the original uncoupled cells are allowed to have different parameter values so the autonomous oscillations in each have different amplitudes and periods. Various forms of response are encountered. From the previous chapter we know to expect quasiperiodicity, entrainment, and chaos in such systems and these are indeed found. Also possible is *phase death*, whereby two cells which oscillate when uncoupled can interact in such a way as to settle on to a stationary state when coupled.

6.8. More than two cells

We have already seen a wealth of complex behaviour displayed by the relatively simple case of two, identical, two-variable coupled cells. A two-cell system then has four variables. If we add further units, the number of variables increases by two at each stage: allowing different conditions in each cell increases the number of parameters to be considered. Not surprisingly, then, most studies of systems with many coupled oscillators tend to be limited to special aspects rather than to generalities. The most popular questions are those relating to the possibilities of various inhomogeneous stationary-state configurations, the conditions under which the whole assembly can become completely synchronized, and the behaviour of arrays comprising weakly coupled units.

As examples of the first of these, Bar-Eli (1985) and Kim and Hlavacek (1986) have considered the behaviour when varying numbers of cells are coupled in a variety of configurations. Bar-Eli used the Brusselator scheme and showed that each unit could, under suitable parameter conditions, choose between either of two possible stable stationary states. Different spatial patterns for the whole system are then constructed by persuading the individual units to select the required state relative to its neighbours. For example, if there are two states a and b and four cells, the arrangements abba and abab represent different 'spatial patterns'. As the number of units increases, the number of possible permutations increases. The whole system may also show multistability, in that two or more 'spatial patterns' may coexist for the same parameter values, with selection between them being determined by the initial conditions. Generally, the ranges of multistability as well as the variety of patterns increase as the number of units increases.

Large arrays of oscillators that are then only weakly coupled (the exchange parameters have small magnitude) are also of wide interest. Many distributed chemical and even living systems can be approximated in such a way. Weak coupling allows only entrainment or quasiperiodicity with small numbers of units, but with large arrays other aspects, such as the propagation of disturbances or 'reaction waves', can be considered.

Ermentrout and Kopell (1983) examined the behaviour of a chain of weakly coupled units, but also imagined that the chain was embedded in

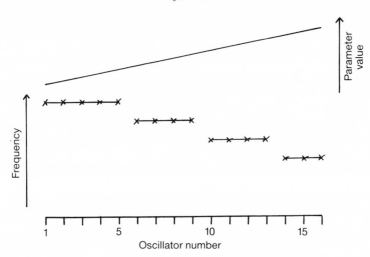

Fɪɢ. 6.18. Schematic representation of frequency plateau responses in a chain of weakly coupled oscillators subject to a parameter gradient.

some 'concentration gradient' so that one of the parameters varied slowly from unit to unit along the chain. With such circumstances, the system responds by forming 'frequency plateaux'. A given section of the chain may become synchronized, so that all of its units have the same oscillatory frequency, but there will then be a discontinuity between the unit at the end of this section and the next along the chain, which begins another section. Thus, as sketched in Fig. 6.18, there are various steps in frequency.

With two-dimensional arrays, mutual entrainment between neighbours over restricted areas within the whole domain can lead to dramatic patterning (see e.g. Kuramoto 1985; Kuramoto and Nishikawa 1988) and to similar spatial chaos to that discussed in Section 4.7.6.

More mathematical aspects of general coupled systems can be found in the extensive article by Aronson *et al.* (1987).

References

Aronson, D. G., Doedel, E. J., and Othmer, H. G. (1987). An analytical and numerical study of the bifurcations in a system of linearly-coupled oscillators. *Physica*, **25D**, 20–104.

Ashkenazi, M. and Othmer, H. G. (1978). Spatial patterns in coupled biochemical oscillators. *J. Math. Biol.*, **5**, 305–50.

Bar-Eli, K. (1985). On the stability of coupled chemical oscillators. *Physica*, **14D**, 242–52.

Boissonade, J. and De Kepper, P. (1980). Transitions from bistability to limit cycle oscillations. Theoretical analysis and experimental evidence in an open chemical system. *J. Phys. Chem.*, **84**, 501–6.

Crowley, M. F. and Epstein, I. R. (1989). Experimental and theoretical studies of a coupled chemical oscillator: phase death, multistability and in-phase and out-of-phase entrainment. *J. Phys. Chem.*, **93**, 2496–502.

Crowley, M. F. and Field, R. J. (1986). Electrically coupled Belousov–Zhabotinskii oscillators. 1. Experiments and simulations. *J. Phys. Chem.*, **90**, 1907–15.

De Kepper, P. and Boissonade, J. (1985). From bistability to sustained oscillations in homogeneous chemical systems in flow reactor mode. In *Oscillations and travelling waves in chemical systems*, (ed. R. J. Field and M. Burger), Ch. 7, pp. 223–56. Wiley, New York.

Ermentrout, B. G. and Kopell, N. (1983). Frequency plateau in a chain of weakly coupled oscillators. *SIAM J. Math. Anal.*, **15**, 215–37.

FitzHugh, R. (1961). Impulses and physiological states in theoretical models of nerve membrane. *Biophys. J.*, **1**, 445–66.

Guckenheimer, J. (1986). Multiple bifurcation problems for chemical reactors. *Physica*, **20D**, 1–20.

Hocker, C. G. and Epstein, I. R. (1989). Analysis of a four-variable model of coupled chemical oscillators. *J. Chem. Phys.*, **90**, 3071–80.

Iooss, G. (1981). *Nonlinear phenomena in chemical dynamics*, (ed. C. Vidal and A. Pacault). Springer, New York.

Kim, S. H. and Hlavacek, V. (1986). On the detailed dynamics of coupled continuous stirred tank reactors. *Chem. Eng. Sci.*, **41**, 2767–77.

Kuramoto, Y. (1985). *Chemical oscillations, waves and turbulence*. Springer, Berlin.

Kuramoto, Y. and Nishikawa, I. (1988). Onset of collective rhythms in large populations of coupled oscillators. *Proc. 2nd Int. Yukawa Seminar on Cooperative dynamics in complex physical systems*. Springer, Berlin.

Langford, W. F. and Iooss, G. (1980). Interactions of Hopf and pitchfork bifurcations. In *Bifurcation problems and their numerical solution*, (ed. H. D. Mittelmann and H. Weber), pp. 103–34. Birkhäuser, Basel.

Leach, J. A., Merkin, J. H., and Scott, S. K. (1991). An analysis of a two-cell coupled non-linear chemical oscillator. *Dyn. Stab. Syst.*, **6**.

Mankin, J. C. and Hudson, J. L. (1986). The dynamics of coupled nonisothermal continuous stirred tank reactors. *Chem. Eng. Sci.*, **41**, 2651–61.

Nagumo, J. S., Arimoto, S., and Yoshizawa, S. (1962). An active pulse transmission line simulating nerve axon. *Proc. IRE*, **50**, 2061–71.

Newhouse, S., Ruelle, D., and Takens, F. (1978). Occurrence of strange axiom A attractors near quasiperiodic flows on T^m, $m \geqslant 3$. *Commun. Math. Phys.*, **64**, 35–40.

Ruelle, D. and Takens, F. (1971). On the nature of turbulence. *Commun. Math. Phys.*, **20**, 167–92.

Schreiber, I. and Marek, M. (1982a). Strange attractors in coupled reaction–diffusion cells. *Physica*, **5D**, 258–72.

Schreiber, I. and Marek, M. (1982b). Transition to chaos via two-torus in coupled reaction–diffusion cells. *Phys. Lett.*, **91**, 263–6.

Sel'kov, E. E. (1968). Self-oscillation in glycolysis. *Eur. J. Biochem.*, **4**, 79–86.

Takens, F. (1974). Singularities of vector fields. *Publ. Math. IHES*, **43**, 47–100.

Tyson, J. J. (1973). Some further studies of nonlinear oscillations in chemical systems. *J. Chem. Phys.*, **58**, 3919–30.

Tyson, J. J. and Kauffman, S. (1975). Control of mitosis by a continuous biochemical oscillation; spatially inhomogeneous oscillations. *J. Math. Biol.*, **1**, 289–310.

Tyson, J. J. and Light, J. C. (1973). Properties of two-component bimolecular and trimolecular chemical reaction schemes. *J. Chem. Phys.*, **59**, 4164–73.

Waller, I. and Kapral, R. (1984). Synchronization and chaos in coupled nonlinear oscillators. *Phys. Lett.*, A **105**, 163–8.

Wang, X.-J. and Nicolis, G. (1987). Bifurcation phenomena in coupled chemical oscillators: normal form analysis and numerical simulations. *Physica*, **26D**, 140–55.

7

Experimental methods

This chapter sets out to link the various theoretical methods described in the previous chapters to their application in experimental situations. In the latter case we must learn to cope with the 'real world' limitations on the availability and precision of our data. The theoretician or computer modeller is generally in an atypically privileged position in this respect. In principle, all species concentrations can be made available at any fixed time interval or other choice of sampling condition. The data can be obtained, and the parameters maintained constant, to almost arbitrary precision (within the confines of a computing budget). Such control over the raw data allows further manipulation such as constructing limit cycles or other attractors. Poincaré sections, etc. to be made with relative ease.

By contrast, the experimentalist can have a much harder time of things. Variables become concentrations or temperatures: parameters become experimental conditions. The data set collectable will almost certainly be restricted in some sense. Typically only one or two species will be monitored directly, and in many cases the raw measurement may have a rather ambiguous relationship to actual concentrations (the potential of a platinum electrode or a total light emission intensity, for instance). The frequency with which the data can be collected will be dictated by practical considerations such as the sampling speed and size of the hard disk on the dedicated microprocessor available in the particular laboratory. The analogue-to-digital (A-to-D) converters employed will impose a finite (and generally rather low) resolution of the data, and there will be the ever-present problem of experimental 'noise'.

As experimentalists we will still wish to reconstruct our attractors and otherwise treat our limited data in some reliable way that will uncover some of the mysteries of our complex bifurcations and hence allow some chemical and dynamical interpretation. We will seek to determine Lyapounov exponents as before, in order to distinguish clearly between complex periodicity and true chaos. Below are described a number of techniques that are currently used to achieve such aims. These frequently require the experimenter to obtain just a single time series record of some response of the system, but do not need a direct measurement of an actual concentration. It is perhaps also worth noting that each method has its limitations and thus it is generally both difficult and dangerous to pass comment on the data from a single experiment, or even repeated experiments for the same experimental conditions. The safest and best conclusions can be drawn following a series of measurements, beginning with relatively simple and recognizable responses, and that then allow the development of complexity as the conditions are changed to be observed and related to one of the characteristic routes discussed in Chapter 4.

7.1. Manipulation of rata data: smoothing, interpolation, and reconstruction

Typically, we will obtain from an experiment a single record of some measureable such as the potential of an electrode pair or the output voltage from a thermocouple. This 'time series' will consist of data collected, usually, at regular time intervals—perhaps every 50 ms. Unfortunately, this may not be the most appropriate form for our purposes. Perhaps we wish to make a Poincaré section, in which case we would have preferred data captured in such a way as to make available the values of suitable quantities each time a further variable attains a specified value. In order to make such manipulations on our real data, we will need some method of interpolation. There are various standard methods for this: a cubic spline fitting of the data is widely employed, and allows some smoothing at the same time. The transformed time series can then be subjected to further tortures, such as numerical differentiation.

An important treatment of the smoothed data set is that of 'reconstructing the attractor'. In this process, we seek to use our single time series to obtain some representation of the limit cycle or other attractor that the system possesses. A popular method for this is that based on a time delay, due to Takens (1981). Let us assume that the observed experimental trace has all the indications of a regular period-1 oscillation that we expect to correspond to a simple (two-dimensional) limit cycle. The time series will consist of a sequence of measurements of some quantity x, the individual elements x_i corresponding to times t_i that will be simple integer multiples of the sampling period Δt.

A two-dimensional plot will require a series of (x, y) pairs that must be constructed from our single data set. For our x coordinate, we can take the direct data $x_i = x(t_i)$. For the corresponding y coordinate, we use the value of x *at some later time* $y_i = x_j = x(t_i + T_d)$ where T_d is some suitably chosen, constant 'delay time'. As an example, consider the data in Table 7.1. This shows the first 10 points of some quantity recorded at 50 ms intervals. If we decide to try a delay time of 200 ms, then the first few (x, y) pairs will be $(51, 17)$, $(37, 29)$, $(24, 42)$, etc. corresponding to $(x(t = 0), x(t = 0 + 200))$, $(x(t = 50), x(t = 50 + 200))$, and so on.

Figure 7.1 shows the reconstructed limit cycle attractor from an oscillatory

TABLE 7.1
Sample raw data for reconstruction

t/ms	0	50	100	150	200	250	300	350	400	450
x	51	37	24	12	17	29	42	57	39	26

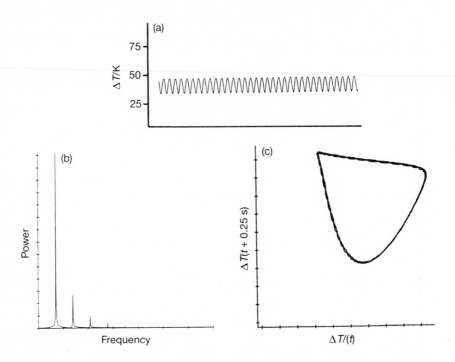

FIG. 7.1. Experimental time series (a). Fourier transform power spectrum (b), and (c) reconstructed limit cycle attractor using delay method for $H_2 + O_2$ reaction. (Courtesy of B. Johnson.)

time series from the gas-phase oxidation reaction of hydrogen (Chapter 9). For this reconstruction a delay period of 250 ms was chosen whilst the oscillatory period is 370 ms. Even with unsmoothed data a convincing limit cycle is obtained. This suggests that the $H_2 + O_2$ system can show dynamics that is basically two-dimensional, even though we will see later that there are many different chemical species participating in the reaction mechanism.

The choice of delay time is somewhat arbitrary, and finding the appropriate value can prove a limitation on this delay technique. Provided the oscillations have a relatively 'nice' waveform (the example above occurs close to the supercritical Hopf point at which oscillations begin and hence has a simple

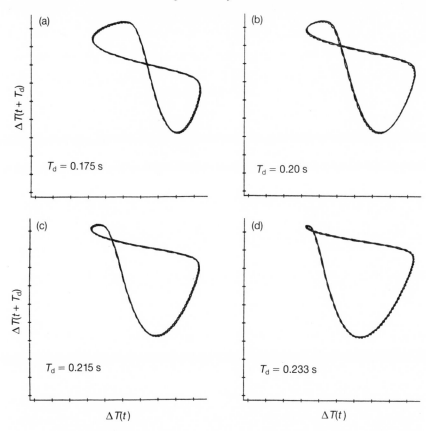

FIG. 7.2. Influence of choice for delay time T_d on reconstructed attractor from Fig. 7.1(c). (Courtesy of B. Johnson.)

power spectrum indicating only a few contributing overtones) the appearance of the attractor can be insensitive to the exact value chosen for T_d. The data from Fig. 7.1 are shown with a selection of delay times in Fig. 7.2. In this case, changing the delay simply seems to be equivalent to altering the direction from which the attractor is being viewed. The apparent crossing of the trajectory is not real, and we can think of the limit cycle lying on a two-dimensional surface in a higher-dimensional phase space.

Of course, we can only expect simple limit cycles to emerge from these (x, y) reconstructions if the time series show period-1 oscillations. If the time evolution is more complex, we know from the previous chapters that the attractor must lie in a phase space that is more than two dimensional. If we wish to reconstruct an attractor for a complex periodic oscillation or a strange attractor we will have to increase the dimension of our plot. To obtain a three-dimensional reconstruction we need to assemble a sequence

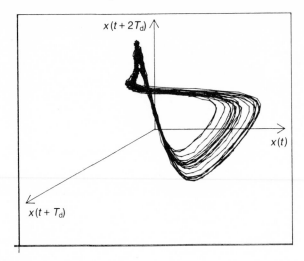

FIG. 7.3. Reconstructed three-dimensional attractor from two-delay method. (Courtesy of P. Ibison.)

of (x, y, z) triples. This is done by applying the delay method twice. Thus with a delay T_d, the cordinates of the ith point will be $(x(t = t_i), x(t = t_i + T_d), x(t = t_i + 2T_d))$. For example, with $T_d = 200$ ms the first two triples from Table 7.1 would be $(51, 17, 39)$ and $(37, 29, 26)$. An example of the reconstructed three-dimensional attractor for a complete time series is shown in Fig. 7.3.

The dimensionality of the phase space that we chose to work with is known as the 'embedding dimension'. It is important to use a dimension higher than that of the attractor, but otherwise we will generally seek the lowest possible.

The time delay method has many similarities to an earlier technique of Packard *et al.* (1980). These authors used phase plots employing the (numerically determined) derivatives of the time series. This idea follows closely the standard procedure from mechanics of plotting position, velocity, acceleration, etc.

More recently, Broomhead and King (1986*a,b*) have adapted 'singular system analysis' (Bertero and Pike 1982; Pike *et al.* 1984) from information theory in order to deal more systematically with typically noisy data and the choice of time delay and embedding dimension. Their approach can be illustrated by considering a time series $x_i(t_i)$ as before. An 'n-window' is constructed in the same way as we would form an n-dimensional vector via the time delay method, except that now we can take the delay time to be simply that between successive points (if we wish)—the 'sampling time' t_s. Then the first n-window, w_1, would be $(x_1, x_2, \ldots, x_{n+1})$, and so on.

If the time series has a total of N_T records, we can construct $N = N_T - (n-1)$ such n-windows. These can then be combined to form an $(N \times n)$ 'trajectory matrix' $\mathbf{X}$

$$\mathbf{X} = N^{-1/2} \begin{pmatrix} w_1 \\ w_2 \\ \vdots \\ w_N \end{pmatrix} = N^{-1/2} \begin{pmatrix} x_1 & x_2 & x_3 & \cdots & x_n \\ x_2 & x_3 & x_4 & \cdots & x_{n+1} \\ \vdots & \vdots & \vdots & & \vdots \\ x_N & x_{N+1} & x_{N+2} & \cdots & x_{N_T} \end{pmatrix}$$

(the factor $N^{-1/2}$ is used to normalize the matrix).

Three more important quantities emerge: a set of singular values σ_i and two sets of singular vectors c_i and s_i of $\mathbf{X}$. These are specified by the conditions

$$\mathbf{XC} = \mathbf{S\Sigma} \qquad \mathbf{X}^\mathrm{T}\mathbf{S} = \mathbf{C\Sigma}$$

where $\mathbf{\Sigma}$ is a diagonal matrix of the singular values, $\mathbf{\Sigma} = \mathrm{diag}(\sigma_1, \sigma_2, \ldots, \sigma_n)$ ordered such that $\sigma_1 \geqslant \sigma_2 \geqslant \cdots \geqslant \sigma_n \geqslant 0$. The columns of $\mathbf{C}$ and $\mathbf{S}$ then give the two sets of eigenvectors associated with each of these singular values: specifically the c_i are the eigenvectors of the covariance matrix of the original n-windows $\mathbf{\Xi} = \mathbf{X}^\mathrm{T}\mathbf{X}$ whilst the s_i are eigenvectors of the matrix $\mathbf{\theta} = \mathbf{XX}^\mathrm{T}$.

An important concept in this procedure is that the n singular values σ_i can frequently be divided into two subsets. Those with largest magnitude, $\sigma_1, \ldots, \sigma_d$, with $d < n$ say, with their associated eigenvectors define a deterministic subspace of dimension d. The remaining $n - d$ singular values will tend to be of low magnitude, corresponding to some noise-dominated 'complementary space'. An inspection of the magnitudes of the σ_i should thus suggest a suitable value for d and we should thus choose $n > d$.

The window length n and the sampling time t_s can also be systematically related by inspecting the Fourier transform power spectrum. If some frequency v^* can be identified above which there is no significant area (power) under the transform curve, then we should choose $nt_sv^* = 1$. The various matrix manipulations required in determining the singular values and vectors are made less onerous by the various symmetries in the trajectory matrix.

Broomhead and King applied this method to data from the Lorenz attractor model, for which a comparison with 'exact' numerical behaviour could be made. For parameter values such that the motion became chaotic they consistently found that the deterministic subspace was three dimensional and were able to identify the corresponding eigenvectors c_1, c_2, and c_3. The projections of the n-window vectors from the trajectory matrix on to these eigenvectors show exactly the same structural features as the directly generated attractors (from the combined x, y, and z time series) as shown in Fig. 7.4. This method has also been applied to genuine experimental data, although not that from a chemical origin.

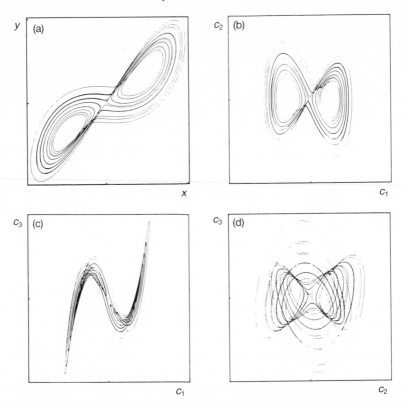

F$_{\text{IG}}$. 7.4. Illustration of the singular spectrum analysis method of attractor reconstruction for the Lorenz model (see Section 4.10.1) with $Pr = 10$, $b = \frac{8}{3}$, and $r = 28$:
(a) projection of strange attractor on to x–y plane from full numerical solutions;
(b)–(d) projections on to the three principal singular vector components c_1–c_3; (b)
shows the same topological structure as (a). (Reproduced with permission from
Broomhead and King (1986*a*), © North-Holland, Amsterdam.)

Of the three reconstruction methods described, that due to Takens is the
most widely used at present, and is typically that incorporated into software
packages such as 'Dynamical Software' (Schaffer *et al.* 1988).

7.2. Further treatment of data

The smoothed and interpolated time series data and the vectors formed for
the reconstructions can be further manipulated in much the same way as
numerical output. Standard library routines for fast Fourier transforms allow
the power spectra to be readily obtained. Poincaré sections, next-return and
next-amplitude, maps, etc. come from the reconstructed attractor. Some

remarkably clear sections and attractors have been derived from experimental results, despite the inevitable accompanying noise, and many examples will be presented in the forthcoming chapters. The significance of experimental noise has been considered in some detail by Schneider and his group (Schneider *et al.* 1989; Freund *et al.* 1986, 1988) who suggest that Lyapounov exponents and dimension of the attractor generally can be reliable measures in distinguishing deterministic chaos from complex periodicity.

7.3. Lyapounov exponents

Lyapounov exponents have been introduced in Chapter 2 (Section 2.6). These quantities are generalized eigenvalues, playing a similar role for any type of attractor to those played particularly for stationary states by the Jacobian eigenvalues or for periodic solutions by Floquet exponents. They are measures of the average exponential decay or growth of infinitesimally small differences in initial conditions or infinitesimal perturbations of the motion along an attractor (or 'of orbits in the phase space'). Viewed in a geometric way, they can be considered to determine the evolution of a small, n-dimensional volume element around the evolving attracting trajectory in the n-dimensional phase space via the growth or decay of a suitably chosen set of axes.

Typically, the dominant evolution will be that which has the 'maximum Lyapounov exponent' (i.e. that which is most positive) and so most interest will be in determining this from the full spectrum. However, we may occasionally wish to find subsequent exponents—perhaps all the positive ones or to help distinguish between an apparently positive but small magnitude result and zero.

For a stable stationary state, all perturbations decay and hence all differences in initial conditions decay. Similarly, the n-dimensional test volume must shrink in all direction to a point. The Lyapounov exponents for such an attractor are all negative. For a stable periodic solution, most disturbances decay except for the special perturbation that corresponds to the direction of motion locally along the one-dimensional limit cycle. This latter mode will not decay on average, but neither will it grow. Thus there is a corresponding (maximum) Lyapounov exponent equal to zero. The remaining exponents will be negative.

With a quasiperiodic solution the attractor is the two-dimensional surface of a torus. Perturbations in either of these two dimensions will persist, giving two zero Lyapounov exponents, with the remainder being negative. For aperiodic attractors, however, there is an exponential divergence from similar but not identical initial conditions—the very sensitivity that characterizes chaos. Thus a chaotic strange attractor will have at least one associated positive Lyapounov exponent (and also will still have one equal to zero). In fact, the possession of a positive Lyapounov exponent can (and will here) be taken as the *definition* of a chaotic system. In some cases there may be multiple positive exponents.

Table 7.1 summarizes the different possibilities for the Lyapounov spectra for three- and four-variable systems. Note that there must always be at least one negative exponent for a bounded flow (i.e. one that does not 'blow up' to infinity) and the sum of the exponents must also be negative.

For model schemes or mechanisms for which the governing differential equations are known and to be integrated, Lyapounov exponents can be evaluated simultaneously with the time series—although there may be practical difficulties of obtaining convergence in a finite number of steps. If the evolution of a system on a particular attractor $\phi(t)$ is governed by the equation (in vector form)

$$d\mathbf{x}/dt = \mathbf{f}(\mathbf{x}) \tag{7.1}$$

where the concentration (and perhaps temperature) variables are $\mathbf{x} = (x_1, \ldots, x_n)^T$ and $\mathbf{f}$ gives the n appropriate rate equations, and if $\mathbf{h}$ represents a small test volume around ϕ, we should allow $\mathbf{h}$ to be sufficiently small so that we can approximate its evolution by a linearized flow

$$d\mathbf{h}/dt = f_{\mathbf{x}}\mathbf{h} \tag{7.2}$$

where $f_{\mathbf{x}}$ is the (time-dependent) Jacbian evaluated as $\mathbf{x}$ flows its solution $\phi(t)$. If eqns (7.1) and (7.2) are integrated along the solution path ϕ for a sufficient time, the Lyapounov exponents evaluated as appropriate vector norms should tend to constant values (Shimada and Nagishima 1979; Seydel 1988). Thus we now have $2n$ equations to integrate. There can be operational difficulties with this direct approach, particularly in finding a suitable initial choice for the text volume $\mathbf{h}(0)$. Also, if any of the directions grow (i.e. the corresponding exponent is positive—the case we are most interested in) the linearization may fail to approximate the evolution adequately and so there is a need for occasional rescaling of the equations.

Clearly the above approach will not be applicable to experimental data. Of more value, therefore, are methods for estimating the Lyapounov spectrum, or at least the most significant elements, from time series. Such methods could then be applied to either experimental or numerical data. At present, the most widely used approached is that derived by Wolf *et al.* (1985). These authors also provided FORTRAN coding to implement their algorithm and this has been incorporated into standard packages (e.g. Dynamical Systems Software).

An initially (n-dimensional) spherical test volume above will deform under the flow to a general n-ellipsoid. The Lyapounov exponents λ_i can be defined in terms of the principal axes p_i of this ellipsoid:

$$\lambda_i = \lim_{t \to \infty} \frac{1}{t} \log_2 \frac{p_i(t)}{p_i(0)} \tag{7.3}$$

where $p_i(0)$ is the initial length of the ith axis.

If the exponents are ordered, the linear extent of the ellipse grows as $2^{\lambda_1 t}$, the area spanned by the two principal axes grows as $2^{(\lambda_1 + \lambda_2)t}$; the volume spanned by the three principal axes grows as $2^{(\lambda_1 + \lambda_2 + \lambda_3)t}$; etc. It is this latter interpretation on which the method of Wolf *et al.* (1985; Wolf 1986) for determining non-negative exponents builds.

The techniques of reconstruction described in the previous sections allow us to obtain a series of discrete points that lie on an attractor in an appropriate phase space from an experimental time series. This attractor will have the same Lyapounov spectrum as the 'real' attractor based in a 'real' variable-state space. If the system contains a strange attractor, we can usefully think of the time-delayed points as forming a series of tracks as the trajectory orbits around the fractal structure. We can also estimate the distance beween any two points from the usual square root of the sum of the squares of the distances in each of the axes components:

$$d = \sqrt{(x^2 + y^2 + z^2 \cdots)}.$$

If two points are chosen initially, that lie relatively close together but on different tracks (separated by at least one mean orbital period), one can be thought of as a small perturbation of the other. If we then follow the growth in distance between these two as we move forwards in time around the points on the attractor, we will gain an estimate of the linear growth of the principal axis. This can be related to the maximum Lyapounov exponent by the definition above (i.e. linear growth $= 2\lambda t$). When implemented, this simple approach can only be followed for a finite time. As the distance between the points grows to be 'finite', so the problems mentioned above for the numerical evaluation arise. This can be circumvented by replacing one of the monitored points with another closer to the first. Thus we perform a 'piecewise' evaluation of the growth in distance.

We begin by choosing a reference point (perhaps the first point in the reconstructed data set) specified in terms of its coordinates $(x(t_0), x(t_0 + T_d)$, $\ldots, x(t_0 + (n - 1)T_d))$ in the n-dimensional space. The next step is to find the corresponding closest point on any other orbit, by calculating the Euclidian distance between our chosen point and all the others, but making sure we find a point on a different orbit. This latter may be achieved by ignoring the first m points ($m \ll n$) in the sequence after our initial point where m corresponds approximately to the number of points per orbit. Having minimized the distance (subject to this constraints) we now follow the evolution along the attractor, i.e. through the sequence of reconstructed

(a)

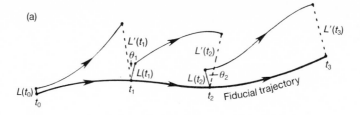

(b)

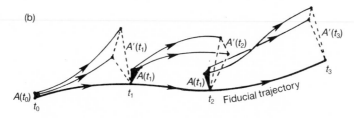

FIG. 7.5. Schematic representation of the Wolf *et al.* method of determining maximum Lyapounov exponents based on the growth of component axes length: (a) a single length scale is required for the largest exponent; the growth of the distance L between two initial points is followed with replacement of one of the points by a closer trajectory at regular intervals; (b) two exponents can be determined if the growth of area spanned by two axes (three points) is followed, again with regular replacement. (Reproduced with permission from Wolf *et al.* (1985), © North-Holland, Amsterdam.)

coordinates. As we move along, we can monitor any growth in the distance between our two evolving points. Let the initial distance be L_0 and that after some time Δt be L'_0. We now 'reorthonormalize' our measurement by discarding the second point and searching again for the nearest neighbour to our first, reference, point (again ensuring any such choice is on a different orbit). If this new point is a distance L_1 away we now proceed as before, monitoring the growth for a time period Δt, when it will have grown to a new distance L'_1. This process is repeated until the end of the data set is reached. At this stage we have a series of M pairs (L_i, L'_i) corresponding to initial and final distances between pairs of points. The maximum Lyapounov exponent can then be estimated as

$$\lambda_1 = \frac{1}{\Delta t} \sum_0^{M-1} \log_2 \frac{L'_k}{L_k}. \tag{7.4}$$

The basis of this method is shown schematically in Fig. 7.5(a), whilst 7.5(b) indicates the extension needed to obtain the second exponent. In the latter case it is necessary to choose three points and monitor the area of the triangle spanned. The same sort of procedures for reorthonormalizing the test area are used, but the search time becomes much longer. The growth in area leads to the sum $\lambda_1 + \lambda_2$ and hence, with λ_1 known, to λ_2. Full details are given in the excellent paper of Wolf *et al.* (1985) including additional considerations such as how to deal with noisy data (briefly, one avoids choosing points that lie too close together even if they are on different orbits).

Wolf *et al.* applied their method to experimental data from the Belousov–Zhabotinskii reaction, which will be discussed in detail in the next chapter, and to the Lorenz model equations, For the former, a value of 0.0054 ± 0.005 was obtained, in agreement with that estimated by other approaches. This is, perhaps, not conclusively differently from zero but also, perhaps, typical of the magnitudes for possibly weakly chaotic experimental data. For the numerical model, the algorithm described above gave a value from a time series within 3 per cent of the 'correct' result $\lambda_1 = 2.16$ determined by direct numerical integration and produced an estimate of $\lambda_2 < 0.1$ (instead of zero).

The units of the exponents, as defined above, are 'bits of information per second'. They thus measure the rate at which significant digits are lost. In a fairly refined experiment with a 16 bit data capture facility (representing optimistically a precision of 1 in 65 000) a result giving $\lambda_1 = 0.5$ indicates that all the initial significance is lost within 32 s.

7.4. Dimensionality of attractors

Another interesting quantity is the dimension of the attractor. A stationary state is a point is a phase space with, therefore, zero dimension. A limit cycle (periodic attractor) is one dimensional, a torus is two dimensional. From previous arguments we expect a chaotic or strange attractor typically to have a non-integer or fractal dimension, $2 < d < 3$ for a three-variable system. Various quantitative measures of attractor dimensionality have been proposed (see Farmer *et al.* 1983). Any choice should reproduce the integer values for the standard attractors just described. A useful definition, related to the Lyapounov spectrum, is that suggested by Kaplan and Yorke (1979): the information dimension d_f

$$d_f = j + \frac{\sum_{i=1}^{j} \lambda_i}{|\lambda_{j+1}|}. \tag{7.5}$$

The integer j in this expression is defined so that the sum of the first j Lyapounov exponents is positive but that of the first $j + 1$ is negative:

$$\sum_{i=1}^{j} \lambda_i > 0 \qquad \sum_{i=1}^{j+1} \lambda_i < 0. \tag{7.6}$$

Thus for the Lorenz system which has $\lambda_1 = 2.16$, $\lambda_2 = 0$, and $\lambda_3 = -32.4$ for a particular set of parameters, the sum over the first two exponents $\lambda_1 + \lambda_2$

is positive, but $\lambda_1 + \lambda_2 + \lambda_3$ is negative, so $j = 2$. Then eqn (7.5) gives $d_f = 2 + (2.16/32.4)$, i.e. a fractal dimension of 2.07.

Another method for evaluating fractal dimensions has been given by Grassberger and Procaccia (1983*a,b*. 1984; Grassberger 1981, 1986). In all cases, this sort of information can be used, for instance, to check that the embedding dimension chosen for the reconstruction is sufficiently large (it must exceed d_f). Similarly, this quantity is conserved throughout data manipulation, so any successful model should predict a dimension equal (within reason) to that determined for the reconstructed attractor.

7.5. Computing Floquet multipliers

As a final numerical note we return to the shooting method for evaluating periodic solutions described briefly in Chapter 3, and give an algorithmic route which allows the important quantities—the Floquet multipliers—to be obtained at no great extra cost. The details of this and various computational path-following methods appropriate to model systems can be found in texts such as Seydel (1988) or Kubicek and Marek (1983).

Again we imagine that our model equations can be expressed in the general form

$$\mathrm{d}\mathbf{x}/\mathrm{d}t = \mathbf{f}(\mathbf{x}, p) \tag{7.7}$$

where p is some parameter (or set of parameters). We are interested in conditions for which eqn (7.7) has a time-dependent solution $\boldsymbol{\phi}(t)$ that has some, probably as yet unknown, period T_p. We can use this period to scale time $\tau = t/T_p$, so the solution $\boldsymbol{\phi}(\tau)$ has period 1 (it does not matter that we do not known T_p yet). It is also useful to add an extra equation giving the evolution of the period T_p. The latter is actually a constant, so we have

$$\mathrm{d}\mathbf{x}/\mathrm{d}\tau = T_p\mathbf{f}(\mathbf{x}, p) \tag{7.8a}$$

$$\mathrm{d}T_p/\mathrm{d}\tau = 0. \tag{7.8b}$$

We will presumably start integrating eqns (7.8) with some chosen initial conditions $\mathbf{x}(0)$ and will compute until $\tau = 1$. If we have chosen the initial conditions properly, so they really lie on the desired periodic solution, then $\mathbf{x}(1) = \mathbf{x}(0)$. We can thus add the following boundary conditions to eqns (7.8):

$$\mathbf{x}(1) - \mathbf{x}(0) = 0. \tag{7.9a}$$

So far, we have $n + 1$ differential equations, with n boundary conditions. For the extra boundary condition, we may use a phase condition, perhaps of the form

$$x_i(0) = x_0 \tag{7.9b}$$

i.e. we choose a value for one of the variables. (In the terms used previously, this extra equation must specify the Poincaré plane: here this is chosen simply to be the $N - 1$ hypersurface $x_i =$ constant.) We now have equations in the form of an $(n + 1)$-variable two-point boundary-value problem, for which standard library routines are readily available. The 'unknowns' for which we are shooting are the $n - 1$ initial values of the variables not specifying the Poincaré plane, corresponding to a point on the limit cycle, and the period (alternatively we may use this method to obtain values for various of the parameters such that a particular limit cycle exists that has given values of two or more of the concentrations).

The library routines will employ an $(n + 1) \times (n + 1)$ iteration matrix, which is usually made available at the end of successful shooting but often not of great interest. In the present case, however, the $n \times n$ leading submatrix $\mathbf{E}$ of this iteration matrix is particularly useful. The 'monodromy matrix' for the system $\mathbf{M}$ is defined by

$$\mathbf{E} = \mathbf{I} - \mathbf{M}$$

where $\mathbf{I}$ is the identity matrix. The eigenvalues of $\mathbf{M}$ are then the Floquet multipliers for the periodic solution.

The transformation of apparently time-dependent equations into two-point boundary-value form is of wide applicability and great benefit. The system of equations can readily be extended by adding various extra conditions that then allow automatic path following, sketching out loci of bifurcations or degenerate bifurcations as required. The various packages PATH (Kaas-Petersen 1987), AUTO (Doedel 1981, 1986), and Dynamical Software mentioned previously generally make use of these approaches.

References

Bertero, M. and Pike, E. R. (1982). Resolution in diffraction-limiting imaging, a singular value analysis. I: the case of coherent illumination. *Opt. Acta*, **29**, 727.

Broomhead, D. S. and King, G. P. (1986*a*). Extracting qualitative dynamics from experimental data. *Physica*, **20D**, 217–26.

Broomhead, D. S. and King, G. P. (1986*b*). On the qualitative analysis of experimental dynamical systems. In *Nonlinear phenomena and chaos*, (ed. S. Sarkar), pp. 113–44. Adam Hilger, Bristol.

Doedel, E. (1981). AUTO: a program for the automatic bifurcation analysis of autonomous systems. *Congress Num.*, **30**, 265–84.

Doedel, E. (1986). AUTO 86 user's manual. *Appl. Math.*, pp. 217–50. California Institute of Technology, Pasadena, CA.

Farmer, J., Ott, E., and Yorke, J. A. (1983). The dimension of chaotic attractors. *Physica*, **7D**, 153–80.

Freund, A. Kruel, Th., and Schneider, F. W. (1986). Distinction between deterministic chaos and amplification of statistical noise in an experimental system. *Ber. Bunsenges. Phys. Chem.*, **90**, 1079–84.

Freund, A., Kruel, Th., and Schneider, F. W. (1988). Distinction between amplified noise and deterministic chaos by the correlation dimension. In *From chemical to biological organization*, (ed. M. Markus, S. C. Mueller, and G. Nicolis). Springer, Berlin.Grassberger, P. (1981). On the Hausdorff dimensions of fractal attractors. *J. Stat. Phys.*, **26**, 173–9.

Grassberger, P. (1986). Estimating the fractal dimension and entropies of strange attractors. In *Chaos*, (ed. A. V. Holden), pp. 291–312. Manchester University Press.

Grassberger, P. and Procaccia, I. (1983*a*). Characterization of strange attractors. *Phys. Rev. Lett.*, **50**, 346–9.

Grassberger, P. and Procaccia, I. (1983*b*). Measuring the strangeness of strange attractors. *Physica*, **9D**, 189–208.

Grassberger, P. and Procaccia, I. (1984). Dimensions and entropies of strange attractors from a fluctuating dynamics approach. *Physics*, **13D**, 34–54.

Kaas-Petersen, C. (1987). *PATH—User's guide*. Centre for Nonlinear Studies, University of Leeds.

Kaplan, J. L. and Yorke, J. A. (1979). Chaotic behaviour in multidimensional difference equations. In *Functional differential equations and the approximation of fixed points*, (ed. H. O. Peitgen and H. O. Walther), Lecture notes in mathematics 730, pp. 218–37. Springer, Berlin.

Kubicek, M. and Marek, M. (1983). *Computational methods in bifurcation theory and dissipative structures*. Springer, New York.

Packard, N. H., Crutchfield, J. P., Farmer, J. D., and Shaw, R. S. (1980). Geometry from a time series. *Phys. Rev. Lett.*, **45**, 712–16.

Pike, E. R., McWhirter, J. G., Bertero, M., and de Mol, C. (1984). Generalized information theory for inverse problems in signal processing. *IEE Proc.*, **131F**, 660.

Schaffer, W. M., Truty, G. J., and Fulmer, S. L. (1988). *Dynamical Software: User's manual and introduction to chaotic systems*. Dynamical Systems Inc., Tucson, AZ.

Schneider, F. W., Kruel, Th., and Freund, A. (1989). How to distinguish between chaos and amplification of statistical fluctuations in a chemical oscillator. In *Structure, coherence and chaos in dynamical systems*, (ed. P. L. Christiansen and R. D. Parmentier). Manchester University Press.

Seydel, R. (1988). *From equilibrium to chaos: practical bifurcation and stability analysis*. Elsevier, New York.

Shimada, I. and Nagashima, T. (1979). A numerical approach to ergodic problems of dissipative dynamical systems. *Prog. Theor. Phys.*, **61**, 1605–16.

Takens, F. (1981). Detecting strange attractors in turbulence. In *Dynamical systems and turbulence*, (ed. D. A. Rand and L.-S. Young), Lecture notes in mathematics, 989, pp. 366–81. Springer, Heidelberg.

Wolf, A. (1986). Quantifying chaos with Lyapounov exponents. In *Chaos*, (ed. A. V. Holden), pp. 273–90. Manchester University Press.

Wolf, A., Swift, J. B., Swinney, H. L., and Vastano, J. A. (1985). Determining Lyapounov exponents from a time series. *Physica*, **16D**, 285–317.

8

The Belousov–Zhabotinskii and other solution-phase reactions

The Belousov–Zhabotinskii reactions is relatively widely known for its long trains of oscillations in closed vessels (Belousov 1951, 1958; Zhabotinskii 1964a,b, 1985; Zhabotinskii 1987; Zhabotinskii et al. 1982). Although there has been much recent debate about some detailed aspects, the basic mechanism appears to be well established. A relatively simple model, the 'Oregonator', has proven remarkably successful at reproducing and predicting the observed behaviour in a wide range of experimental situations, even with independently determined rate data. We will return to mechanistic and modelling questions below, after introducing the experimental phenomena: comprehensive accounts of such issues as bromide ion control, the role of the catalyst, values of the rate constants, etc. can be found in the multiauthor work edited by Field and Burger (1985) or the recent papers by Field and Försterling (1986), Noyes (1986), Ruoff et al. (1988), and Noyes et al. (1989). Here, we concentrate on the behaviour in well-stirred flow reactors (CSTRs).

8.1. Transient oscillations in a closed vessel

A typical initial mixture for the Belousov–Zhabotinskii reaction in a well-stirred beaker might have the composition: malonic acid, 0.275 M; potassium bromate, 6.25×10^{-2} M; ammonium ceric nitrate 2×10^{-3} M; in 500 ml of 1M H_2SO_4 at 20 °C, with 1–2 ml of 0.025 M ferrion (1.485 g of 1,10 phenanthroline and 0.695 g $FeSO_4$ in 100 ml H_2O) added as an indicator. After a short induction period, the reaction enters its oscillatory phase. The colour alternates between magenta and blue and there are corresponding oscillations in the potential of a bromide ion sensitive electrode or a platinum electrode (with a calomel reference in each case). The waveform is shown in Fig. 8.1.

The amplitude and period of the platinum electrode, which responds primarily to the transition metal catalyst couple (Ce(III):Ce(IV)), soon become regular, so each oscillation is virtually indistinguishable from the next. For the bromide ion concentration there is a significant additional feature: the maximum achieved on each successive cycle increases slowly, but continuously, during the course of the reaction.

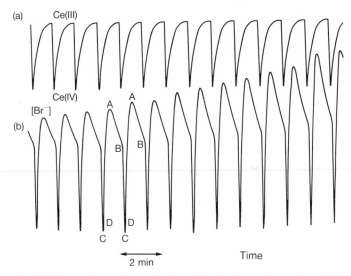

FIG. 8.1. Typical experimental records from (a) platinum electrode and (b) bromide ion sensitive electrode for the Belousov–Zhabotinskii reaction in a closed system. In each case the reference electrode is calomel.

Oscillatory behaviour may last for several hours, with perhaps 200 pulses. However, reactant consumption and the approach to chemical equilibrium must eventually take their toll: oscillations can only be a transient phenomenon in a thermodynamically closed system and must eventually die out.

Further details of the oscillatory waveform can be discussed alongside their mechanistic interpretation. The traces in Fig. 8.1 have 'relaxation oscillation' character. Concentrating particularly on the bromide ion record, each completely oscillatory pulse can be divided into two sections of relatively slow evolutions AB and CD, separated by rapid jumps BC and DA. This division is perhaps less clear in the Ce(III):Ce(IV) record, which has a slow increase to maximum (in the Ce(III) concentration) followed by a sharp drop. High Ce(III) corresponds to the indicator ferroin also being in its reduced state, Fe(II), and for which it has the magenta colour. (If more ferroin is used, the colour in this state becomes a clearer red: red = reduced is then particularly satisfying). Thus the Ce record indicates the sharp change from red/magenta to blue and the slow recovery of the former colour.

Most interpretations are based on the work of Field, Körös, and Noyes (FKN) (1972), although some refinements of the 'organic' reactions involving malonic acid are now favoured. The simplest, skeleton FKN scheme is shown in Table 8.1 and consists of three sets of three reaction steps. These three triplets are frequently referred to as processes A, B, and C. We will see other variations on this scheme in due course, but it serves here as a guide through the significant features of the closed vessel oscillations.

TABLE 8.1
Skeleton FKN scheme for Belousov–Zhabotinskii reaction

(R1)	$Br^- + HOBr + H^+ \rightleftharpoons Br_2 + H_2O$
(R2)	$HBrO_2 + Br^- + H^+ \rightleftharpoons 2HOBr$
(R3)	$BrO_3^- + Br^- + 2H^+ \rightleftharpoons HBrO_2 + HOBr$
(R4)	$HBrO_2 + HBrO_2 \rightleftharpoons HOBr + BrO_3^- + H^+$
(R5)	$BrO_3^- + HBrO_2 + H^+ \rightleftharpoons 2BrO_2^\cdot + H_2O$
(R6)	$Ce(III) + BrO_2^\cdot + H^+ \rightleftharpoons Ce(IV) + HBrO_2$
(R7)	$Ce(IV) + BrO_2^\cdot + H_2O \rightleftharpoons Ce(III) + BrO_3^- + 2H^+$
(R8)	$Br_2 + CH_2(CO_2H)_2 \rightarrow BrCH(CO_2H)_2 + Br^- + H^+$
(R9)	$6Ce(IV) + CH_2(CO_2H)_2 + 2H_2O \rightarrow 6Ce(III) + HCO_2H + 2CO_2 + 6H^+$
(R10)	$4Ce(IV) + BrCH(CO_2H)_2 + 2H_2O \rightarrow$
	$\qquad 4Ce(III) + HCO_2H + Br^- + 2CO_2 + 5H^+$

Starting at point A on the cycle in Fig. 8.1, the reaction has a relatively high concentration of bromide ion and most of the cerium catalyst is in the reduced Ce(III) state. The mixture is red/magenta. The important reactions under these conditions are those which constitute process A:

(R3)	$BrO_3^- + Br^- + H^+ \rightarrow HBrO_2 + HOBr$
(R2)	$HBrO_2 + Br^- + H^+ \rightarrow 2HOBr$
(R1)	$HOBr + Br^- + H^+ \rightarrow Br_2 + H_2O.$

All of these steps remove bromide ion in the process of reducing bromate to bromine. The overall stoichiometry of process A is obtained from (R3) + (R2) + 3(R1), to give

$$BrO_3^- + 5Br^- + 6H^+ \rightarrow 3Br_2 + 3H_2O. \qquad (A)$$

As the bromide ion concentration decreases, so the rate of all the reaction steps in process A decreases. As the rates of (R3) and (R2) slow down, the intermediate $HBrO_2$ can begin to complete for reaction with bromate. This takes us into process B:

(R5)	$BrO_3^- + HBrO_2 + H^+ \rightarrow 2BrO_2 + H_2O$
(R6)	$BrO_2 + Ce(III) + H^+ \rightarrow HBrO_2 + Ce(IV)$
(R4)	$2HBrO_2 \rightarrow BrO_3^- + HOBr + H^+.$

Reaction (R5) produces two molecules of the radical species BrO_2 (perhaps first as the dimer Br_2O_4 which then rapidly equilibrates with the monomer form). Each of these then react rapidly, principally via step (R6), returning an $HBrO_2$ molecule and oxidizing the catalyst. The stoichiometry (R5) + 2(R6) thus leads to an increase in the concentration of $HBrO_2$:

$$BrO_3^- + HBrO_2 + 2Ce(III) + 3H^+ \rightarrow 2HBrO_2 + 2Ce(IV) + H_2O \qquad (B)$$

The slow part of this sequence (the rate-determining step) is reaction (R5). This allows autocatalysis, as the rate of increase of $[HBrO_2]$ becomes proportional to its own concentration, as discussed in Chapter 1. The characteristic feature of autocatalysis is the rapid acceleration of the overall rate, so once process B is 'switched on' by the decrease in the bromide ion concentration, so it soon proceeds rapidly. The onset is marked by point B in the figure and leads to a rapid increase in the concentration of the oxidized form of the catalyst and hence to a sharp colour change from red to blue. There is also a sharp, further drop in the bromide ion concentration through reaction (R2). The exponential growth in the reaction rate is limited by step (R4) and by consumption of the reduced form of the catalyst.

Processes A and B thus remove bromide ion and oxidize the catalyst. The other half of the oscillatory process must, therefore, reset the conditions by producing bromide ion and reducing the catalyst. This occurs through process C and involves the organic species malonic acid. As stated prevously, this part is probably not so well understood and is still being debated. In the original form, it was suggested that malonic acid may enolize and then react with Br_2 to form bromomalonic acid BrMA and a bromide ion. If all three Br_2 formed in process A undergo this reaction, three Br^- are returned, but we still have a net loss of two, which remain bound in BrMA. These latter, and that from the original bromate, can be released if bromomalonic acid is oxidized by the Ce(IV), which also serves to reduce the catalyst. This overall change is often represented in the form

$$2Ce(IV) + BrMA + MA \rightarrow fBr^- + 2Ce(III) + \text{other products}. \qquad (C)$$

As this part of the process proceeds, so the solution changes colour back to the reduced red/magenta. The rate of process C is taken to be first order in the oxidized catalyst concentration $[Ce(IV)]$ and in a quantity $[Org]$ that represents the total concentration of organic species. The 'stoichiometric factor' f is to some extent an adjustable parameter. It represents the number of bromide ions produced as two Ce(IV) ions are reduced. If the oxidized catalyst were to react exclusively with BrMA and not with MA or any other species, then $f = 2$. For $f > \frac{2}{3}$ there is a net increase in the bromide ion concentration through a complete cycle.

The ambiguity associated with process C has allowed scope for a number of different 'improvements' that have been particularly concerned with modelling the more complex phenomena to be described shortly. Fairly recent work (Varga *et al.* 1985) has suggested that the route through bromomalonic acid is not of prime significance. An alternative suggestion is that an organic species RH (such as MA) reduces Ce(IV) and forms a radical R. This latter may then react with HOBr, producing ROH and a bromine atom Br: the bromine atom then forms a chain by reacting with another RH to form Br^-, H^+, and R. The chain nature here allows many bromide ions to be formed from the first Ce(IV) reduction, so that f may have a relatively high value even when there are many competing reactions for the oxidized catalyst.

One other important aspect of the BZ modelling which needs to be borne in mind when comparing different approaches is the value used for the reaction rate constants—for steps (R2), (R4), and (R5) in particular. Two sets of kinetic constants are commonly used—referred to as the 'Hi' and 'Lo' sets.

The original 'Hi' set were part derived and part estimated by FKN (see Noyes (1986) for a full historical discussion). In putting together the jigsaw of fragmentary data that existed at the time, these authors were required to estimate the ionization constant for $HBrO_2$. Following Pauling, they chose $pK \approx 2$. In subsequent years, evidence began to accrue that suggested this might be too low a value, leading to corresponding overestimates in the assigned reaction rate constants. It is particularly impressive that an important role in this process of re-evaluation was played by Tyson's analytical investigations of the Oregonator model derived from the FKN scheme (Tyson 1979, 1985). Field and Försterling (FF) have been mainly responsible for a programme in which the individual rate constants have been measured independently of the oscillatory reaction (Field and Försterling 1986). This has provided the 'Lo' set which now seems almost universally accepted in principle, although there is still a rather annoying tendency amongst authors to continue with the older 'Hi' set—'to allow easier comparison with earlier work'.

Table 8.2 lists the two sets of forward and reverse rate constants for reactions (R1)–(R6).

The simplest models derived from the FKN scheme are the two- and three-variable Oregonators discussed in Section 3.12.3. For these the mechanism is reduced by ignoring reverse reactions, taking (R5) + 2(R6) as a single step determined by the rate of (R5) and representing process C by the single step above. This then gives the model scheme

(O1)	$A + Y \rightarrow X + P$	rate $= k_3 ay$
(O2)	$X + Y \rightarrow 2P$	rate $= k_2 xy$
(O3)	$A + X \rightarrow 2X + 2Z$	rate $= k_5 ax$
(O4)	$2X \rightarrow A + P$	rate $= k_4 x^2$
(O5)	$B + Z \rightarrow \frac{1}{2} f Y$	rate $= k_0 bz$.

The identity and concentrations of the various species in Oregonator models

TABLE 8.2
Rate constants for FKN mechanism of BZ reaction

		'Hi'	'Lo'
(R1)	$k_1/\mathrm{M}^{-2}\,\mathrm{s}^{-1}$	8×10^9	8×10^9
	k_{-1}/s^{-1}	110	110
(R2)	$k_2/\mathrm{M}^{-2}\,\mathrm{s}^{-1}$	2×10^9	3×10^6
	$k_{-2}/\mathrm{M}^{-1}\,\mathrm{s}^{-1}$	5×10^{-5}	2×10^{-5}
(R3)	$k_3/\mathrm{M}^{-3}\,\mathrm{s}^{-1}$	2.1	2
	$k_{-3}/\mathrm{M}^{-1}\,\mathrm{s}^{-1}$	1×10^4	3.2
(R4)	$k_4/\mathrm{M}^{-1}\,\mathrm{s}^{-1}$	4×10^7	3×10^3
	$k_{-4}/\mathrm{M}^{-2}\,\mathrm{s}^{-1}$	2×10^{-10}	1×10^{-8}
(R5)	$k_5/\mathrm{M}^{-2}\,\mathrm{s}^{-1}$	1×10	42
	$k_{-5}/\mathrm{M}^{-1}\,\mathrm{s}^{-1}$	2×10^7	4.2×10^7
(R6)	$k_6/\mathrm{M}^{-2}\,\mathrm{s}^{-1}$	6.5×10^5	8×10^4
	$k_{-6}/\mathrm{M}^{-1}\,\mathrm{s}^{-1}$	2.4×10^7	8.9×10^3
(R7)	$k_7/\mathrm{M}^{-1}\,\mathrm{s}^{-1}$	9.6	0
	$k_{-7}/\mathrm{M}^{-3}\,\mathrm{s}^{-1}$	1.3×10^{-4}	0

tend to be translated into single letters A, B, etc. as above. Table 8.3 gives the cypher for this code as used throughout this book and most other works.

8.2. Open systems

In the past 15 years, closed vessels (batch reactors) have slowly tended to be replaced by open systems (flow reactors) for serious study of the BZ reaction. A flow reactor is provided with continuous, and often separate, inflows of fresh reactants: usually BrO_3^-, Ce(III) or some other catalyst, MA, and H^+ at certain, known concentrations, as shown in Fig. 8.2.

If the inflow concentrations correspond, after allowing for dilution on mixing, to a composition that shows long-lived oscillatory behaviour in batch, then at low flow rates the flow system may exhibit fully sustained oscillations. These will tend to have large amplitude and relatively long period, i.e. generally similar form and character to the relaxation oscillations typical of batch reactors. In the flow system, however, transient behaviour decays to leave a true limit cycle. Each excursion becomes exactly the same as its predecessor.

TABLE 8.3
Identification of symbols for Oregonator models

A	BrO_3^-
B	All oxidizable organic species
C	Ce(III)—reduced form of catalyst
H	H^+
M	Malonic acid
P	HOBr
W	$BrO_2^{\cdot}$
X	$HBrO_2$
Y	Br^-
Z	Ce(IV)-oxidized form of catalyst

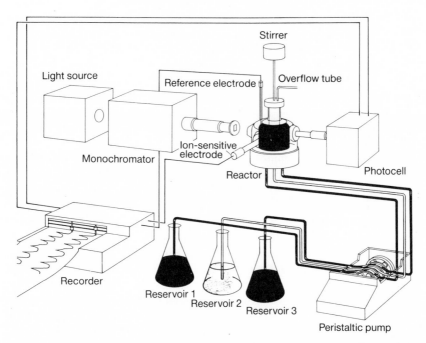

FIG. 8.2. Schematic representation of a typical flow reactor system for studying solution-phase oscillatory reactions. (Reprinted with permission from Epstein *et al.* (1983), *Sci. Am.*, **248** (March), 96–108, © Scientific American Inc.)

8.2.1. Bursting patterns

The first report of more complex oscillatory patterns in an open system was made by Sørensen (1974, 1979) who recorded a typical 'bursting pattern' as

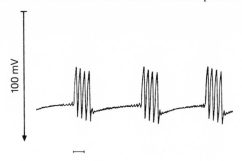

FIG. 8.3. Original 'bursting' patterns for the Belousov–Zhabotinskii reaction in a CSTR: potential of platinum electrode. (Reprinted with permission from Sørensen (1974). *Faraday Symp. Chem. Soc.*, **9**, 88. © Royal Society of Chemistry.)

shown in Fig. 8.3. The characteristic bursting waveform has short trains of perhaps four large excursions separated by periods of non-oscillatory or at least much smaller amplitude evolution. Such behaviour is typically found at higher flow rates (shorter residence times) than the simple, large period-1 relaxation oscillations.

Marek and Svobodova (1975) showed experimentally how different initial compositions approached the same period-1 limit cycle and investigated the influence of the flow rate on the amplitude, period, and waveform. They also observed similar bursting patterns to Sørensen, as well as coupling two flow reactors (to which we return below).

A thorough study of complex oscillations in the BZ reaction by Hudson. Schmitz, and coworkers began with the investigation of Graziani *et al.* (1976). A 25.1 ml CSTR with four separate inflows of bromate, malonic acid, sulphuric acid, and either cerous ion or ferroin respectively were maintained at 25 ($\pm$0.1) °C with post-mixing inflow concentrations of 0.3 M MA, 0.14 M BrO_3^-, 0.2 M H_2SO_4, and 10^{-3} M Ce(III) or 1.25×10^{-4} ferroin. The corresponding batch system showed (transient) oscillations.

With a flow rate of 1.0 ml min^{-1} (a mean residence time of 25 min), simple, large, period-1 waveforms were found (Fig. 8.4(a)). For higher flow rates, however, a bursting pattern with one large excursion per cycle was observed. Figure 8.4(b) shows the waveform for 4.7 ml min^{-1} ($t_{res} = 5.34$ min). For flow rates in excess of 4.8 ml min^{-1} the reaction settled into stationary-state operation.

Other, early, isolated reports of complex behaviour, each with a tentative and mainly geometric interpretation, are due to Wegmann and Rössler (1978; Rössler and Wegmann 1978; Rössler 1981) and to Maselko (1980*a,b*).

8.2.2. Chaos

The variations in waveform with changing flow rate were investigated in

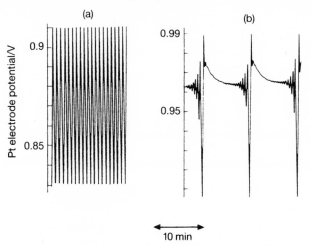

FIG. 8.4. Period-1 and complex oscillations for the Belousov–Zhabotinskii reaction in a CSTR: (a) flowrate = 1.0 ml min^{-1}; (b) flowrate = 4.7 ml min^{-1}. (Reprinted with permission from Graziani *et al.* (1976). *Chem. Eng. J.*, **12**, 9–21. © Pergamon Press.)

more detail (Schmitz *et al.*, 1977). In this latter study, ferroin was used as the catalyst. The inflow concentration of bromide (from the bromate stock) was estimated as 3.5 μM. The range of flow rates k_f studied was 1.0 to 7.0 ml min^{-1}: above this range the stationary-state reaction is stable. Figure 8.5 shows five typical time series from a bromide ion sensitive electrode. At low flow rates the oscillations are of large amplitude, or large amplitude interspersed with small peaks. At higher flow rates there are small amplitude, high-frequency oscillations which die out at a (supercritical) Hopf bifurcation at $k_f = 7$ ml min^{-1}. In this paper, Schmitz *et al.* also reported the observation of 'non-periodic' states or chaos.

Hudson and coworkers (Hudson *et al.* 1979, 1981; Hudson and Mankin 1981; Baier *et al.* 1989) returned to the question of chaotic behaviour in this system (also returning to a Ce(III) rather than ferroin inflow). The transition from simple large oscillations at low flow rate to simple small-amplitude oscillations at high flow rate over the range 2.91–5.42 ml min^{-1} was shown to involve many different waveform patterns.

It will be convenient to refer to the resulting waveforms using the L^s notation seen earlier, where L is the number of large-amplitude excursions and s the number of small peaks respectively in one complete period.

With $k_f = 2.91$ ml min^{-1} ($t_{res} = 8.63$ min), there is simple large period-1 (Fig. 8.6(a))—a 1^0 pattern. At $k_f = 4.06$, the oscillation is 1^1 with one large and one small peak per cycle (Fig. 8.6(c)). In between these two forms, at $k_f = 3.76$ ml min^{-1}, the observed response appears as a mixture of these 'parent states', i.e. a 2^1 pattern (Fig. 8.6(b)).

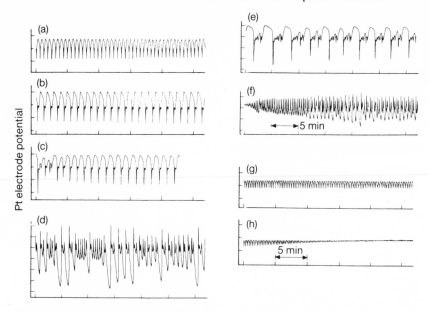

FIG. 8.5. Periodic and aperiodic responses for the Belousov–Zhabotinskii reaction in a CSTR, flowrates/ml min^{-1}: (a) 1.0, period-1; (b) 2.0, period-2; (c) 3.0, period-2 established after complex transient; (d) 3.5, chaotic; (e) 4.0, chaotic; (f) 4.25, chaotic; (g) 4.5, period-1; and (h) 7.0, emerging steady state. (Reprinted with permission from Schmitz *et al.* (1977). *J. Chem. Phys.*, **67**, 3040–4. © American Institute of Physics.)

The response at $k_f = 4.34$ ml min^{-1} is a 1^2: so going through the sequence (a) → (c) → (f) in Fig. 8.6, the number of small peaks per cycle is simply increased by one. Just as a mixed state could be found between the 1^0 and 1^1, so similar daughters are born from the 1^1 and 1^2 parents. Figure 8.6(d) shows the 2^3 pattern for $k_f = 4.11$ ml min^{-1}. All of the responses mentioned so far are strictly period, despite the increasing complexity. In Fig 8.6(e), the 'mixing' of the 1^1 and 1^2 parents for $k_f = 4.31$ ml min^{-1} is not regular but aperiodic.

A 1^3 pattern exists for $k_f = 4.62$ ml min^{-1} (Fig. 8.6(h)) and again there is a mixed state 2^5 between this and the 1^2. Above the 1^3 there is a chaotic mixing of 1^3 and 1^4 (Fig. 8.6(i)) before the 1^4 state emerges for $k_f = 4.81$ ml min^{-1} (Fig. 8.6(j)). The time series shown in Fig. 8.6(k) for $k_f = 5.37$ ml min^{-1} has a 1^{many} waveform, paving the way to the 1^∞ or 0^1 of the simple small-amplitude pattern of Fig. 8.6(l).

The apparently chaotic mixing between the 1^1 and 1^2 states has been tested further using many of the experimental methods discussed earlier (Hudson and Mankin 1981). The Fourier transform power spectrum for the aperiodic state with $k_f = 4.2$ ml min^{-1} shows a much broader band structure than those for the periodic parents (1^1 at $k_f = 4.0$ ml min^{-1} and 1^2 at $k_f = 4.4$ ml min^{-1}) as shown in Fig. 8.7. The phase-plane reconstructions

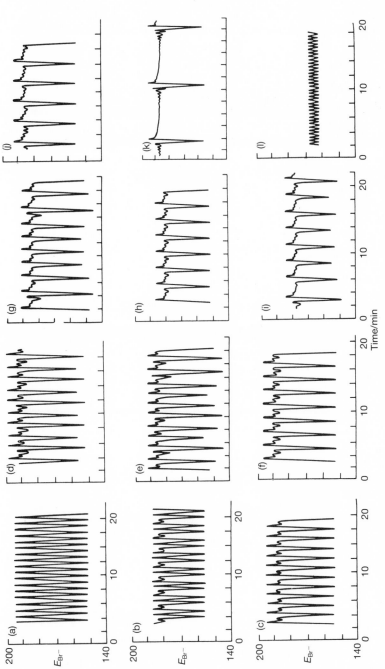

Fig. 8.6. Complex oscillations and chaos for the Belousov–Zhabotinskii reaction in a 25.4 ml CSTR at various flow rates k_f (quoted in ml min^{-1}): (a) 1^0, $k_f = 2.91$; (b) $1^0 1^1$, $k_f = 3.76$; (c) 1^1, $k_f = 4.06$; (d) $1^1 1^2$, $k_f = 4.11$; (e) chaos, $k_f = 4.31$; (f) 1^2, $k_f = 4.34$; (g) chaos, $k_f = 4.51$; (h) 1^3, $k_f = 4.62$; (i) chaos, $k_f = 4.76$; (j) 1^4, $k_f = 4.81$; (k) 1^{many}, $k_f = 5.37$; (l) 0^1, $k_f = 5.42$. (Reprinted with permission from Hudson *et al.* (1979). *J. Chem. Phys.*, **71**, 1601–6. © American Institute of Physics.)

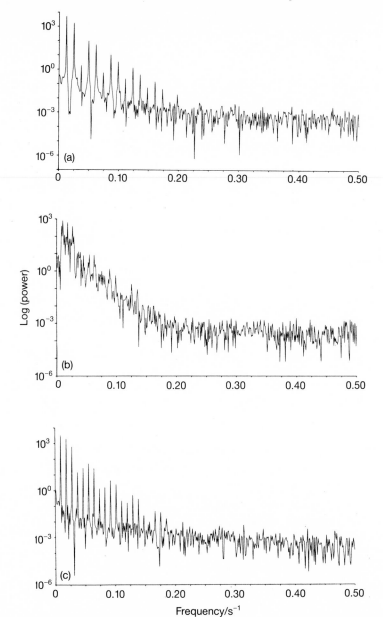

Fig. 8.7. Power spectra for periodic and chaotic states in the Belousov–Zhabotinskii reaction: (a) flowrate = 3.95 ml min^{-1} appropriate to 1^1 oscillation from Fig. 8.6(c); (b) flowrate = 4.18 appropriate to chaos from Fig. 8.6(d); (c) flowrate = 4.42 appropriate to 1^2 from Fig. 8.6(e). (Reprinted with permission from Hudson and Mankin (1981). *J. Chem. Phys.*, **74**, 6171–7. © American Institute of Physics.)

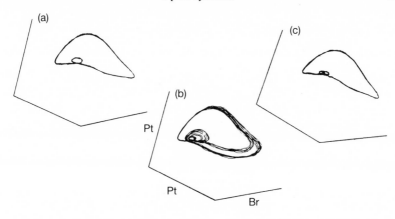

FIG. 8.8. Reconstructed attractors for periodic and chaotic states: details as in Fig. 8.7. (Reprinted with permission from Hudson and Mankin (1981). *J. Chem. Phys.*, **74**, 6171–7. © American Institute of Physics.)

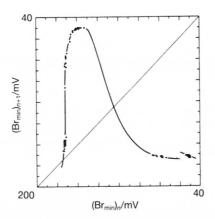

FIG. 8.9. One-dimensional next-return map for chaotic state in Fig. 8.6(d) constructed from successive minima in the potential of a bromide ion sensitive electrode. (Reprinted with permission from Hudson and Mankin (1981). *J. Chem. Phys.*, **74**, 6171–7. © American Institute of Physics.)

(Fig. 8.8) suggest a more complex attractor for the middle time series and the corresponding next-return map from the bromide ion sensitive electrode gives a single-humped maximum curve (Fig. 8.9) similar to the cubic mapping of Chapter 2. The Lyapounov exponent evaluated by fitting a curve to this map also appeared to be healthily positive, although no reliable quantitative result could be determined.

8.3. Concatenations and Farey arithmetic

Another substantial investigation of large- and small-amplitude interactions in the Belousov–Zhabotinskii reaction has been carried out by Maselko and Swinney (Maselko 1980*a, b*; Swinney and Maselko 1985; Maselko and Swinney 1986). These authors employed a 30 ml reactor thermostated at 28 (± 0.1) °C with three separate inflows and used Mn^{2+} as the catalyst. Typical inflow concentrations were $[BrO_3^-]_0 = 0.033$ M and $[H_2SO_4]_0 = 1.5$ M (a comparison of the experimental conditions in various studies is given in Table 8.4). The residence time and inflow concentration of malonic acid were varied systematically over relatively wide ranges in the course of a sequence of experiments. In reward for this effort, the reaction exhibited many interesting features including a 5^6 state, concentrations (mixing) of two parent states (such as that described above) and also of three or four parents, and birhythmicity.

8.3.1. Concatenations: parents and daughters

Figure 8.10 shows some divisions of the malonic acid–residence time parameter plane corresponding to different waveforms. Two regions are marked with cross-hatching. Just below the lower of these, the reaction exhibits a 1^2. If the residence time is further decreased (increasing flow rate), we find 1^3, then 1^4, etc. In between the cross-hatched regions. the waveform is 1^1 whilst immediately above the upper region is the 2^1 pattern. Moving to higher residence time brings in 3^1, 4^1, etc. The lower and upper cross-hatched regions correspond to conditions for which the 1^2 and 1^1 or the 1^1 and 2^1 mix, respectively. These will be similar regions between, say, the 1^2 and the 1^3 etc. but these are not shown in the figure.

An example of the patterns observed in a typical mixing region is shown in Fig. 8.11. The parent states are the 1^2 at the higher residence time end and 1^3 at shorter t_{res}. The mixed states are the 1^2 at the higher residence time end and 1^3 at shorter t_{res}. The mixed states in between are the concatenations of these two parents. The simplest mixing is the 1:1 combination $1^2 1^3$ as shown in Fig. 8.11(c). This is, in fact, not the first observed as a $(1^2)^2 1^3$ is found between this and the 1^2 parent (Fig. 8.11(b)): this can be interpreted as the daughter of the 1^2 and the $1^2 1^3$ states.

As the residence time is decreased below that for the $1^2 1^3$ state, so a succession of waveforms are traversed, each with an extra 1^3 unit being added. A $1^2(1^3)^5$ state, shown in Fig. 8.11(g), was distinguished before the parent 1^3 emerges.

TABLE 8.4
Typical inflow concentrations for BZ studies in a CSTR

	MA	BrO_3^-	10^3 cat	H_2SO_4	t_{res}	$T/°C$
Hudson, Schmitz	0.3	0.14	(CF) 1.0	0.2	4.9–9	25
Bordeaux-1 (a)	0.08	0.036	0.25	1.5	3.8–5.5	39
(b)	0.056	0.018	0.58	1.5	7–100	39.6
Texas	0.25	0.14	(C) 0.83	0.2	52–136	28.3
Hourai *et al.*	0.056	0.039	0.25	0.05	1–200	24–55
(a)	0.066	0.0022	0.17	0.63	15–60	43
(b)	0.174	0.012	0.083	1.0	20–36	43
Bordeaux-2 (c)	0.165	0.012	0.083	1.0	3–6	40
(d)	0.051	0.025	0.17	1.0	25	43
(e)	0.033	0.0042	0.17	1.5	12–33	43
Maselko, Swinney	0.033	0.033	(M) 0.4	1.5	4.8–8.4	28
Leeds	0.3	0.14	(C) 1.0	0.23	5–170	25

Concentrations quoted in mol dm^{-3}, residence time in minutes: different catalysts are denoted as C = Ce(III), F = ferroin, M = Mn^{2+}.

8.3.2. Firing numbers, Farey sequences, and trees

A convenient way of at least cataloguing such a sequence is to use the Farey arithmetic discussed in Chapter 4. Each waveform can be assigned a firing number F defined by

$$F = s/(L + s) \qquad (8.1)$$

where, as above, L is the number of large-amplitude excursions and s the number of small peaks in a complete period. The parent states in the above sequence then have $F = \frac{2}{3}$ and $\frac{3}{4}$ respectively. When the parents mix, they produce a new state whose firing number is given by the Farey sum of those for the parents. This sum for two firing numbers p_1/q_1 and p_2/q_2 is formed by the rule

$$\frac{p_1}{q_1} \oplus \frac{p_2}{q_2} = \frac{p_1 + p_2}{q_1 + q_2}. \qquad (8.2)$$

Thus the 1^2 and 1^3 parents of Fig. 8.11(a) and (h) give a daughter with a firing number of $(2 + 3)/(3 + 4)$ or $\frac{5}{7}$, which is the $1^2 1^3$ state of Fig. 8.11(c). One can relatively quickly confirm that each pattern of Fig. 8.11 has a firing number given by the Farey sum of the firing number for the two states immediately on either side.

The sequence can also be represented in terms of a Farey tree, as shown in Fig. 8.12. A more impressive tree is formed by considering a larger range

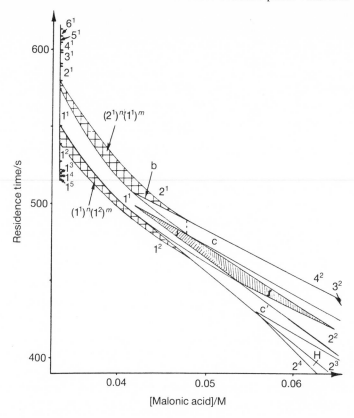

Fig. 8.10. 'Phase diagram' indicating regions of different periodic responses for the Belousov–Zhabotinskii reaction from the studies of Maselko and Swinney. (Reprinted with permission from Maselko and Swinney (1986). *J. Chem. Phys.*, **85**, 6430–41. © American Institute of Physics.)

of residence time, sweeping through a larger number of concatenations along the vertical line corresponding to [MA] = 0.033 M close to the residence time axis in Fig. 8.10. This produces a specimen with numerous branches and trunks (Fig. 8.13) going from the 1^0 to the 0^1 states.

8.3.3. The Devil's staircase

The same basic data—i.e. how the waveform and its associated firing number vary with some experimental parameter, such as the residence time—can be plotted as shown in Fig. 8.14. This gives the *Devil's staircase* (see Section 4.8.3). The states forming the treads at the top and bottom of figure are the 1^5 ($F = \frac{5}{6}$) and the 6^1 ($F = \frac{1}{7}$) respectively, although the floors connected by the staircase are the 1^0 (lower) and 0^1 or 1^∞ (higher). The length of individual

(a) $1^2[\frac{2}{3} = 0.667$ (b) $(1^2)^2 1^3 \, [\frac{7}{10} = 0.700]$ (c) $1^2 1^3 \, [\frac{5}{7} = 0.714]$

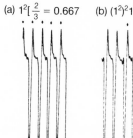
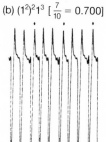
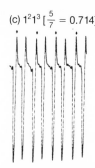

(d) $1^2(1^3)^2 \, [\frac{8}{11} = 0.727$ (e) $1^2(1^3)^3 \, [\frac{11}{15} = 0.733]$ (f) $1^2(1^3)^4 \, [\frac{14}{19} = 0.737]$

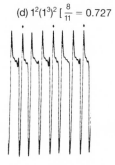
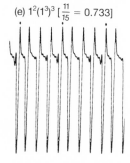

(g) $1^2(1^3)^5 \, [\frac{17}{23} = 0.739$ (h) $1^3 \, [\frac{3}{4} = 0.750]$

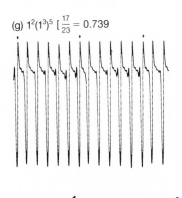

$\longleftrightarrow$
20 min

FIG. 8.11. Concatenated Farey sequence between the 1^2 and 1^3 parent states. (Reprinted with permission from Maselko and Swinney (1986). *J. Chem. Phys.*, **85**, 6430–41. © American Institute of Physics.)

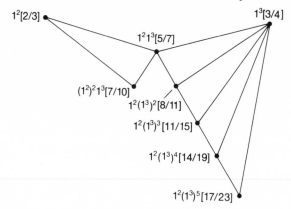

FIG. 8.12. Farey tree for the concatenation in Fig. 8.11. (Reprinted with permission from Maselko and Swinney (1986). *J. Chem. Phys.*, **85**, 6430–41. © American Institute of Physics.)

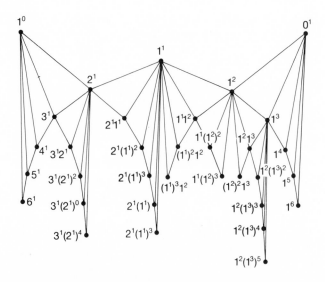

FIG. 8.13. Farey tree for complete concatenated sequence from 1^1 to 0^1 states. (Reprinted with permission from Maselko and Swinney (1986). *J. Chem. Phys.*, **85**, 6430–41. © American Institute of Physics.)

steps is not equal, the strongest resonances such as 1^1, 1^1, and 1^2 having greater measure than more complex periodicities. The exact form of the staircase would, however, vary if the other parameters, such as inflow concentration of MA were altered.

In the present case, the treads on the staircase appear to have full measure, i.e. there are no gaps corresponding to residence times for which the

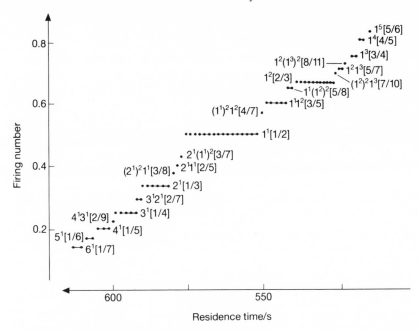

FIG. 8.14. The Devil's staircase for Farey sequence from Maselko and Swinney. (Reprinted with permission from Maselko and Swinney (1986). *J. Chem. Phys.*, **85**, 6430–41. © American Institute of Physics.)

waveform has no firing number. Such gaps might have existed if the reaction had displayed either quasiperiodicity or aperiodicity over some ranges, but these responses were not encountered by Masleko and Swinney in their study. These authors also noted that, given a greater degree of control over experimental conditions, more concatenations and hence more steps on the staircase would have been located: some transient evidence could be obtained by careful probing near the change from one step to the next in many cases.

It is also possible to assess the dimension of the staircase, which turns out to be fractal. Typical dimensions lie between $\frac{2}{3}$ and 1 (although there is reason to believe that a more accurate upper bound is 0.9240 . . .). Maselko and Swinney conclude that their results strongly suggest that the complex oscillations in this system correspond to a frequency-locked motion on a two torus. The non-linear coupling between the two oscillatory modes in this case must, then, be sufficient to ensure locking persists over all residence times, so quasiperiodicity is avoided.

8.2.4. *Birhythmicity of states with same firing number*

At higher inflow concentrations of malonic acid, to the right in Fig. 8.10,

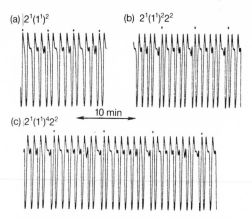

(a) $|2^1(1^1)^2$
(b) $2^1(1^1)^22^2$
(c) $|2^1(1^1)^42^2$

10 min

FIG. 8.15. Concatenation of three parent states. (Reprinted with permission from Maselko and Swinney (1986). *J. Chem. Phys.*, **85**, 6430–41. © American Institute of Physics.)

the oscillatory waveform can show other patterns. In particular, Maselko and Swinney obtained each of the 1^1, 2^2, 3^3, and 6^6 states (the latter also required an adjustment of the bromate inflow). The regions of parameter space for the 1^1 and 2^2 are indicated in Fig. 8.10: 'between' them is a shaded region. This latter is a region of birhythmicity or hysteresis. Both the 1^1 and 2^2 can be found for identical conditions within this region, and the system can be perturbed from one to the other by momentarily stopping the stirrer.

8.3.5. *Concatenations of three or more basic states*

As well as finding concatenation sequences of two states (ultimately the 1^0 and 0^1 and then their daughters), Maselko and Swinney also discovered complex waveforms and sequences between them that could be interpreted as involving three or even four basic states. These may indicate the possibility of frequency-locked motions on three- and four-tori respectively. Figure 8.15 shows some of the patterns containing various numbers of the three basic units 1^1, 2^1, and 2^2.

To apply Farey arithmetic to these concatenations we need to define the triple (p, q, r) for a given waveform. If a given response has the form $(2^1)^a(1^1)^b(2^2)^c$ then a suitable choice is $p = b + c$, $q = a + b$, and $r = a + b + c$. A three torus is formed from two pairs of irrational frequencies (a two torus has one pair): these pairs can be approximated by p/r and q/r. Two parent states with (p_1, q_1, r_1) and (p_2, q_2, r_2) then couple under the Farey sum

$$(p, q_1, r_1) \oplus (p_2, q_2, r_2) = (p_1 + p_2, q_1 + q_2, r_1 + r_2) \qquad (8.3)$$

Rather than a Farey tree, we now need a Farey triangle, as shown in

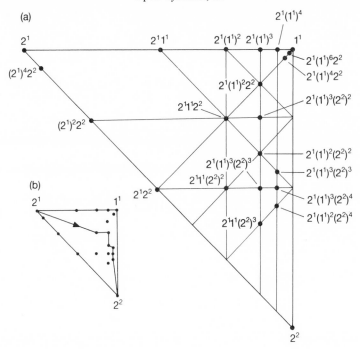

FIG. 8.16. Farey triangle for concatenations of three parent states. (Reprinted with permission from Maselko and Swinney (1986). *J. Chem. Phys.*, **85**, 6430–41. © American Institute of Physics.)

Fig. 8.16. The apexes correspond to the parent states 2^1, 1^1, and 2^2 and have the triples $(0, 1, 1)$, $(1, 1, 1)$, and $(1, 0, 1)$ respectively. Figure 8.16 (a) then shows some of the (infinitely many) possible concatenated states from these parents, whilst 8.16(b) shows the actual path realized in the experiments of Maselko and Swinney.

The concentrations of four states observed were built on the parents 1^1, 1^2, 2^2, and 2^3 and required a quadruple (p, q, r, s) and representation through a Farey pyramid.

8.4. Open systems, II

A second programme with similar experimental conditions to those of Hudson *et al.* has been carried out mainly in Texas (Turner *et al.* 1981; Roux and Swinney 1981; Simoyi *et al.* 1982; Roux *et al.* 1982, 1983; Swinney and Roux 1984). The main difference between these and the Hudson studies is the range of residence times investigated. The Texas work concentrated on the long residence time (low flow rate) behaviour. We can imagine starting in the region the large period-1 relaxation oscillations observed at the lowest flow rates in the Hudson experiments. As the flow rate is now decreased further, in a sense moving closer towards the closed-vessel situation, it is perhaps surprising that more changes in waveform are to be encountered. In fact, we should expect that at very low flow rates, oscillations will cease and a stable stationary state will emerge as we approach sufficiently close to the thermodynamic limit. The ultimate loss of oscillations may well be through a Hopf bifurcation, and if this is supercritical we expect small-amplitude near-sinusoidal traces just before this.

Within this framework, the Texas experiments reveal a structure that has almost a 'mirror-image' relationship to the Hudson and Schmitz work. As the residence time is increased, the waveform changes from 1^0 through 1^1, 1^2, etc. to the very long residence time 0^1 just above the Hopf point. In the Hudson case, of course, this sequence accompanied an increase in flow rate and so a decrease in residence time.

Turner *et al.* (1981) also report that between each of the periodic modes there is a range of aperiodicity, as shown in Fig. 8.17. The waveform in a

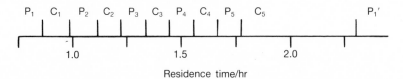

FIG. 8.17. Summary of periodic–chaotic sequences observed in Belousov–Zhabotinskii experiments. (Reprinted with permission from Turner *et al.* (1981). *Phys. Lett.*, A **85**, 9–12.)

given 'aperiodic window' appears in some sense to be a chaotic mixing of the parent states immediately on either side. The experimental control of such low pumping rates would probably not have been sufficient to allow complex concatentations to be identified: none are reported. Figure 8.18 shows a selection of the time series from these studies, along with their power spectra and reconstructed attractors (formed using the time delay method from the bromide ion sensitive electrode record). Again, the characteristic shape of a strange attractor for the aperiodic series confirms the broad band structure of the power spectra. The next-return map for one of the reconstructed strange attractors has a single-humped maximum (Fig. 8.19)

Coffman *et al.* (1987) have collected together a convincing sequence of experimental records which reveal a period-doubling cascade as residence time is increased, converging to chaotic behaviour. From the chaotic traces a one-dimensional next-amplitude map was constructed (from successive minima in the bromide ion sensitive electrode potential) as shown in Fig. 8.20.

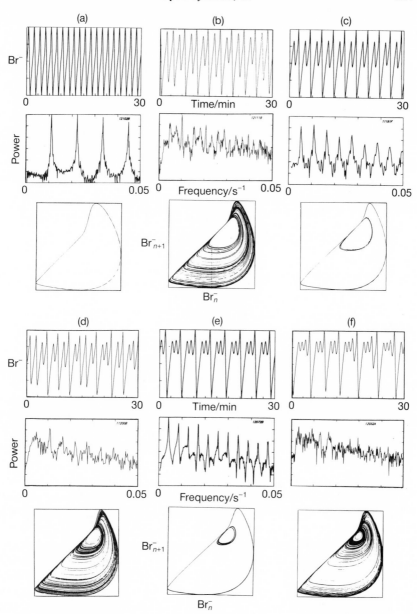

FIG. 8.18. Bromide ion time series, power spectra, and reconstructed attractors from periodic–chaotic sequence: (a) 1^0 for $t_{res} = 0.49$ h; (b) chaos for $t_{res} = 0.9$ h; (c) 1^1 for $t_{res} = 1.03$ h; (d) chaos for $t_{res} = 1.11$ h; (e) 1^2 for $t_{res} = 1.25$ h; (f) chaos for $t_{res} = 1.38$ h. (Reprinted with permission from Roux *et al.* (1982). *Nonlinear problems: present and future*, (ed. A. R. Bishop, D. K. Campbell, and B. Nickolaenko), pp. 409–22. North-Holland, Amsterdam.)

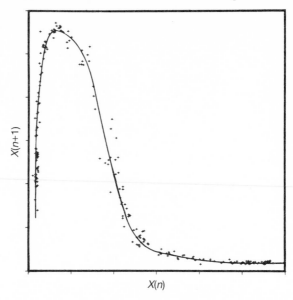

FIG. 8.19. One-dimensional next-return map constructed via the Poincaré section indicated in Fig. 8.18(b). (Reprinted with permission from Roux *et al.* (1982). *Nonlinear problems: present and future*, (ed. A. R. Bishop, D. K. Campbell, and B. Nickolaenko), pp. 409–22. North-Holland, Amsterdam.)

The chaotic response is interrupted as the residence time varied further, with a series of periodic windows. The order in which particular periodicities appear corresponds to the classic universal U sequence appropriate to smooth, single-humped one-dimensional maps, with period-6, period-5, and period-3 (plus subsequent period doubling from this latter state) being found first. Of the states up to period-6, only two from the universal sequence were not located experimentally. Different waveforms with the same periodicity but different ordering of the large and small peaks were observed as suggested by the results from simple mappings. Figure 8.20 also shows some of the experimental records and sequences between them from this study. As we will discuss later, Coffman *et al.* were able to correlate their observed results with relatively simple exponential maps. Such an interpretation also allows them to reconcile other sequences observed with slightly different inflow conditions (mainly they appear due to a change in the manufacturing supplier of the malonic acid used) for which some responses were found to exist in more than one range of residence time.

   Hourai *et al.* (1985) also performed experimental investigations which they then interpreted through one-dimensional maps. Using relatively dilute solutions, these authors studied the BZ system over a wide range of both residence time, involving a total of seven reactors of differing size, and

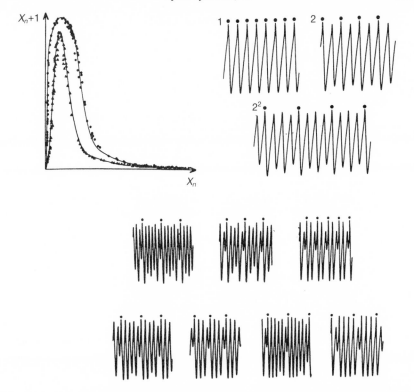

FIG. 8.20. One-dimensional map (upper curve) and associated period doubling and periodic window traces observed on increasing the residence time in experiments of Coffman *et al.*: the lower curve on the one-dimensional map gives rise to multiplicity of periodic states. (Reprinted with permission from Coffman *et al.* (1987). *J. Chem. Phys.*, **86**, 119–29. © American Institute of Physics.)

temperature. The observed responses are summarized in Fig. 8.21 with example time series in Fig. 8.22. At the longest residence times there is a stable stationary state: this loses stability on decreasing the residence time or the temperature, via a supercritical Hopf bifurcation giving rise to small-amplitude period-1 oscillations. At lower residence times, occasional large-amplitude excursions also appear in the waveform; Fig. 8.22(b) shows the 1^9 response, which is followed by a 1^5 and a 1^3. Between the 1^3 and the 1^1 oscillations is a range of experimental conditions corresponding to a chaotic behaviour which is approximately an aperiodic mixing of 1^2 and 1^1. Further alternations between periodic and chaotic responses, in particular a 1^2 or period-3 pattern, occur before a large-amplitude period-1 appropriate to high flow rates.

Whilst the analysis of the experimental data in the various studies discussed so far provides many indications that chaos is a fundamental

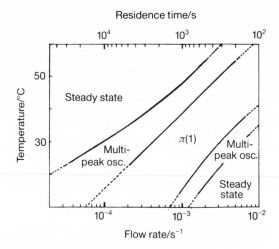

Fig. 8.21. 'Phase diagram' for complex oscillations and chaos in experiments of Hourai *et al.* (Reprinted with permission from Hourai *et al.* (1985). *J. Phys. Chem.*, **89**, 1760–4. © American Chemical Society.)

feature of the BZ system, some qualifications may still be in order. Some of these will need to wait until the corresponding numerical investigations have been discussed.

8.5. Open systems, III

Two groups of experiments can be ascribed to the Bordeaux school. Pioneering studies of the BZ system in a flow reactor (De Kepper *et al.* 1976) independently confirmed the existence of sustained period-1 oscillations and other features such as bistability. Next, Vidal *et al.* (1979, 1982) and Roux *et al.* (1980; Roux 1983; Roux and Rossi 1984), Bordeaux-1 in Table 8.4, employed inflow concentrations etc. similar to those of the (later) investigations of Maselko and Swinney (Section 8.3). On increasing flow rate, the system passes from 1^0 to 0^1 oscillations (and ultimately a supercritical Hopf bifurcation to a stable stationary state) via a sequence alternating between periodic and reportedly chaotic states. The periodic states all have relatively simple waveforms—1^0, 3^1, 2^1, 3^2, 2^3, etc.—whilst the chaotic states are 'mixtures' of the neighbouring periodicities.

One might suspect, from the Maselko and Swinney results, that given sufficient experimental control many of these 'aperiodicities' (and those of the Texas experiments described in the previous section) may have been resolvable into concatenated periodic states governed by Farey sequences etc. and hence are not the unequivocal evidence for chemical chaos that they have sometimes been presented as.

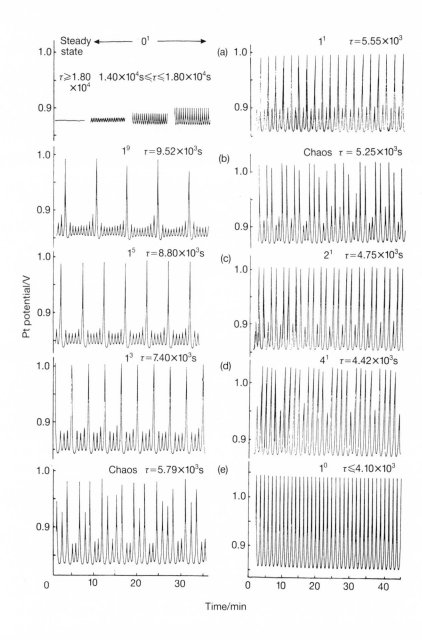

FIG. 8.22. Example periodic and chaotic states for different regions in Fig. 8.21. (Reprinted with permission from Hourai *et al.* (1985). *J. Phys. Chem.*, **89**, 1760–4. © American Chemical Society.)

One significant extra pattern reported in this Bordeaux-1 work is that of a quasiperiodic trace. Vidal *et al.* (1979) show a two-frequency power spectrum for residence times between their period-1 oscillations and a 'chaotic' response. In Farey terms, this suggests that the corresponding Devil's staircase would not have full measure, i.e. there are gaps at the residence times for quasiperiodicity. This change in behaviour may be due to the slight variations in experimental parameters (different temperature and form of the catalyst) between Bordeaux-1 and Maselko and Swinney.

Roux *et al.* (1981), Bordeaux-1(b) in Table 8.4, examined the changes in waveform on increasing the residence time (decreasing flow rate) from the 1^0 at $t_{res} = 7$ min (a large-amplitude limit cycle) surrounding three stationary-state singularities) through to the 0^1 at $t_{res} = 100$ min (a small limit cycle surrounding only a single stationary state on the thermodynamic branch). At residence times below the small-amplitude period-1 (0^1) there is some homoclinic bifurcation which leads to the appearance of responses that are approximately 1^{20}. The reported traces are not quite regular, but have varying numbers of small peaks between each large (single) excursion. Roux *et al.* interpret this as 'intermittency', with the system trying to remain close to the small 0^1 limit cycle, but this has just lost stability so there are occasional excursions away from its vicinity before some reinjection process. Also observed were the coexistence of different periodic solutions at various residence times and the onset of quasiperiodicity at shorter residence times from the 1^0 pattern: the quasiperiodic torus appears to coexist with periodicities of the form M^n with $M, n \gg 1$.

With this background, attention then turns to the role of tori and their quasiperiodic responses in the onset of higher-order dynamic behaviour in the Belousov–Zhabotinskii reaction, in a sequence of experiments referred to as Bordeaux-2 in Table 8.4 (Argoul and Roux 1985; Argoul *et al.* 1987*a,b*; Richetti *et al.* 1987*a,b*).

In a thoughtful and painstaking series of experiments, Argoul *et al.* (1984, 1987*b*) followed the evolution of the observed attractor in the BZ system as the flow rate is increased. Period-1 oscillations emerge from a stable stationary state via a supercritical, primary Hopf birfurcation: in the immediate vicinity beyond this bifurcation point the amplitude of the waveform grows as the square root of the increase in flow rate whilst the period increases linearly. As the flow rate is further varied, the next (secondary) bifurcation, now from the periodic solution, is to a torus. Figure 8.23 shows the evolution of the quasiperiodic waveform: between (a) and (b) a second frequency, corresponding to the long period modulation of the oscillatory amplitude, appears in the Fourier power spectrum. The reconstructed attractor in a three-dimensional phase plane, obtained via the time delay method, are shown in Fig. 8.24. (The secondary Hopf bifurcation in this system does not appear to show the classical scaling of amplitude and period with flow rate such as that seen at the primary Hopf point.

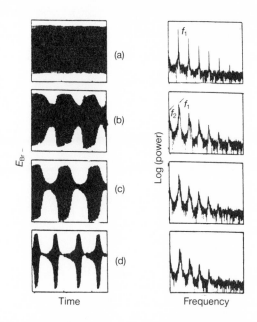

FIG. 8.23. Quasiperiodic behaviour and corresponding power spectra showing two fundamental frequencies. (Reprinted with permission from Argoul *et al.* (1987*b*). *J. Chem. Phys.*, **86**, 3325–38. © American Institute of Physics.)

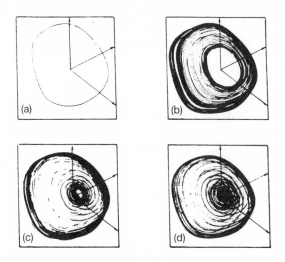

FIG. 8.24. .Growth of reconstructed torus corresponding to Fig. 8.23(a)–(d). (Reprinted with permission from Argoul *et al.* (1987*b*). *J. Chem. Phys.*, **86**, 3325–38. © American Institute of Physics.)

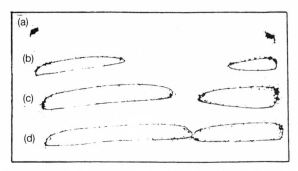

FIG. 8.25. Poincaré sections through tori in Fig. 8.24. (Reprinted with permission from Argoul *et al.* (1987*b*). *J. Chem. Phys.*, **86**, 3325–38. © American Institute of Physics.)

Argoul *et al.* suggests that this may be a degenerate bifurcation with amplitude growing as the fourth root of the increase in flow rate. From the earlier discussions about structural stability, we cannot expect this degeneracy, even if it exists, to survive over any finte range of experimental parameters. Also any future evolution of the torus will reflect non-local behaviour from this secondary bifurcation. Nevertheless, the proximity of some degeneracy may be significant and can be used to test possible models.)

As the flow rate is further increased the torus evolves: two typical scenarios are reported by Argoul *et al.* (1987*b*). For the inflow and reactor conditions of Fig. 8.23 and 8.24, the minimum amplitude during the quasiperiodic trace decreases as the flow rate increases. This corresponds to a narrowing of the inner hole of the torus. This is also shown in the Poincaré sections of the attractor (Fig. 8.25). As this part of the attractor narrows, so the time spent by the system in its vicinity increases with longer periods of almost non-oscillatory evolution separating bursts of large-amplitude excursions around the remainder of the torus. The ultimate conclusion of this process is that ratio of the two fundamental frequencies in the power spectrum tends to zero and the period between bursts tends to infinity: quasiperiodicity gives way to a stationary state (but one on a different branch from the original, low flow rate state).

With different inflow concentrations, and a slightly different type of reactor, the high flow rate behaviour of the torus is significantly altered. Now, the torus does not remain simple and smooth as it grows. Instead its surface becomes increasingly folded, until a fractal strange attractor is obtained. The resulting Poincaré section reveals this folding (Fig. 8.26).

The distinction between the regular two torus and the fractal attractor can also be seen to be in a 'circle map' from the Poincaré section. Rather than plotting the value of a given variable as it passes through a particular line in the plane against the value of the same parameter next time round

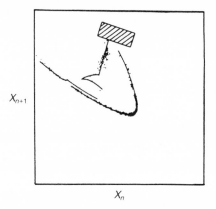

FIG. 8.26. Poincaré section of the BZ torus: the response is far from a simple circle and allows stretching and folding corresponding to aperiodic behaviour. The shaded region indicates the approximate location of the 'hole' at the centre of the torus. (Reprinted with permission from Argoul *et al.* (1987*b*). *J. Chem. Phys.*, **86**, 3325–38. © American Institute of Physics.)

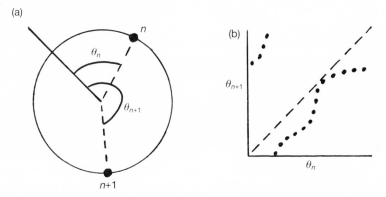

FIG. 8.27. An idealized representation of a circle map. (a) The Poincaré map of a torus, shown here as a circle, consists of a set of points comprising the returns to the Poincaré section. Two consecutive points n and $n + 1$ are indicated on the circle. The angles θ_n and θ_{n+1} formed by the rays from the centre to these points relative to some arbitrary reference zero are the raw data for the circle map (b). In the case shown, there is no intersection of the map with the identity line, so there can be no invariant point on the Poincaré circle. The mapping curve approaches the identity line indicating a possible tangency (tangent bifurcation) for some parameter values close by.

(the next-return map), the circle map concentrates on one of the two 'circular' cross-sections of the torus in the Poincaré plane and plots the angle made by the radial line from the centre to a given point relative to some chosen reference direction against the angle of the next return to the sectional plane, as illustrated in Fig. 8.27. For a regular torus, the circle map is single

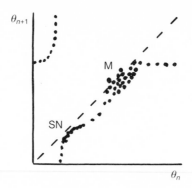

Fig. 8.28. The circle map for the BZ reaction under the experimental conditions of Argoul *et al.* showing a thickening (non-invertibility) close to the point M. Also shown is the dashed line $\theta_n = \theta_{n+1}$ which shows how the mapping curve is nearing a tangent bifurcation at the point SN. (Reprinted with permission from Argoul *et al.* (1987*b*). *J. Chem. Phys.*, **86**, 3325–38. © American Institute of Physics.)

valued and hence invertible (we can run the mapping forwards or backwards): for a fractal torus the map is non-invertible and has thickened noticeably in some parts (Fig. 8.28).

Next we must consider how the fractal torus develops. One of the two areas in which the points accumulate in the circle map is labelled SN. As the flow rate is increased a pair of invariant points appears in this map as a saddle–node bifurcation. The resulting nodal point corresponds to an accessible stationary-state solution to which the system moves from the torus.

Finally, in this set, Argoul *et al.* report a sequence for yet another combination of inflow concentrations. In this case a periodic–chaotic alternation is observed, with no apparent quasiperiodicity. The toroidal character of the attractor is now suppressed, but still underlies the observed behaviour and allows its interpretation. A selection of time series is shown in Fig. 8.29 along with power spectra and reconstructed attractors. Argoul *et al.* use a slightly different notation from that employed above: P_i^j identifies a periodic solution with i large and j small peaks—an i^j waveform.. In between two different periodic states there exists an aperiodic window: this appears as a chaotic mixing of the parent periodicities, and is denoted as $C_i^{j,j+1}$ if the parents are P_i^j and P_i^{j+1}.

The small-amplitude peaks correspond to motion through the inner hole of the torus. If this passage is slow, there will be many of these small oscillations. The amplitude of the last of these is particularly sensitive to the motion and there is some variation in the waveform here. This, in turn, gives rise to variability in the number of large-amplitude excursions as the trajectory moves over the outer surface of the torus.

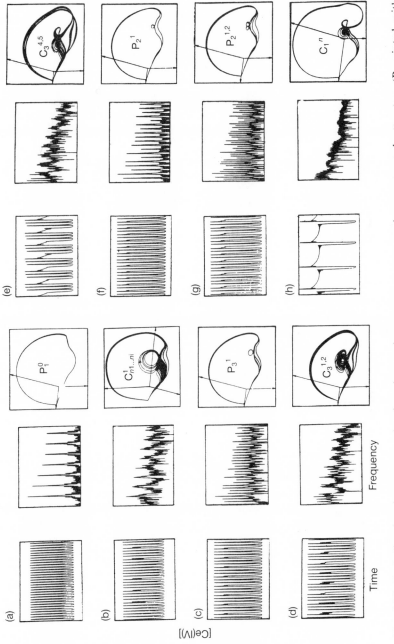

Fig. 8.29. Periodic–chaotic sequence showing time series, power spectra, and reconstructed attractor. (Reprinted with permission from Argoul *et al.* (1987*b*). *J. Chem. Phys.*, **86**, 3325–38. © American Institute of Physics.)

8.6. Other open-system studies and general comments

Few studies have attempted to catalogue in detail the full range of dynamical responses to be found between the stable stationary state at the longest residence times and that at high flow rates. Between these extremes, the steady-state response loses stability. Somewhere in the middle of this range, there are the typically large-amplitude period-1 relaxation oscillations. Hudson and coworkers have examined most closely the transition from these period-1 states to the high flow rate stationary state, revealing the mixed-mode and chaotic sequences described above, whilst the Texas and Bordeaux studies have tended to work in the opposite direction, to longer residence times. Ibison and Scott (1990) have attempted a relatively thorough scan across the whole residence time range for a particular set of inflow concentrations, and their results are summarized in Fig. 8.30.

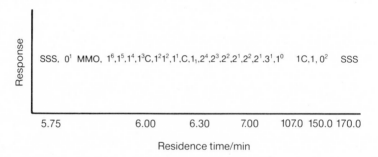

FIG. 8.30. Sequence of bifurcations and responses observed as a function of residence time by Ibison and Scott: SSS, stable stationary state; MMO, various mixed-mode oscillatory waveforms; I, intermittency; C, chaos.

For these studies, there is again a supercritical Hopf bifurcation from the high flow rate, stable stationary state as the residence time is increased. This gives rise to small-amplitude 0^1 waveforms that then develop through a concatenated mixed-mode sequence. Chaotic windows are observed. In addition to the Farey sequence with varying numbers of small excursions separating single large peaks, 2^n and 3^n sequences were also found, culminating in the emergence of the 1^0 state. This is similar in spirit to the observations of Hudson *et al.* and of Maselko and Swinney. On further increasing the residence time, the period-1 large oscillations appear to lose stability through an intermittency bifurcation. Again chaotic rates emerge, before a return to small-amplitude period-1 oscillations and another supercritical Hopf bifurcation at the longest residence times. Some remarkable time series have been obtained, including a 1^{48} response shown in Fig. 8.31—perhaps the experimental record, although recent computations (Peng and Showalter 1990)

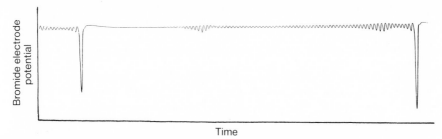

FIG. 8.31. The 1^{48} trace observed by Ibison and Scott.

have gone well beyond this with a 1^{180} pattern (see later).

The mixed-mode oscillatory waveforms reported in earlier studies tend to have an 'hourglass' shape: the train of small excursions between the large peaks often shows decreasing size, perhaps to almost zero amplitude, before a slow growth that then 'triggers' the large oscillation. In the current work, many traces showed more intricate variations in the amplitude of the small peaks—in some cases almost a quasiperiodic form. The Ibison study made no attempt at a full catalogue and it seems clear from earlier work that the harder one looks the greater the detail to be found—with perhaps an infinitely detailed structure. Parts of the apparent regions of chaos may also resolve into complex but periodic responses if examined more closely.

Argoul *et al.* recognized that better experimental resolution might have revealed less aperiodicity and more Farey concatenations in their study. This is something of a general point. Despite their theoretical advantages, flow reactors are much easier to recommend and envisage than they are to construct and operate successfully. This is particularly so with a sensitive non-linear system such as the Belousov–Zhabotinskii reaction that is influenced by factors such as dissolved oxygen, stirring rate, and production of CO_2 bubbles. An important consideration is the method of pumping the solutions through the reactor. Peristaltic pumps are frequently used, but by their action these do impose a small periodic perturbation on the system: we have already seen that forcing a non-linear oscillator can have drastic consequences, so too might imperfect mixing of separate inflows. It is also important to allow the system long enough before making an observation such that all transients have decayed. Even under simple circumstances such as stationary-state operation this may typically mean waiting for three residence times. For residence times of 100 min, therefore, stable operation must be maintained for perhaps 5 hours before data can even begin to be collected.

With all these considerations it is clear that much commendable dedication has gone into all the investigations described in this and the preceding sections. Nevertheless, it seems fair to expect that there will be some ambiguities and some questions necessarily not fully answered by experimental study. The tell-tale marks of chaos are universal, i.e. the same for a chemical and physical and biological and any other system, so their identification in the BZ reaction does not give unequivocal confirmation of 'chemical chaos' (chaos which would arise from the non-linear kinetics even given perfect experimental control). Chaos is a generic response of non-linear systems, and chemical reactions have non-linear kinetics. To help decide whether a particular chemical system with a particular set of experimental parameter values should be aperiodic we can hope for additional guidance from numerical computations if the reaction rate laws are sufficiently well known.

We will move on to consider the various numerical studies in the next section. In the meantime, Table 8.5 gives a selective summary of experimental investigations of the BZ system in flow reactors.

TABLE 8.5
Selected chronology of BZ CSTR studies

1964, 1973	Zhabotinskii, Vavilin, Zaikin
1974	Sørensen
1975	Marek and Svobodova, Marek and Stuchl
1976	De Kepper *et al.*
1976–81	Graziani, Hudson, Schmitz *et al.*
1978	Rössler and Wegmann
1980	Maselko
1980–2	Bordeaux-1
1981–3	Texas/Roux
1985	Hourai *et al.*
1985–7	Bordeaux-2
1986	Maselko and Swinney

8.7. Numerical studies, I: mechanisms

Table 8.6 lists those computational investigations which have taken the most obvious approach: direct numerical integration of the reaction rate equations corresponding to an assumed kinetic scheme. Various different schemes, sets of rate constants, and experimental parameters have been used, so different investigators frequently obtain differing results and apply differing interpretations.

The first entries in Table 8.6 (1972–5) are applicable to closed-vessel studies, and are of interest here primarily because they saw the initial development of the basic FKN mechanisms and the Oregonator model. For our local purposes the important analysis papers of Tyson (1979, 1982, 1984) are not included in the table.

TABLE 8.6
Mechanistic and computational studies of BZ reaction

1972	FKN mechanism
1974	Field and Noyes—Oregonator
1975	Field—reversible Oregonator
1975–9	Edelson et al.—computations
1978	SNBE computations
	Rössler and Wegmann—mapping
1980	Janz et al.—P feedback
	Pikovsky—mapping
1982	Ganapathisubramanian and Noyes
	Rinzel and Troy—analysis
1983	De Kepper and Bar-Eli
1984	Rinzel and Schwartz, Ringland and Turner—mapping
1985	Richetti and Arneodo
1987	Barkley et al., Richetti et al.
1988	Bar-Eli and Noyes
1988–9	Field and Györgyi

The first set of computations specifically designed to model the complex oscillations and chaotic patterns observed by Hudson *et al.* is that of Showalter *et al.* (SNBE) (1978). These authors employed an Oregonator scheme (Section 8.1) and followed Field (1975) in allowing reversibility for all the 'inorganic' steps. They also split the autocatalytic process (O3) into two steps, much as in the original FKN scheme. Thus, they arrived at the model

(S1) $$A + Y \rightleftharpoons X + P$$

(S2) $$X + Y \rightleftharpoons 2P$$

(S3) $$A + X \rightleftharpoons 2W$$

(S4) $$C + W \rightleftharpoons X + Z$$

(S5) $$2X \rightleftharpoons A + P$$

(S6) $$Z \rightarrow \tfrac{1}{2}fY + C.$$

(Showalter *et al.* use the symbol $g = \tfrac{1}{2}f$ for the stoichiometric factor.) The reaction rate equations are written in the form appropriate to a well-stirred flow reactor—ordinary differential equations with inflow and outflow terms. No attempt to reduce the set of seven equations by stoichiometric relationships was made. The inflow concentrations of bromate and cerium(III) were taken to match the Hudson experiments, $A_0 = 0.14$ M and $C_0 = 1.25 \times 10^{-4}$ M. The rate constants taken were the original FKN set (the FF revision had not been made at that stage), appropriate to a constant concentration of

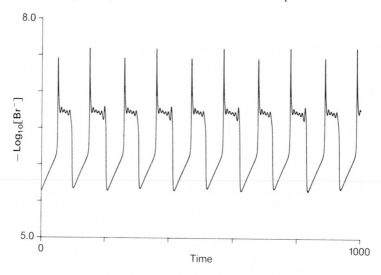

FIG. 8.32. Complex, but strictly periodic, $1^4 1^5$ trace from SNBE model of Belousov–Zhabotinskii reaction in a CSTR.

$H^+ = 0.2$ M: $k_1 = 0.084$, $k_{-1} = 10^4$, $k_2 = 4 \times 10^8$, $k_{-2} = 5 \times 10^{-5}$, $k_3 = 2 \times 10^3$, $k_{-3} = 2 \times 10^7$, $k_4 = 1.3 \times 10^5$, $k_{-4} = 2.4 \times 10^7$, $k_5 = 4 \times 10^7$, and $k_{-5} = 4 \times 10^{-11}$ in $M^{-n} s^{-1}$ units where the numbers correspond to the scheme above. The flow rate k_f, the inflow concentration of bromide ion Y_0, the rate constant k_6, and the stoichiometric factor f were varied for different integrations. We may notice that the organic species concentration B does not enter this model explicitly. The reversibility of the reaction steps crucially allows the species HOBr (P) to play a kinetic role.

Some important conclusions were drawn from this seminal study. First, the reversible Oregonator form used was indeed capable of supporting oscillations more complex than simple period-1. Many of the computed traces bore a striking resemblance to the experimental results. Figure 8.32 shows a $1^4 1^5$ pattern for a particular choice of the four 'adjustable' parameters: the experimental records do not show the sharp spike on the large-amplitude excursion but otherwise the comparison is impressive. The power spectrum for this time series shows two fundamental frequencies and their overtones: the two frequencies are not irrational and so the resulting evolution appears to be frequency locked rather than quasiperiodic.

Close examination of the small peaks within the cycle reveals another feature: the amplitude within a given train between neighbouring large excursions at first decreases to a minimum and then increases, leading to a typical 'hourglass' shape to this section of the waveform. The detail of the limit cycle is shown in Fig. 8.33: the small peaks occur in a region of the phase space apparently close to the unstable stationary-state solution.

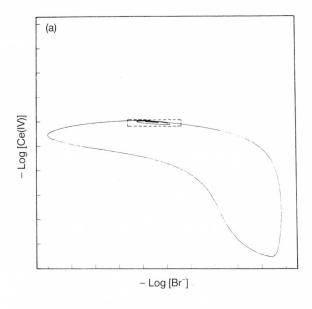

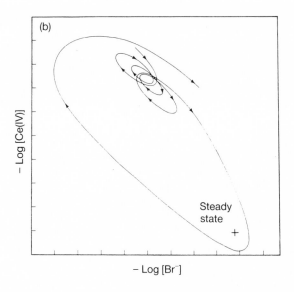

Fig. 8.33. Limit cycle for complex oscillations in SNBE model showing (a) full cycle and (b) detail of small excursions close to unstable steady state. (Reprinted with permission from Showalter *et al.* (1978). *J. Chem. Phys.*, **69**, 2514–24. © American Institute of Physics.

Perhaps most significant, however, is that Showalter *et al.* only observed periodic responses. High degrees of complexity were located, but these always contained a strictly repeating unit. No aperiodic mixing of the forms reported by Hudson *et al.* emerged in the computations.

The major problem associated with this reversible Oregonator model, and some variations to be discussed below, is that the assumed rate constants allow unrealistically high vaues for the concentration of species P (HOBr). With these, the major source of bromide then becomes not the organic steps but simply the reverse of reaction (S1). This seems unsatisfactory, but most attempts to alleviate this also stop the model oscillating.

Complex, but still periodic, oscillations were also observed by Janz *et al.* (1980) with particular reference to the experimental results of Marek and Svobodova (1975). Here, the kinetic mechanism was again based on an Oregonator form but in a quite different adaptation from that of Showalter *et al.* above. Janz *et al.* retained the irreversible five-reaction scheme of Section 8.1. Feedback from the species HOBr (P) was introduced by allowing the stoichiometric factor f to vary with the instantaneous composition of the reaction mixture rather than remaining constant. This is quite a satisfying approach as one would expect the number of bromide ions produced per Ce(IV) ion reduced through process C to vary with the composition for reasons discussed above. (In fact, Janz *et al.* identify P as the brominated form of the organic species rather than as HOBr, but there is no fate for the latter other than participation in process C: we may also note that in their notation $Z = 2[Ce(IV)]$.)

The model leads to a set of five coupled rate equations for A, X, Y, Z, and P, although for the last of these the equation does not strictly follow the law of mass action. The rate constants are again those of FKN with a constant H^+ concentration of 3 M. The inflow concentration of bromate is $A_0 = 0.07$ M.

The dependence of f on P is incorporated via the relationship

$$f = \frac{FP^2}{K + P^2} \tag{8.4}$$

where F and K are two new adjustable parameters. Typically $F = 4$ and $K = 0.0005$. This form leads the stoichiometric factor to increase with increasing concentration of the brominating (or brominated) species P, initially as the second power but ultimately saturating to a maximum value of F at large P.

For the related system with constant f, the above parameter values, and a flow rate of 4 ml min^{-1} through a 200 ml reactor ($t_{res} = 50$ min) the system has a Hopf bifurcation at $f = f^* \approx 1.5$. This is a subcritical Hopf point: the stationary state is unstable for $f < f^*$ and stable for $f > f^*$. An unstable limit cycle emerges as f increases, but this is also surrounded by a stable limit

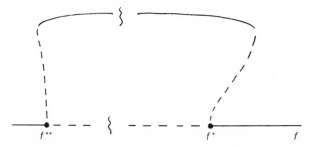

FIG. 8.34. Character of Hopf bifurcations and limit cycle development for basic Oregonator model as function of parameter f.

cycle which exists for lower f, as shown in Fig. 8.34. For the present P-feedback model, the stoichiometric factor given by eqn (8.4) varies slowly, moving backwards and forwards across the Hopf bifurcation point. The resulting behaviour has also been determined analytically for a reduced version of this scheme by Rinzel and Troy (1982).

The important feature of the present model is the coexistence of a stable stationary state and a stable limit cycle for some finite range of f above f^* and the consequent hysteresis between these two modes. For the stationary-state reaction with $f > f^*$ there is a net consumption of the intermediate P, so f given by eqn (8.4) falls. As f passes through the Hopf point, so the system moves away from the now unstable stationary state on to the stable limit cycle solution. During oscillatory reaction, there is a net production of P, so f now increases again. The stable limit cycle continues to exist for some $f > f^*$ because of the subcritical nature of the Hopf point. Only when the stable and unstable limit cycles merge must the system move back to steady reaction, with f then failing again.

In this qualitative way, we can thus expect periods of oscillation interspersed by periods of steady reaction, i.e. classic bursting patterns. (Actually, the quiescent periods would not be completely steady; the stationary state is a stable focus so there would be a damped oscillatory approach—and an initial oscillatory divergence when it loses stability.) Figure 8.35 shows such waveforms evolving after some initial transients for a full computation of this model. These are bursts of oxidizing chemical reaction. Rinzel and Troy (1982) also showed that bursting could occur in a reducing system, around the lower Hopf bifurcation point which occurs for $f = f^{**} \approx 0.5$. This is again a critical Hopf point, with the stationary state being unstable and surrounded by a stable limit cycle for $f > f^{**}$ and a stable stationary state and stable limit cycle coexisting and separated by an unstable limit cycle for $f < f^{**}$. The hysteresis below f^{**} allows the same basic mechanisms to produce bursting, with now the stationary-state reaction producing P and the oscillatory mode consuming it.

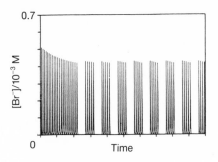 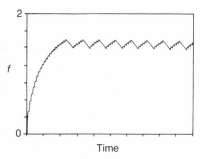

FIG. 8.35. Bursting oscillations from modified 'P-feedback' Oregonator model. (Reprinted with permission from Janz *et al.* (1980). *J. Chem. Phys.*, **73**, 3132–8. © American Institute of Physics.)

Ganapathisubramanian and Noyes (1982), meanwhile, examined the Showalter *et al.* (1978) scheme over a wider range of parameter values, again corresponding to the Hudson experiments (Hudson and Mankin 1981). Again, a large variety of complex periodic waveforms of the general L^s pattern were obtained, but again no aperiodic solutions could be found. One feature that did emerge is the very small changes in the flow rate parameter that are sometimes needed to change the response from one pattern to the next: sometimes this variation need only be as small as one part in 10^6. This high sensitivity, whilst not corresponding to chaos itself, does suggest that experimental studies would have extreme difficulty in stabilizing a single response in such a region and that what is observed is somehow a sampling of many states. Such arguments have some appeal, but it is not then easy to account for some of the clear evidence of strange attractors etc. from experiments.

De Kepper and Bar-Eli (1983) took what is effectively the full FKN scheme with the FKN rate constants. For completeness, the representation of this model in terms of A, C, etc, becomes

(B1) $A + Y + 2H \rightleftharpoons X + P$

(B2) $X + Y + H \rightleftharpoons 2P$

(B3) $A + X + H \rightleftharpoons 2W$

(B4) $C + W + H \rightleftharpoons Z + X$

(B5) $2X \rightleftharpoons P + A + H$

(B6) $Z + M \rightarrow \frac{1}{2}fY + C$

(B7) $Z + W \rightleftharpoons C + A.$

In fact, reaction (B7) was found to be of little significance, increasing the similarity of this model to that of Showalter *et al.* (1978). Because of the rather elementary representation of process C retained in this scheme via step (B6), it was not found possible to include the extra step

$$H^+ + Br^- + HOBr \rightleftharpoons Br_2 + H_2O$$

without losing all the oscillatory and multiple stationary-state behaviour. This shortcoming can probably only be avoided by a more detailed model for the organic chemistry steps in the reaction, beyond that available at present.

De Kepper and Bar-Eli allowed the concentration of H^+ and malonic acid to appear as normal variables in their model, rather than assuming them to be constant as in Showalter *et al.* (1978). They were, however, also able to reduce their system to just four independent rate equations by invoking various mass-conservation conditions on the individual atomic species. The main interest in the 1983 paper was a comparison of the full model with the simplest three-variable Oregonator scheme with regard to matching experimental results concerning multiple stationary states and simple period-1 relaxation oscillations. The authors did, however, also report the experimental observation of some complex and possible chaotic forms. More significant here, however, is later work with this model.

Bar-Eli and Noyes (1988) computed virtually the same set of equations (they also included H_2O as a variable) and still with the FKN rate constants, for inflow concentrations and a residence time intended to allow comparison with the experiments of Maselko and Swinney (1986). They took $[BrO_3^-]_0 = 0.03$ M, $[Ce(III)]_0 = 0.003$ M, $[H^+]_0 = 1.5$ M, and $t_{res} = 255$ s, varying $[MA]_0$ and $[Br^-]_0$. Two of their traverses across the parameter plane are of particular interest.

With $[MA]_0 = 0.1$ M and varying $[Br^-]_0$ between 22.75 and 23.15 μM the system exhibits complex oscillations following Farey concatenations. In one sequence the parent states are 2^2 and 2^3, whilst elsewhere in the range the basic states are 1^1 and 1^2. The qualitative comparison with Maselko and Swinney is claimed by the authors, with some justification to be 'spectacular'.

A similar traverse in $[Br^-]_0$ with $[MA]_0 = 1.0$ M also produced Farey sequences, generally with higher numbers of large and small peaks: from 9^4 with $[Br^-]_0 = 15.15 \mu$M to 3^8 for $[Br^-]_0 = 17.05 \mu$M. Concatenations based on more than two parents and birhythmicity amongst different states were also found, again in 'spectacular' qualitative agreement with experiment. Bar-Eli and Noyes also mention, in passing, that they observed chaotic patterns, e.g. at $[Br^-]_0 = 15.60524 \mu$M, but do not show any computed traces. Figure 8.36 shows an example time trace from this work.

Richetti *et al.* (1987*a,b*) employed yet another modification of the FKN

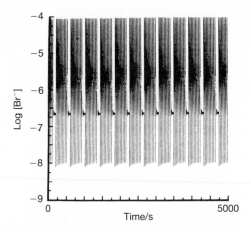

F_{IG}. 8.36. Complex (9^4) patterns from the De Kepper–Bar-Eli model that also shows chaotic responses. (Reprinted with permission from Bar-Eli and Noyes (1988). *J. Chem. Phys.*, **88**, 3646–54. © American Institute of Physics.)

scheme. In terms of the present symbols, their model (M2) becomes

(R1) $A + Y + 2H \rightarrow X + P$

(R2) $X + Y + H \rightarrow 2P$

(R3) $P + Y + H \rightarrow Br_2 + H_2O$

(R4) $A + X + H \rightleftharpoons 2W$

(R5) $2X \rightarrow A + P + H$

(R6) $W + C + H \rightarrow Z + X$

(R7) $P + M \rightarrow BM$

(R8) $BM + Z \rightarrow Y + R + C + H$

(R9) $R + Z \rightarrow C.$

The first six steps in this scheme are relatively familiar as the inorganic processes A and B. Only one of the steps is considered to be reversible. The formation of Br_2 in step (R3) is included in this scheme, but is assumed to be irreversible. This last point has been criticized by Györgyi and Field (1989*a,b*) who argue that the reverse step, i.e. bromine hydrolysis, would be far from negligible. Reactions (R7)–(R9) are an attempt to represent the organic process C without recourse to a stoichiometric factor and are an adaptation of the model of Richetti and Arneodo (1985) proposed for so-called catalyst-free BZ systems (Orban and Köros 1978). The brominating species is assumed to be HOBr (P) and bromomalonic acid (BM) is then oxidized in step (R8) by cerium(IV). This latter step also produces an organic radical R which competes for the oxidized catalyst through step (R9). The combination of steps (R8) and (R9) is roughly equivalently to having a constant stoichiometric factor $f \approx 1$ (for the particular values of the FKN rate constants and inflow concentrations used by Richetti *et al.*) and an effective rate constant for process C which depends on the concentration of the brominating agent P (approximately $k \sim P^{1/2}$).

Richetti *et al.* further assumed that the concentrations of bromate A, H^+, the reduced catalyst C, and malonic acid M could be considered as constants, leaving seven variable species concentrations: P, W, X, Y, Z, BM, and R. Once the values for the constant concentrations have been specified, the remaining parameters are the inflow concentration of bromide ion Y_0 and the flow rate (or residence time).

In the first set of computations reported, with parameter values to correspond with the Texas experiments, a period-1 solution emerges with increasing flow rate via a subcritical Hopf bifurcation (hard excitation). As the flow rate is further increased, so there is a period-doubling sequence based on the 1^0 state which eventually converges to a chaotic response. The chaotic region is interspersed with periodic windows. The period doubling and intermediate periodicities are similar to those observed in the Coffman experiments. With high flow rates there is a reverse period-doubling cascade: the system regains periodicity with a 1^1 waveform. With the resolution available to these numerical studies it is possible to identify a period doubling based on the 1^1 state leading to a second region of chaos, which also has periodic windows. Another reverse cascade brings in the 1^2 periodic state: this period-doubles to chaos; a reverse sequence gives a 1^3. With only limited resolution, as is likely in all but the most sophisticated experiments, one would see only the gross features of such a scenario—an alternation of periodic and chaotic states: 1^0, C_1, 1^1, C_2, 1^2, C_3, etc. with the chaotic states appearing as 'mixing' of the neighbouring periodic forms. This is just as reported in the Texas and Bordeaux-1 investigations and also, perhaps, those of Hudson *et al.*

Figure 8.37 from Richetti *et al.* (1987*b*) compares experimental and numerical records, showing compelling qualitative similarities, although quantitatively there are discrepancies (compare the time axes, for instance). One should not expect more than qualitative agreement; the rate constants etc. are not known to anything better than ± 10 per cent even in the best cases and there is no allowance for variation of these with the different reaction temperatures employed in the studies. It is, perhaps, less comforting that the experiments and computations have these sequences as the flow rate is varied in opposite directions (decreased in the experiments, increased in the computations), but the similarity may at least point to the same underlying mechanism in each case: and that mechanism might be easier to find in the study of a model than in an experiment.

The second set of computations, simulation B, performed by Richetti *et al.*

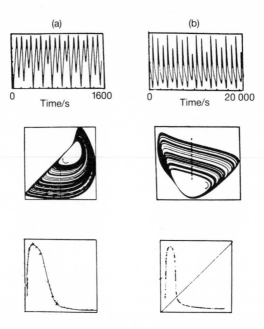

FIG. 8.37. Comparison of (a) experimental and (b) computed time series, attractor, and one-dimensional map from Richetti *et al.* (Reprinted with permission from Richetti *et al.* (1987*b*). *J. Chem. Phys.*, **86**, 3339–56. © American Institute of Physics.)

(1987*b*) correspond to the different scenario leading to chaos appropriate to the Bordeaux-2 experiments (Argoul *et al.* 1987*b*). Here, there appear to be two underlying frequencies in the observed waveforms, leading to frequency locking or quasiperiodicity on a torus. Richetti *et al.* also include the Maselko and Swinney results in their interpretation, although the latter reported no chaotic states.

A typical variation of waveform with flow rate might give the following sequence:

$$1^0, C_0, 2^1, 2^2, 2^3, \ldots, 2^7, 1^1, 1^1 1^2, C_1, 1^2, C_2 \ldots.$$

Here, the chaotic states are not simply mixings of the neighbouring periodicities. The three C_0–C_2 listed above characterize the different types of chaotic response observed. Response C_0 involves single peaks of varying amplitude. Although the amplitude varies, all the peaks may be identified as large excursions. Type C_1 chaos involves both large- and small-amplitude excursions and hence are of mixed-mode form. Typically the time series consists of single large excursions separated by an aperiodic selection of one, two, or three small peaks—a chaotic mixture of 1^1, 1^2, and 1^3. However, greater variety is allowed within this framework: for example, there may be single small peaks separating trains of different numbers of large excursions. The third type C_2 is a 'small-scale' chaos: at first sight the time series appears quite regular, a 1^3 say. On close inspection, however, the amplitude of the small peak preceding the large excursion varies aperiodically. Within experimental measurements such irregularity could be almost undetectable and also probably of no practical significance.

More complex forms such as $4^6 4^5 3^1 3^7$ can also be found. Rather than using the firing number defined by Maselko and Swinney, Richetti *et al.* prefer a winding number W. This number is related to properties of the corresponding circle map (see Section 8.5) and its approximate value can be computed from

$$W = t/(L + s) \qquad (8.5)$$

where L and s are the number of large- and small-amplitude oscillations per cycle as before and t is the number of times there is a change from a small to a large excursion. Thus for the $4^6 4^5 3^1 3^7$ pattern $L = 14$, $s = 19$, and there are four L^s units, so $W = 4/(14 + 19) = \frac{4}{33}$. If W is a simple rational number, this implies that the corresponding motion in phase space is a frequency-locked motion on a (two-dimensional) torus. An irrational winding number corresponds to quasiperiodic motion.

Plotting the winding number against their bifurcation parameter (the flow rate), Richetti *et al.* found the Devil's staircase governed by Farey arithmetic in much the same way as Maselko and Swinney. In some cases, the steps of the staircase overlap. In these cases, the corresponding circle map in not invertible (we can run the map forwards in time but not backwards). There are also gaps in the staircase in which the system is chaotic and hence W is not defined. Various other features such as the fractal dimension could be determined.

As well as integrating their particular Oregenator equations, Richetti *et al.* showed that a qualitatively similar scenario is exhibited by a 'normal form' set of three equations—suggesting that the above responses can justifiably be interpreted as motion in a low-dimensional phase space.

Another version on the Oregonator theme is Turner's (1984) four-variable model. This takes the classic three-variable FKN scheme, but with reverse reactions, and allows for reaction and outflow of the species Br^-, $HBrO_2$, $HOBr$, and $Ce(IV)$. By varying the organic reaction rate constants and the flow rate various mixed-mode, bursting, and chaotic time series can be obtained. The assumption that the organic reaction can be treated as a reversible process, even in the simplified Oregonator form, has, however, been criticized recently (Györgyi and Field 1988).

Quasiperiodic solutions were computed in a study by Barkley *et al.* (1987) from the SNBE model. The original 'Hi' inorganic reaction rate data for steps (S1)–(S5) were employed, with $k_6 = 2.9 \, \text{s}^{-1}$, $g = 0.42$, a flow rate

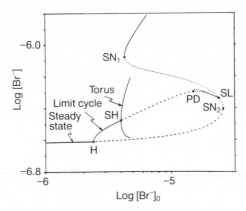

FIG. 8.38. Bifurcation diagram showing stationary-state multiplicity, Hopf bifurcation, and the growth of a stable limit cycle from the lower branch at point H. At SH there is a secondary Hopf bifurcation giving rise to a stable torus. (Reproduced with permission from Barkley *et al.* (1987). *J. Chem. Phys.*, **87**, 3812–20. © American Institute of Physics.)

$k_0 = 7.8 \times 10^{-3}$ s and an inflow concentration of bromate given by $A_0 = 0.14$ M. These parameter values appear targeted at the experimental studies of Hudson and Schmitz (Schmitz *et al.* 1977; Hudson *et al.* 1979; etc.), but here the response of the system is followed as the inflow concentration of either bromide or Ce(III) is varied, rather than as a function of flow rate. Figure 8.38 shows the bifurcation diagram as a function of $[Br^-]_0$ with $[Ce(III)]_0 = 1.25 \times 10^{-4}$ (typical of the Hudson experiments). The stationary-state bromide ion concentration shows an S-shaped hysteresis loop. Along the lower branch, the first bifurcation accompanying increasing $[Br^-]_0$ is a primary, supercritical Hopf point giving birth to a stable, period-1 limit cycle. At higher inflow concentrations the period-1 solution also becomes unstable, via a secondary Hopf bifurcation. The Floquet multipliers leave the unit circle as a complex pair and so a quasiperiodic state with two incommensurate oscillatory frequencies emerges. An example of such a quasiperiodic solution and the corresponding torus in phase space is shown in Fig. 8.39. The torus disappears at larger $[Br^-]_0$, perhaps by coalescence with an unstable torus: the system then moves to the upper branch along which the stationary state is stable.

A slightly different scenario is observed with the inflow of bromide ion fixed ($[Br^-]_0 = 4.2 \times 10^{-6}$ M) and that of the catalyst varied. Again the system shows primary and secondary Hopf bifurcations to a quasiperiodic state. At higher $[Ce(III)]_0$, the observed behaviour becomes that of mixed-mode oscillations, with many small oscillations between each large peak. The details of the transition between quasiperiodicity and mixed-mode waveforms could not be determined completely (hence the gap in Fig. 8.40)

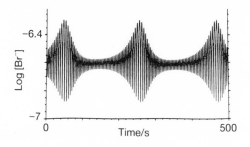

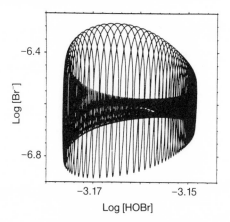

FIG. 8.39. Quasiperiodic solution and its associated torus from the modified SNBE scheme. (Reprinted with permission from Barkley *et al.* (1987). *J. Chem. Phys.*, **87**, 3812–20. © American Institute of Physics.)

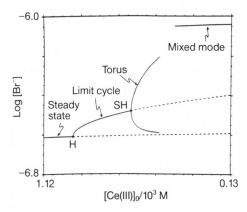

FIG. 8.40. Bifurcation diagram indicating some features of the route from period-1 oscillation to mixed-mode sequences via a T^2 torus in the modified SNBE scheme. (Reprinted with permission from Barkley *et al.* (1987). *J. Chem. Phys.*, **87**, 3812–20. © American Institute of Physics.)

but two features did emerge. On increasing $[Ce(III)]_0$ the quasiperiodic torus undergoes a number of deformations, wrinkling and phase locking. At the lower end of the mixed-mode oscillation range the small-amplitude excursions correspond to a spiralling in the phase plane which appears, at least in some projections, to come very close to the stationary state singularity. This suggests the possibility of a Shilnikov homoclinicity on decreasing $[Ce(III)]_0$. There are thus signs at each end of this gap that might indicate the possibility of chaos in this model as formulated.

An interesting common point, albeit rather mathematical, of the two studies of quasiperiodic states discussed so far (Richetti *et al.* 1987b; Barkley *et al.* 1987) which use different versions of the Oregonator model is their comparison of the observed behaviour with that of the 'hysteresis–Hopf' normal form. The simplest set of equations that can be used to discuss the interactions between stationary-state and the periodic solutions which arise as a hysteresis loop unfolds close to a point of Hopf bifurcation can be written as

$$\frac{dx}{dt} = x(z - \beta) - \omega y$$

$$\frac{dy}{dt} = \omega x + y(z - \beta)$$

$$\frac{dz}{dt} = \lambda + \alpha z - \tfrac{1}{3}z^3 - b(x^2 + y^2).$$

The stationary-state locus $z(\lambda)$ has a hysteresis loop for $\alpha > 0$ which unfolds for $\alpha = 0$. The system has a Hopf bifurcation when $z_{ss} = \beta$. For the special case $\alpha = \beta = 0$, the Hopf point moves to the inflection point ($\lambda = 0$) of the unfolding hysteresis loop. At this point the stationary-state $x_{ss} = y_{ss} = z_{ss} = 0$ has a zero and two imaginary eigenvalues. When the parameters α and β are varied (perturbed) slightly from this degenerate case, the stationary state and Hopf bifurcations typically interact to form secondary bifurcations at which a torus emerges in a similar way to that described above. Moreover, the torus may terminate either via a heteroclinic loop (connecting the lowest stationary state to the intermediate saddle and which corresponds to the hole in the centre of the torus shrinking to zero) or by collision with an unstable torus (the latter born at a heteroclinic connection).

The behaviour of this normal form model can be further adjusted by including higher-order non-linearities in the governing equations. In particular, if the underlying symmetry of the equations is broken the heteroclinic loop becomes atypical (the torus no longer typically reaches two different stationary states simultaneously). Now the torus can be expected, typically, to develop phase-locked orbits or to wrinkle and become fractal. Richetti *et al.* (1987b) especially have shown the strong qualitative comparisons between computations from an Oregonator form and the normal form.

Some features of fine detail in the experimental traces and the Oregonator computations cannot be produced from this normal form, however. Barkley (1988*a*) has commented particularly on the 'hourglass' shape between large excursions, first noted by Showalter *et al.* (1978). Such a waveform does not arise for the three-variable normal form. Barkley constructed an abstract four-variable model which is in effect two coupled two-variable subschemes: the first subscheme is driven backwards and forwards across its simple Hopf bifurcation by the second, which is constructed to show large relaxation oscillations. In fact the second subsystem is chosen to be excitable, as discussed in Chapter 3, and it is the growing small-amplitude oscillations from the first subscheme that push the second across the excitability threshold and into its large excursion. This idea, however, does not lead naturally to quasiperiodicity, nor to the waveforms with successive large peaks.

Another mathematical point has been addressed by Barkley (1988*b*) through his four-variable model. In many of the experimental studies described, most notably Maselko and Swinney (1986), the Devil's staircase is found to be complete (no gaps of quasiperiodicity between the treads). One might expect a complete staircase to be a degenerate case separating those with gaps from those with overlapping treads (the latter being non-invertible and allowing chaotic responses). In the spirit of structural stability, one would not then expect completeness to survive as a second parameter, such as when an inflow concentration is varied (or even to be observed in any 'real' experiment). Despite such arguments, complete staircases do appear over whole regions of parameter space in experiments and in computations (e.g. the Bar-Eli and Noyes results mentioned earlier and, particularly, those of Showalter *et al.*). Barkley has rationalized this behaviour, showing that apparently complete, and structurally stable, staircases may result if the corresponding maps have certain properties. We shall discuss mappings applied to realistic models of the BZ system in the next section.

If we were to summarize the situation at this stage, we might be tempted to conclude that even if direct computations of Oregonator-type mechanisms have not provided a complete picture, they do seem to present a picture of chemical chaos very similar to the experimental studies. That view, however, has been severely challenged recently in a series of papers by Györgyi and Field (1988, 1989*a,b*). These authors have critically assessed many of the above studies in terms of the 'chemical reasonableness' of the different Oregonator forms. They have also repeated some of the computations. Particular attention was paid to the apparently successful model of Richetti *et al.* (1987*a,b*; Richetti and Arneodo 1985). Györgyi and Field suggest that the aperiodicity observed by the latter in their computations is a numerical artefact: by changing the form of the error control from an absolute error to a relative error, strict periodicity appears to emerge. Similar comments about the role of error control and step length have been made by Sheppard (1990).

Györgyi and Field also provide the first attempts at modelling this system with the Field and Fösterling rate constants. They use an irreversible Oregonàtor with the stoichiometric factor f given as a function of the instantaneous HOBr concentration, as discussed earlier—eqn (8.4). This model has also been carefully re-examined, with the FKN rate constants, by Peng and Showalter (1990). In each case, periodicities of the form L^s and Farey sequences are the order of the day, with no aperiodicity observed despite a fine coverage of the parameter space. Györgyi and Field (1989*a,b*) also showed that aperiodic traces similar to experiments can arise of two Oregonators are coupled or if one is forced (see Chapters 5 and 6). From this they suggest that the experimental observations of chaos need not reflect a purely chemical origin (hence, the 'failure' of the most stringent integrations of the Oregonator models to go beyond periodicity) but may come out of experimental imperfections. The idea that a pulsed peristaltic inflow may be significant has been mentioned earlier, although some studies with syringe pumps also reveal aperiodicity, Györgyi and Field postulate that in the vicinity of the inflow ports the stirring will never be sufficient to keep the local composition identical to that in the main part of the reactor. To model such imperfect mixing they imagine the whole system as represented by two coupled reactors—a small one corresponding to the inlet region and a larger one for the well-mixed bulk of the CSTR. The Oregonator scheme used was chosen so that the high concentrations of HOBr that are predicted here and in the work of Barkley *et al.* (1987) do not influence the dynamics: high [HOBr] is not realized in experiments.

The concentrations of the major reactants (MA, BrO_3^-, Ce(III), and H^+) are assumed to be constant, and the parameters chosen such that the whole system has a stable limit cycle at zero flow rate. (If the coupling is 'turned off' computationally, the bulk reactor oscillates but the inflow subsystem has a stable station state.) As the flow rate in the model is increased, two features develop: a second unstable limit cycle emerges whilst the original limit cycle undergoes a period-doubling cascade to chaos. Later, the unstable cycle becomes stable via a secondary Hopf bifurcation to a torus.

This controversy between chemical and physical interpretations of the chaos which undoubtedly has been observed experimentally can be expected to continue for some time. We will return to coupled and forced BZ systems later. There has, however, been another approach to modelling the BZ reaction which we discuss now.

8.8. Numerical studies, II: mappings

Rössler and Wegmann (1978) and Pikovsky (1981) have both used a simple mathematical model quite similar to the hysteresis–Hopf normal form equations above, to mimic experimental observations. Pikovsky also used his (numerical) results to construct a one-dimensional next-maximum map. His equations have the form

$$\frac{dx}{dt} = hx + y - 0.1z$$

$$\frac{dy}{dt} = -x$$

$$\varepsilon \frac{dz}{dt} = \tanh[100(1 + 4z - 16x)] - 4(z + x + x^3).$$

The 'variables' here are clearly not obviously related to any BZ chemistry. For $\varepsilon = 0.1$, the behaviour depends on h. The corresponding one-dimensional map formed from successive maxima in z typically has a maximum and a minimum, as shown in Fig. 8.41. The first increasing section of the map is almost linear and the decreasing portion is almost a discontinuity whilst the final increase is smooth again. Intersections of the map with the line $z_{n+1} = z_n$ give 'stationary' or invariant states of the map. On the increasing sections these are stable, corresponding to simple period-1 oscillations (each iteration returns to the same point indicating that successive maxima have the same amplitude) with either small peaks (on the left-hand, linear branch) or large excursions (on the upper branch beyond the discontinuity). On the other hand, intersections with the decreasing section, which has a high absolute gradient, are unstable. The system visits the stable branches in turn, corresponding to a mixed-mode waveform of small and large excursions.

Varying h from 0.188 (stable, large period-1) to 0.1004 (stable, small period-1) gives rise to a whole set of periodic mixed-mode and also to aperiodic responses with the same qualitative map: for $h = 0.08$ a stable stationary state appears in the model equations.

Rinzel and Schwartz (1984) used the SNBE Oregonator model (Showalter *et al.* 1978) to derive a one-dimensional map. They defined a Poincaré section by specifying a particular value of the bromide ion concentration ($Y = \hat{Y}$ in the notation of the model) with $dY/dt < 0$. By integrating the equations for one return to this section for various initial values of the HOBr concentration, the first return map $P_1(P_0)$ is constructed. Over a wide range of [HOBr], the map appears as a series of steps, with each tread curving inwards just before an almost discontinuous step down, as shown in Fig. 8.42. For any given parameter set, in particular for any given flow rate k_0, the only relevant portion of the map is that close to the intersection with the identity line $P_1 = P_0$. An example is shown in Fig. 8.43. Each tread can be identified with a particular subunit of the total waveform. In Fig. 8.43 the upper tread corresponds to a 1^2 pattern, the lower one to the 1^1. As the map is iterated for this example an invariant cycle of six intersections is found. This has five

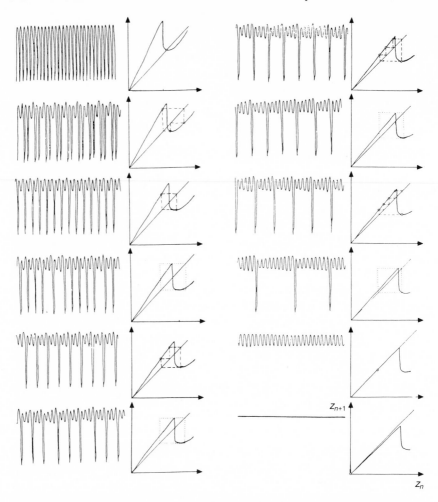

FIG. 8.41. A mapping approach to modelling the complex oscillations and chaos in the Belousov–Zhabotinskii reaction. These traces can be compared with the experimental time series shown in Fig. 8.6. (Reprinted with permission from Pikovsky (1981). *Phys. Lett.*, A **85**, 13–16.)

intersections on the upper (1^2) branch and one on the lower (1^1) branch, giving rise to a total $(1^2)^5(1^1)$ response. The same waveform appears if the original Oregonator differential equations are integrated directly.

If the flow rate is varied, the mapping curve moves with respect to the identity line, shifting down and slightly to the right. As this happens, so the upper tread moves closer to the identity line. Eventually these may meet tangentially, but before this the difference becomes very small creating a narrow channel through which the iterated mapping must pass. The

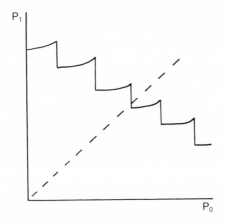

FIG. 8.42. Mapping approach to modelling the BZ reaction from the SNBE scheme: the construction of the next-return map is described in the text and leads to the 'stepped' locus. Intersections with the identity line (dashed) correspond to invariant fixed points, but these will not be stable if the gradient there is large (as shown).

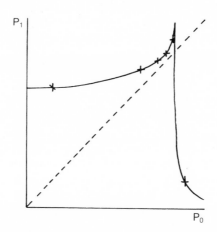

FIG. 8.43. Detail close to typical intersection from Fig. 8.42. The mapping gives five intersections ($+$) on the upper tread (the 1^2 branch) and one intersection on the lower (1^1) tread. (Reprinted with permission from Schwartz (1984). *Phys. Lett.*, A **102**, 25–31. © Elsevier Science Publishers.)

narrower this channel, the more iterations intersect on the upper tread, i.e. the more 1^2 units appear in the waveform. (If the minimum distance between the mapping curve and the identity line is m^{-2} for large m, there will be approximately $m\,1^2$ units between each 1^1, giving a $(1^2)^m(1^1)$ waveform. With large m there are thus only occasional, or 'intermittent', 1^1 units interrupting the 'laminar' 1^2 pattern.)

After the 'tangent bifurcation' a stable fixed point corresponding to the pure 1^2 response exists and the previous mixed state is unstable. We stay with this new simple state until the next tread of the staircase, corresponding to the 1^3 pattern, comes into the picture. The 1^2 fixed point can then become unstable, giving rise to a stable state that visits both 1^2 and 1^3 branches. We then go through the 1^2, 1^3 concatenations as the flow rate is varied further, until a stable 1^3 solution emerges through its tangent bifurcation.

This scenario clearly mirrors the Farey sequences of Hudson *et al.* or Maselko and Swinney. The mapping approach does not introduce any new features beyond the direct computations of the same model: like Showalter *et al.*, Rinzel and Schwartz observed only periodic states. Only two steps are involved at any stage, so there will be no counterparts of Farey triangles etc. where three or more parents concatenate—although such behaviour may emerge for different parameter values or different Oregonator models.

In a development of this work, Schwartz (1984) showed how chaotic mixing of parent states could be generated in a way similar to that observed in experiments, by adding a stochastic forcing term to one of the parameters of the map.

Hudson and Mankin (1981) used a mapping function fitted to their experimental next-maximum maps in an attempt to help convince sceptical chemists that their traces really are chaotic. Despite some experimental uncertainty in the resulting coefficients, the map generally produced positive values for the Lyapounov exponent following the procedure outlined in Chapter 2, giving evidence for the divergence of neighbouring trajectories.

Ringland and Turner (1984) also constructed maps from an Oregonator-type model—the four-variable Turner (1984) scheme discussed earlier. The interest there was to match the period-doubing sequences seen at very long residence times in the Texas experiments rather than those of Hudson *et al.* They obtained smooth, single maximum maps which then, not surprisingly, gave rise to the 'universal' period doubling and periodic windowing typical of such systems (see Chapter 2).

As a variation, Coffman *et al.* (1987) used successive minima in the bromide ion potential to construct their map from chaotic experimental traces. The resulting curves have something of the cubic shape seen in Chapter 2. Again this will support a 'universal' sequence, and Coffman *et al.* showed that not only do the periodic windows occur in the correct order from the map to coincide with their experimental observations, finer detail such as distinguishing between different period-5 or period-6 states in symbolic dynamic terms is also successful. Thus if iterations of the map are denoted l or r if they occur to the left or right respectively of the intersection with the identity line, the three period-5 states are lrlrr, lrllr, and lrlll and they occur in this order as the parameter is increased for both the map and the experiment.

By allowing the simple cubic form to become distorted to an indented trapezoid, higher-order complications in their results could also be 'reproduced'. There are four parameters required to describe the map as shown: the coordinates a, b, c, and h. Coffman *et al.* took $a = 0.13$, $b = 0.16$, and $c = 0.25$. With $0.62 < h < 0.893$, the map is not qualitatively different from a cubic and so sticks with the normal U sequence. For greater extends of indentation (smaller h) the map allows some responses to occur for more than one range of parameter values, i.e. gives the multiplicity observed experimentally.

Mappings often appear rather arbitrary and disconnected from the chemical systems they are being used to represent (and, indeed, there is no certainty that the form of a mapping should arise solely from chemical interactions). However, there does seem at present firmer progress along this route than the rather disputed conclusions from direct numerical computation.

8.9. Forced BZ reactors

Various investigators have used the BZ reaction to study the effects of periodic perturbations on non-linear chemical systems. The earliest work appears to be that of Vavilin *et al.* (1968) and of Zaikin and Zhabotinskii (1973), who perturbed the cerium-catalysed reaction in a CSTR with periodic ultraviolet illumination. This produced notable quasiperiodic responses and also the first complex oscillations. Marek and coworkers (Marek 1979, 1985; Marek *et al.* 1986; Dolnik *et al.* 1984, 1986, 1989) perturbed their CSTR system with periodic additions of bromide or ceric ions. A similar procedure has been employed by Lamba and Hudson (1987) and Hudson *et al.* (1986). Nagashima (1982) worked with a batch reactor and periodically altered the stirring rate. The variation of the inflow rate (or residence time), in the spirit of many of the theoretical studies discussed in Chapter 5, has been exploited by Schneider and coworkers (Buchholtz and Schneider 1983; Freund *et al.* 1985, 1986; Schneider 1985).

Marek's work was conducted by introducing 1 ml additions of bromide ion solution of known concentration periodically at time t_f or, occasionally, by a 'square-wave' perturbation of the inflow concentration (see below). Typical results from these studies show either a 'chaotic' response or an entrainment of the observed and forcing frequencies. Examples of the latter include the $\frac{7}{3}$ and $\frac{12}{5}$ states, as well as the less exotic $\frac{5}{2}$, $\frac{2}{1}$, and $\frac{3}{2}$ phase locking as the ratio of the forcing and natural frequencies is varied from 2.51 to 1.18. These authors also constructed a partial Devil's staircase.

Hudson *et al.* (1986; Lamba and Hudson 1987) also used a 'square-wave' forcing by periodically switching the inlet bromate between stock reservoirs containing low and high bromide ion concentration. The low bromide supply was estimated to have 4×10^{-6} M Br$^-$ (the impurity concentration), whilst

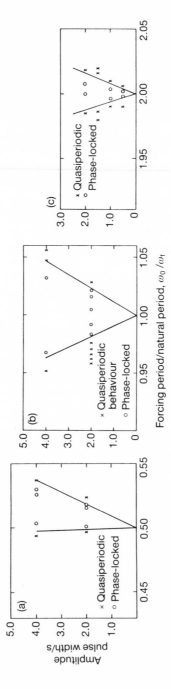

FIG. 8.44. Experimental excitation diagrams showing the regions surrounding (a) 1:2, (b) 1:1, and (c) 2:1 resonance horns for a forced Belousov–Zhabotinskii system. (Reprinted with permission from Hudson *et al.* (1986). *J. Phys. Chem.*, **90**, 3430–4. © American Chemical Society.)

the higher concentration was 6.85×10^{-5} M. The natural unforced period T_0 of their BZ system was about 57 s and they forced this with frequencies corresponding to forcing period close to $\frac{1}{2}T_0$, T_0, and $2T_0$. During a forcing period, the high bromide supply is provided for typically 0.5–4 s and then the inflow is switched to the low bromide stream. Varying this 'pulse width' and the forcing period allows the construction of the entrainment bands in the 'amplitude–period' excitation diagram, as shown in Fig. 8.44. Quasi-periodic solutions outside the entrainment bands are also observed and give rise to closed curves in the stroboscopic map (Fig. 8.45).

With higher forcing amplitudes, period doubling cascades, as represented by the Feigenbaum 'tree' in Fig. 8.46. Deterministic chaos, suggested by such a cascade, is further confirmed by stroboscopic maps, reconstructed attractors, and broad-band power spectra.

Buchholtz and Schneider (1983; Buchholtz *et al.* 1985; Schneider 1985) obtained quasiperiodicity and 1:1, subharmonic, and superharmonic entrainment by sinusoidally forcing the inflow at various multiples of the natural frequency. They also produced a tentative excitation diagram (Fig. 8.47).

Computations by Bar-Eli (1985) based on a variation of the FKN scheme (De Kepper and Bar-Eli 1983; see Section 8.7) also indicated that forcing, in his imagined situation due to the peristaltic effect of pumps, should have a significant quantitative and qualitative effect on the oscillatory waveform in the BZ system. (We should note that the above workers tended to use syringe pumps that are not prone to this effect.)

More recently, Dolnik *et al.* (1989) have returned to forcing BZ systems that do not oscillate in their unforced state, but which have the property of being excitable (see Chapter 3). The excitation diagrams shown in this work have the usual Arnol'd tongue corresponding to regions of entrainment, Dolnik *et al.* ascribe the regions between the tongues to that of aperiodicity, although these would normally be expected to support quasiperiodic behaviour. In their modelling based on an 'expanded' four-variable Oregonator (Ruoff and Noyes 1986), a more conventional excitation diagram was obtained, with entrainment tongues embedded in a quasiperiodic background, and only small regions of aperiodicity (Fig. 8.48).

8.10. Coupled BZ reactors

Again, the BZ reaction has seemed a natural system through which to study coupled oscillators experimentally. Whilst the possibility of coupling by the exchange of mass between, say, two reactors is easily envisaged, it is less easy to achieve in such a way as to allow control or systematic variation of the exchange coefficient. Some progress towards the latter has been made in recent years in a remarkable development by the groups in Bordeaux and Texas (Ouyang *et al.* 1989; Tam *et al.* 1988; Vastano 1988). There, the

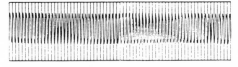

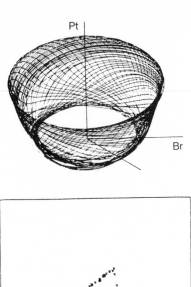

FIG. 8.45. Quasiperiodic time series, torus, and stroboscopic portrait for forced Belousov–Zhabotinskii system. (Reprinted with permission from Hudson *et al.* (1986). *J. Phys. Chem.*, **90**, 3430–4. © American Chemical Society.)

reactors are coupled among the annulus between two concentric cylinders: by rotating the inner cylinder a turbulent (Taylor–Couette) fluid motion is achieved and this provides transport of material between the two cells. The rate of exchange is determined primarily in this case by the rate of rotation and hence can be varied (within certain limits). In these studies interest has centred not just on the time dependence of solutions but also on spatial patterning along the connecting cylinder. Some of these features are sketched in Fig. 8.49 for the BZ and other reaction systems. Tam *et al.* report a 'complete' bifurcation sequence from a stable spatial pattern through periodicity, quasiperiodicity, entrainment, and spatio-temporal chaos.

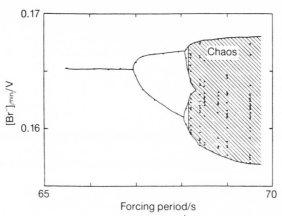

FIG. 8.46. Variation of stroboscopic intersections with forcing period for forced Belousov–Zhabotinskii system showing period doubling to chaos. (Reprinted with permission from Lamba and Hudson (1987). *Chem. Eng. Sci.*, **42**, 1–8.)

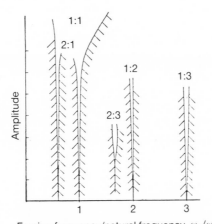

FIG. 8.47. Tentative excitation diagram showing resonance horns for sinusoidal forcing of inflow in Belousov–Zhabotinskii reaction. (Reprinted with permission from Buchholtz *et al.* (1985). *Temporal order*, pp. 116–21. Springer, Berlin.)

Earlier work has generally allowed direct mass exchange by constructing holes that can be opened or closed in some more-or-less controlled manner between two or more cells: a typical arrangement has been shown in Fig. 6.1.

Again, Marek and his group have made some of the earliest contributions to this part of the field (Marek and Stuchl 1975; Marek and Svobodova 1975) with a coupled pair of reactors separated by three removable windows. By thermostating the reactors at different temperatures, the uncoupled

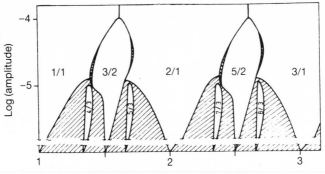

Forcing period/natural period

FIG. 8.48. Computed excitation diagram for forced excitable Belousov–Zhabotinskii reaction. (Reprinted with permission from Dolnik *et al.* (1989). *J. Phys. Chem.*, **93**, 2764–77. © American Chemical Society.)

systems (windows closed) may be maintained at different oscillatory frequencies. When coupling is then allowed, the two cells may become synchronous, adjusting to oscillations with a common frequency—usually close to the higher of the two uncoupled frequencies. A 'harmonic' synchronization is also observed, if one of the uncoupled cells has a natural frequency close to a simple multiple of the other: on coupling, the slower cell may adopt a frequency that is a harmonic of the faster (driving) cell. Marek also observed 'rhythm splitting' giving rise to mixed-mode-type oscillations and a complex synchronization reminiscent of quasiperiodicity, for which the resulting coupled oscillations contain contributions from the driving frequency and multiples of the slower. Fujii and Sawada (1978) and Nakajima and Sawada (1980) provided similar evidence for phase ocking, synchronization, etc. in closed, coupled reactors. They also provided a claim for chaotic behaviour.

Stuchl and Marek (1982) coupled up to seven identical reactors in a linear array, as shown in Fig. 8.50. The natural (uncoupled) state of each cell did not correspond to oscillatory behaviour, but to a region of bistability so that each could operate at a steady state with either high or low $[Ce^{4+}]/[Ce^{3+}]$ respectively. By suitable perturbations, the coupled system could be adjusted into a variety of 'dissipative structures' or spatially inhomogeneous states, so that neighbouring cells might have different $[Ce^{4+}]/[Ce^{3+}]$ values.

Bar-Eli (1984*a,b*) first predicted and then confirmed experimentally (Bar-Eli and Reuveni 1985) that coupling two oscillating BZ systems together could under certain circumstances create an overall steady state. Crowley and Epstein (1989) also obtained such results with the apparatus shown in Fig. 6.1 and also found 1:1 entrainment with both cells either in phase or out of phase. They successfully matched their experimental traces with

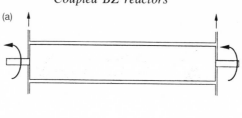

(a)

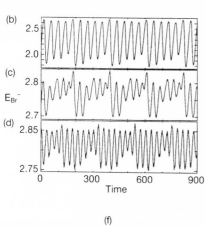

(b)

(c)

(d)

E_{Br^-}

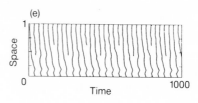

(e)

Space

Time

(f)   (g)   (h)

Fig. 8.49. Schematic representation of 'Couette' reactor and possible spatial or spatiotemporal patterns. (a) Reactor consisting of two concentric cylinders, the inner of which is rotated. At the ends of the cylinders are 'reservoirs' that may be conventional CSTRs or simple flows of some of the reactants. (b)–(d) Experimental records of bromide ion potential at points 20 per cent, 40 per cent, and 67 per cent along reactor: the flows at each end are of oxidant and reductant respectively and these particular conditions correspond to a $\frac{7}{5}$ entrainment. (e) Variation of position of the maxima in bromide ion potential with time for entrained state. (f)–(h) Representation of fronts observed when reservoirs contain bistable system in different states (characterized by different colours). ((a)–(e) Reprinted with permission from Tam *et al.* (1988). *Phys. Rev. Lett.*, **61**, 2163–6. © American Physical Society. (f)–(h) After Ouyang *et al.* (1989). *Phys. Lett.*, A **134**, 282–6.).

computations based on Tyson's scaling (1982) of the original three-variable Oregonator—although apparently without inflow and outflow terms. The model also produced some mixed-mode sequences, particularly the 1^1 pattern.

A noticeably (and imaginatively) different form of coupling has been

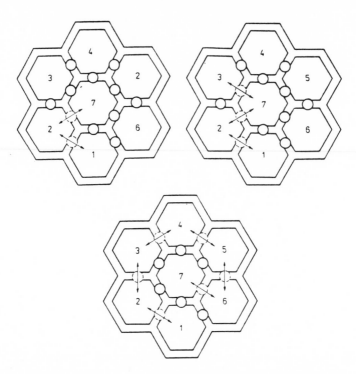

FIG. 8.50. Different numbers of linearly coupled Belousov–Zhabotinskii reaction cells employed by Stuchl and Marek. (Reprinted with permission from Stuchl and Marek (1982). *J. Chem. Phys.*, **77**, 2956–63. © American Institute of Physics.)

employed by Crowley and Field (1981, 1986). They arranged a large-area platinum electrode in each of the cells, connecting these through an external circuit and an internal ion bridge. Electrical coupling arises if the [Ce(IV)]/[Ce(III)] ratio varies in the two cells, the generated potential difference driving a current through the system, with the magnitude and direction of the current determined by the chemistry. The CSTR cells were of rather small volume (2.5 ml) for technical reasons and malonic acid, which generates CO_2, was relaced by an alternative organic substrate, acetylacetone. The cells could be effectively decoupled by introducing a large resistance into the external electrical circuit, and the coupling controlled by adjusting this resistance. Despite apparently identical inflows, the isolated cells showed quite different responses, and variations of those responses with flow rate.

Weak coupling (relatively high resistance in the external circuit) of two oscillatory states with similar (uncoupled) frequencies may lead to entrainment (in phase or out of phase) of quenching of the oscillations (with the

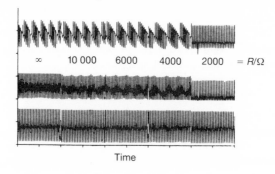

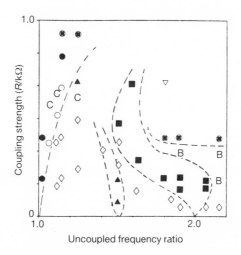

FIG. 8.51. Electrically coupled oscillators showing different possible responses and excitation diagram with regions of quasiperiodicity $\diamond$, 1:1 entrainment (synchronized ● or out-of-phase ○), B bursting, C chaos, 2:1 entrainment ■, 3:2 entrainment ▲, or with one or both oscillations quenched ⊗, ▽. (Reprinted with permission from Crowley and Field (1986). *J. Phys. Chem.*, **90**, 1907.15. © American Chemical Society.)

two cells perhaps then sitting in different stationary states). If the isolated cells have a large difference in frequency then at low coupling the slower oscillation is significantly perturbed by the faster, but not vice versa: entrainment at some intermediate frequency emerges at higher degrees of coupling. With markedly dissimilar uncoupled states, subharmonic entrainment (2:1, 3:1, 3:2, etc.), quasiperiodicity, complex (bursting) oscillations, and chaos emerge, as well as the possibility of quenching to steady reaction. Figure 8.51 shows some typical time series and the tentative parameter plane (excitation diagram) constructed by Crowley and Field.

Finally, we may mention more work by Dolnik and Marek (Dolnik *et al.* 1987; Dolnik and Marek 1988) in which they both forced and coupled their reaction cells. Again, the main patterns of response are entrainment and more complex behaviour including chaos.

8.11. Other solution-phase reactions

The BZ system has a (growing) number of oscillating cousins in CSTRs, related through variations amongst the halogen atoms in the reactants. The simplest of all is the so-called minimal bromate oscillator (Geiseler and Föllner 1977; Bar-Eli and Noyes 1977, 1978; Geiseler and Bar-Eli 1981; Bar-Eli 1981; Geiseler 1982; Orban *et al.* 1982*a,b*; Bar-Eli and Geiseler 1983): this is the BZ mixture without the organic species, i.e. bromate, bromide, Ce(III), and H^+. The clock-resetting role of process C is now played by the inflow of reduced catalyst and Br^-. Without the organic component, this system is somewhat simpler in its range of response than the BZ, exhibiting multistability (bistability) and oscillations (plus the associated phenomena of Hopf bifurcation, homoclinic orbits, and double zero eigenvalue solutions). This system has also been successfully modelled using the new FF set of rate constants (Sasaki 1988).

The same range of behaviour, bistability, and oscillation is seen if bromide ion is replaced by iodide (Alamgir *et al.* 1983) and without the metal ion catalyst. In the presence of the metal ion catalyst, the $BrO_3^- - Mn^{2+} - I^-$ system, which generates bromide ions in the course of the reaction, can be described in some sense as a 'coupling' of the minimal bromate and bromate–iodide systems in the same reactor (Alamgir and Epstein 1983, 1984). Each of the two 'subsystems' gives rise to a classic 'cross-shaped diagram', with apparently connected but non-overlapping regions of bistability and oscillations meeting at a point. For the 'coupled' system, these two features can still be recognized (Fig. 8.52(c)) but the regions of bistability overlap, creating a set of conditions for which the reaction shows tristability (three stable stationary states, implying a total of five stationary states in all, two of which will be unstable saddles).

Another system showing bistability and oscillations with a cross-shaped diagram form is the chlorite–iodide reaction (Dateo *et al.* 1982). If this is conducted in a CSTR which also has a bromate inflow, there is again the possibility of a 'coupling' between two subsystems ($ClO_2^- - I^-$ and $BrO_3^- - I^-$). Here, the interaction appears to be stronger and hence have more consequences (Alamgir and Epstein 1984) than for the 'coupling' described above. Two different types of oscillation were observed (Alamgir and Epstein 1983): a small-amplitude waveform with a relatively high mean potential (measured by a platinum electrode versus $Hg/HgSO_4$ reference) reminiscent of those

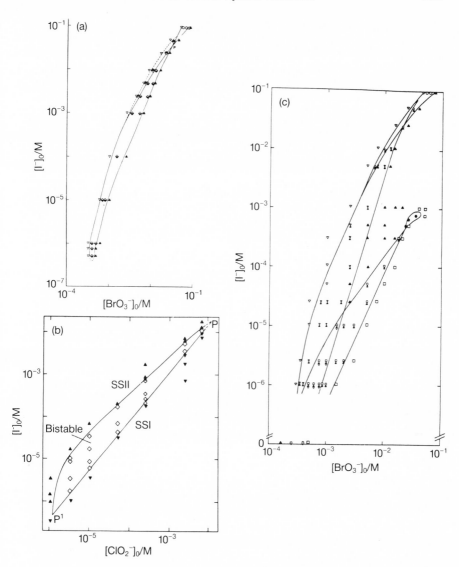

FIG. 8.52. Phase diagrams for (a) $BrO_3^--I^-$, (b) $ClO_2^--I^-$, subsystems and (c) $BrO_3^--ClO_2^--I^-$ mixed systems. (Reprinted with permission from Alamgir *et al.* (1983). *J. Am. Chem. Soc.*, **105**, 2641–3; Dateo *et al.* (1982). *J. Am. Chem. Soc.*, **104**, 505–9; Alamgir and Epstein (1984). *J. Phys. Chem.*, **88**, 2848–51. © American Chemical Society.)

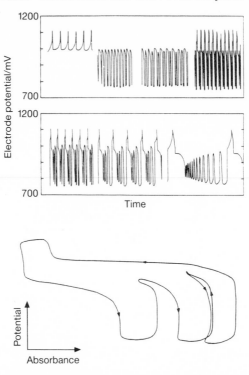

FIG. 8.53. Example complex oscillations and limit cycle for $BrO_3^- - ClO_2^- - I^-$ mixed system. (Reprinted with permission from Alamgir and Epstein (1984). *J. Phys. Chem.*, **88**, 2848–51. © American Chemical Society.)

found in the $BrO_3^- - I^-$ system and a large-amplitude oscillation about a lower mean typical of the $ClO_2 - I^-$ subsystem—the minimum in the small-amplitude trace is generally higher than the maximum for the large-amplitude excursions. For some conditions, these two waveforms coexist, giving rise to birhythmicity. Alamgir and Epstein also discovered 'compound' and 'complex' oscillations, for which the waveform appears as some combination of the small and large peaks as shown in Fig. 8.53: a compound oscillation has the character of one large + one small; complex oscillations, labelled as C_i in the original paper, consists of a compound oscillation followed by i large excursions. (These may also be thought of as one small followed by $i + 1$ large.) Periodic solutions up to C_{13} have been observed. Additionally, there can be concatenations of various neighbouring states, and these may lead to aperiodic waveforms. Figure 8.54 shows a chaotic mixing of C_1 and C_2.

There is, in fact, a whole family of chlorite-based oscillators (Orban *et al.* 1982*b*) of which perhaps the most interesting is one of the simplest in terms

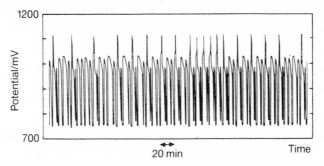

FIG. 8.54. Aperiodic behaviour in the $BrO_3^- - ClO_2^- - I^-$ mixed system. (Reprinted with permission from Alamgir and Epstein (1984). *J. Phys. Chem.*, **88**, 2848–51. © American Chemical Society.)

of ingredients—the chlorite–thiosulphate system, $ClO_2^- - S_2O_3^{2-}$ (Orban *et al.* 1982*c*; Orban and Epstein 1982; Maselko and Epstein 1984). Orban and Epstein report on a series of experiments in which the flow rate is varied for constant inflow concentrations. The reaction typically exhibits bistability: on reducing the flow rate (increasing t_{res}), the low-potential state undergoes a Hopf bifurcation from which relatively large, simple period-1 oscillations emerge (whether this is a subcritical Hopf point or a supercritical point with rapid growth in amplitude would be almost impossible to determine). As the flow rate k_f decreases further, the reaction shows mixed-mode oscillations with large and small excursions in each complete cycle, much like those described earlier for the BZ system. The responses reported are all of the 1^s form with the number of small peaks per cycle s increasing as k_f decreases. The record here seems to be a 1^{16} state, before the oscillation becomes a simple period-1 small-amplitude signal. Some examples are shown in Fig. 8.55. Concatenation of the 1^s and 1^{s+1} states may also be expected, although Orban and Epstein observed only aperiodic mixing of various parents: the $1^0 - 1^1$, $1^1 - 1^2$, $1^8 - 1^9$, and three-parent $1^3 - 1^4 - 1^5$ traces are shown in Fig. 8.56.

Maselko and Epstein (1984) showed that the chaotic response in the chlorite–thiosulphate system gives rise to a 'cubic' next-amplitude map (constructed from successive minima in the potential of a platinum electrode). With relatively low inflow concentrations of chlorite ion, the cubic map is supposed to be intersected only once by the identity line. The mixed-mode oscillation scenario, with increasing s, can then be interpreted in much the same way as that of Pikovsky or the hysteresis–Hopf arguments applied to the BZ system. At higher $[ClO_2^-]_0$, the cubic map can have three intersections with the identity line, as shown in Fig. 8.57. Maselko and Epstein use this to account for a different transition from periodicity to chaos, via almost quasiperiodic waveforms.

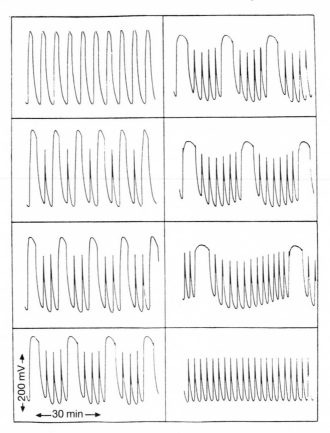

FIG. 8.55. Complex periodic oscillations in the chlorite–thiosulphate system. (Reprinted with permission from Orban and Epstein (1982). *J. Phys. Chem.*, **86**, 3907–10. © American Chemical Society.)

The reaction between iodate and arsenite or sulphite ions is known to follow a cubic autocatalytic rate law similar to that discussed in many of the earlier chapters. Ganapathisubramanian and Showalter (1984*a*,*b*, 1986) have obtained experimentally many of the stationary-state patterns (isola, mushroom, etc.) predicted by these analyses, as well as some of the time-dependent behaviour. When the reactants are mixed with ferrocyanide, a typical cross-shaped diagram for bistability and large-amplitude relaxation oscillation emerges (Edblom *et al.* 1986, 1987; Gaspar and Showalter 1987, 1990; Luo and Epstein 1989) but no complex oscillations have yet been observed.

Finally, we may mention examples of multistability and oscillations in inorganic solution-phase reactions that do not involve halogen species. The

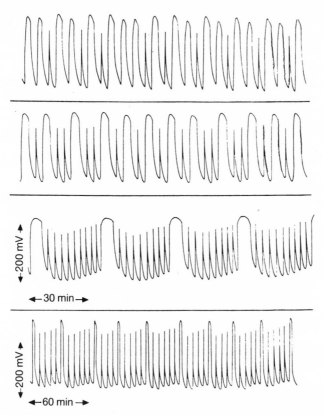

FIG. 8.56. Aperiodic behaviour in the chlorite–thiosulphate system. (Reprinted with permission from Orban and Epstein (1982). *J. Phys. Chem.*, **86**, 3907–10. © American Chemical Society.)

autocatalytic oxidation reactions involving permanganate ion MnO_4^- with various substrates are common in analytical chemistry, but also exhibit non-linear kinetic behaviour (Reckley and Showalter 1981; De Kepper *et al.* 1984; Nagy and Treindl 1986).

Orban and Epstein report bistability and oscillations in a number of reactions involving hydrogen peroxide, with sulphide (Orban and Epstein 1985), thiosulphate (Orban and Epstein 1987), and thiocyanate ion (Orban 1986)—the latter two catalysed by Cu(II). The first two of these peroxide systems appear to be driven by autocatalytic changes in the pH, much like the iodate–sulphite–ferrocyanide and the peroxide–sulphite–ferrocyanide (Rabai *et al.* 1989) oscillators. A mechanistic interpretation based on a 30 reaction scheme has been suggested by Luo *et al.* (1989).

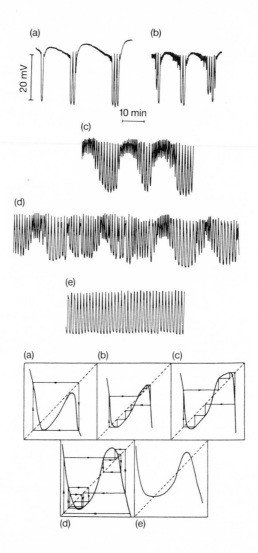

Fɪɢ. 8.57. The transition to chaos on increasing residence time in the chlorite–thiosulphate reaction with associated one-dimensional maps. (Reprinted with permission from Maselko and Epstein (1984). *J. Chem. Phys.*, **80**, 3175–8. © American Institute of Physics.)

References

Alamgir, M. and Epstein, I. R. (1983). Birhythmicity and compound oscillation in coupled chemical oscillators: chlorite–bromate–iodide system. *J. Am. Chem. Soc.*, **105**, 2500–2.

Alamgir, M. and Epstein, I. R. (1984). Experimental study of complex dynamical behaviour in coupled chemical oscillators. *J. Phys. Chem.*, **88**, 2848–51.

Alamgir, M., Orban, M., De Kepper, P., and Epstein, I. R. (1983). A new type of bromate oscillator: the bromate–iodide reaction in a stirred-flow reactor. *J. Am. Chem. Soc.*, **105**, 2641–3.

Argoul, F. and Roux, J. C. (1985). Quasiperiodicity in chemistry: an experimental path in the neighbourhood of a codimension-two bifurcation. *Phys. Lett*, A **108**, 426–30.

Argoul, F. Richetti, P., and Arneodo, A. (1984). Nonlinear interactions between instabilities leading to chaos in the Belousov–Zhabotinskii reaction. In *Nonequilibrium dynamics in chemical systems*, (ed. C. Vidal and A. Pacault), pp. Springer, Berlin.

Argoul, F., Arneodo, A., Richetti, P., Roux, J. C., and Swinney, H. L. (1987a). Chemical chaos: from hints to confirmation. *Acc. Chem. Res.*, **20**, 436–42.

Argoul, F., Arneodo, A., Richetti, P., and Roux, J. C. (1987b). From quasiperiodicity to chaos in the Belousov–Zhabotinskii reaction. I. Experiment. *J. Chem. Phys.*, **86**, 3325–38.

Baier, G., Wegmann, K., and Hudson, J. L. (1989). An intermittent type of chaos in the Belousov–Zhabotinskii reaction. *Phys. Lett*, A **141**, 340–5.

Bar-Eli, K. (1981). The behavior of multistable chemical systems near the critical point. In *Nonlinear phenomena in chemical dynamics*, (ed. C. Vidal and A. Pacault), pp. 228–34. Springer, Berlin.

Bar-Eli, K. (1984a). Coupling chemical oscillators. *J. Phys. Chem.*, **88**, 3616–22.

Bar-Eli, K. (1984b). The dynamics of coupled chemical oscillators. *J. Phys. Chem.*, **88**, 6174–7.

Bar-Eli, K. (1985). The peristaltic effect on chemical oscillations. *J. Phys. Chem.*, **89**, 2852–5.

Bar-Eli, K. and Geiseler, W. (1983). Oscillations in the bromate–bromide–cerous system. The simplest chemical oscillator. *J. Phys. Chem.*, **87**, 3769–74.

Bar-Eli, K. and Noyes, R. M. (1977). Model calculations describing bistability for the stirred flow oxidation of cerous ions by bromate. *J. Phys. Chem.*, **81**, 1988–90.

Bar-Eli, K. and Noyes, R. M. (1978). Detailed calculations of multiple steady states during oxidation of cerous ions by bromate in a stirred flow reactor. *J. Phys. Chem.*, **82**, 1352–9.

Bar-Eli, K. and Noyes, R. M. (1988). Computations simulating experimental observations of complex bursting patterns in the Belousov–Zhabotinskii system. *J. Chem. Phys.*, **88**, 3646–54.

Bar-Eli, K. and Reuveni, S. (1985). Stable stationary-states of coupled chemical oscillators. Experimental evidence. *J. Phys. Chem.*, **89**, 1329–30.

Barkley, D. (1988a). Slow manifolds and mixed-mode oscillations in the Belousov–Zhabotinskii reaction. *J. Chem. Phys.*, **89**, 5547–59.

Barkley, D. (1988b). Near critical behaviour for one-parameter families of circle maps. *Phys. Lett.*, A **129**, 219–22.

Barkley, D., Ringland, J., and Turner, J. S. (1987). Observations of a torus in a model of the Belousov–Zhabotinskii reaction. *J. Chem. Phys.*, **87**, 3812–20.

Belousov, B. P. (1951). A periodic reaction and its mechanism. *Personal Archives.* (Published in *Autowave processes in sysems with diffusion*, (ed. M. T. Grecova), 1981. USSR Academy of Sciences, Gorky and also in *Oscllations and traveling waves in chemical systems*, (ed. R. J. Field and M. Burger), 1985, pp. 605–13. Wiley, New York.

Belousov, B. P. (1958). A periodic reaction and its mechanism. *Sb. Ref. Radiat. Med.*, p. 145. Medzig, Moscow.

Buchholtz, F. and Schneider, F. W. (1983). First experimental demonstration of chemical resonance in an open system. *J. Am. Chem. Soc.*, **105**, 7450–2.

Buchholtz, F., Freund, A., and Schneider, F. W. (1985). Periodic perturbation of the BZ-reaction in a CSTR: chemical resonance, entrainment and quasi-periodic behavior. In *Temporal order*, (ed. L. Rensing and N. I. Jaeger), pp. 116–21. Springer, Berlin.

Coffman, K. G., McCormick, W. D., Noszticzius, Z., Simoyi, R., and Swinney, H. L. (1987). Universality, multiplicity and the effect of iron impurities in the Belousov–Zhabotinskii reaction. *J. Chem. Phys.*, **86**, 119–29.

Crowley, M. F. and Epstein, I. R. (1989). Experimental and theoretical studies of a coupled chemical oscillator: phase death, multistability and in-phase and out-of-phase entrainment. *J. Phys. Chem.*, **93**, 2496–502.

Crowley, M. F. and Field, R. J. (1981). Electrically coupled Belousov–Zhabotinskii oscillators: a potential chaos generator. In *Nonlinear phenomena in chemical dynamics*, (ed. C. Vidal and A. Pacault), pp. 147–54. Springer, Berlin.

Crowley, M. F. and Field, R. J. (1986). Electrically coupled Belousov–Zhabotinskii oscillators. 1. Experiments and simulations. *J. Phys. Chem.*, **90**, 1907–15.

Dateo, C. E., Orban, M., De Kepper, P., and Epstein, I. R. (1982). Bistability and oscillations in the autocatalytic chlorite–iodide reaction in a sitrred-flow reactor. *J. Am. Chem. Soc.*, **104**, 505–9.

De Kepper, P. and Bar-Eli, L. (1983). Dynamical properties of the Belousov–Zhabotinskii reaction in a flow system. Theoretical and experimental analysis. *J. Phys. Chem.*, **87**, 480–8.

De Kepper, P., Rossi, A., and Pacault, A. (1976). Etude experimental d'une reaction chimique periodique. Diagramme d'état de la reaction de Belousov–Zhabotinskii. *CR Acad. Sci.*, **C283**, 371–5.

De Kepper, P., Ouyang, Q., and Dulos, E. (1984). Bistability in the reduction of permanganate by hydrogen peroxide in a stirred tank flow reactor. In *Nonequilibrium dynamics in chemical systems*, (ed. C. Vidal and A. Pacault), pp. 44–9. Springer, Berlin.

Dolnik, M. and Marek, M. (1988). Extinction of oscillations in forced and coupled reaction cells. *J. Phys. Chem.*, **92**, 2452–5.

Dolnik, M., Schreiber, I., and Marek, M. (1984). Experimental observations of periodic and chaotic regimes in a forced chemical oscillator. *Phys. Lett.*, **100A**, 316–19.

Dolnik, M., Schreiber, I., and Marek, M. (1986). Dynamic regimes in a periodically forced reaction cell with oscillatory chemical reaction. *Physica*, **21D**, 78–92.

Dolnik, M., Padusakova, E., and Marek, M. (1987). Periodic and aperiodic regimes in coupled reaction cells with pulse forcing. *J. Phys. Chem.*, **91**, 4407–10.

Dolnik, M., Finkeova, J., Schreiber, I., and Marek, M. (1989). Dynamics of forced excitable and oscillatory chemical reaction systems. *J. Phys. Chem.*, **93**, 2764–74.

Edblom, E. C., Orban, M., and Epstein, I. R. (1986). A new iodate oscillator: the Landolt reaction with ferrocyanide in a CSTR. *J. Am. Chem. Soc.*, **108**, 2826–30.

Edblom, E. C., Györgyi, L., Orban, M., and Epstein, I. R. (1987). A mechanism for dynamical behavior in the Landolt reaction with ferrocyanide. *J. Am. Chem. Soc.*, **109**, 4876–80.

Edelson, D., Field, R. J., and Noyes, R. M. (1975). Mechanistic details of the Belousov–Zhabotinskii oscillations. *Int. J. Chem. Kinet.*, **7**, 417–32.

Edelson, D., Noyes, R. M., and Field, R. J. (1979). Mechanistic details of the Belousov–Zhabotinskii oscillations. II. The organic reaction subset. *Int. J. Chem. Kinet.*, **11**, 155–64.

Epstein, I. R., Kustin, K., De Kepper, P., and Orban, M. (1983). Oscillatory chemical reactions. *Sci. Am.*, **248**, 96–108.

Field, R. J. (1975). Limit cycle oscillations in the reversible Oregonator. *J. Chem. Phys.*, **63**, 2289–96.

Field, R. J. and Burger, M. (1985). *Oscillations and traveling waves in chemical systems*. Wiley, New York.

Field, R. J. and Försterling, H.-D. (1986). On the oxybromine chemistry rate constants with cerium ions in the Field–Körös–Noyes mechanism of the Belousov–Zhabotinskii reaction: the equilibrium $HBrO_2 + BrO_3^- + H^+ = 2BrO_2$. *J. Phys. Chem.*, **90**, 5400–7.

Field, R. J. and Noyes, R. M. (1974). Oscillations in chemical systems. IV. Limit cycle behavior in a model of a real chemical reaction. *J. Chem. Phys.*, **60**, 1877–84.

Field, R. J., Körös, E., and Noyes, R. M. (1972). Oscillations in chemical systems. II. Thorough analysis of temporal oscillation in the bromate–cerium–malonic acid system. *J. Am. Chem. Soc.*, **94**, 8649–64.

Freund, A., Buchholtz, F., and Schneider, F. W. (1985). Slow fluctuations berween attractors in a forced chemical oscillator: the Belousov–Zhabotinskii reaction in a CSTR. *Ber. Bunsenges. Phys. Chem.*, **89**, 637–41.

Freund, A., Kruel, Th., and Schneider, F. W. (1986). Distinction between deterministic chaos and amplification of statistical noise in an experimental system. *Ber. Bunsenges. Phys. Chem.*, **90**, 1079–84.

Fujii, H. and Sawada, Y. (1978). Phase-difference locking of coupled oscillating chemical systems. *J. Chem. Phys.*, **69**, 3830–2.

Ganapathisubramanian, N. and Noyes, R. M. (1982). A discrepancy between experimental and computations: evidence for chaos. *J. Chem. Phys.*, **76**, 1770–4.

Ganapathisubramanian, N. and Showalter, K. (1984a). Washout effects in pumped tak reactors. *J. Am. Chem. Soc.*, **106**, 816–7.

Ganapathisubramanian, N. and Showalter, K. (1984b). Bistability, mushrooms and isolas. *J. Chem. Phys.*, **80**, 4177–84.

Ganapathisubramanian, N. and Showalter, K. (1986). Relaxation behavior in a bistable chemical system near the critical point and hysteresis limits. *J. Chem. Phys.*, **84**, 5427–36.

Gaspar, V. and Showalter, K. (1987). The oscillatory Landolt reaction. Empirical rate-law model and detailed mechanism. *J. Am. Chem. Soc.*, **109**, 4869–76.

Gaspar, V. and Showalter, K. (1990). A simple model for the oscillatory iodate ferrocyanide. *J. Phys. Chem.*, **94**, 4973–9.

Geiseler, W. (1982). Multiplicity, stabilty and oscillations in the stirred flow oxidation of manganese(II) by acid bromate. *J. Phys. Chem.*, **86**, 4394–9.

Geiseler, W. and Bar-Eli, K. (1981). Bistability of the oxidation of cerous ions by bromate in a stirred flow reactor. *J. Phys. Chem.*, **85**, 908–14.

Geiseler, W. and Föllner, H. H. (1977). Three steady-state situation in an open chemical reaction system. Part 1. *Biophys. Chem.*, **6**, 107–15.

Graziani, K. R., Hudson, J. L., and Schmitz, R. A. (1976). The Belousov–Zhabotinskii reaction in a continuous flow reactor. *Chem. Eng. J.*, **12**, 9–21.

Györgyi, L. and Field, R. J. (1988). Aperiodicity resulting from external and internal two-cycle coupling in the Belousov–Zhabotinskii reaction. *J. Phys. Chem.*, **92**, 7079–88.

Györgyi, L. and Field, R. J. (1989*a*). Aperiodicity resulting from external and internal two-cycle coupling in the Belousov–Zhabotinskii reaction. 2. Modelling of the effect of dead spaces at the input ports of a continuous-flow stirred tank reactor. *J. Phys. Chem.*, **93**, 2865–67.

Györgyi, L. and Field, R. J. (1989*b*). Aperiodicity resulting from external and internal two-cycle coupling in the Belousov–Zhabotinskii reaction. 3. Analysis of a model of the effects of spatial inhomogeneities at the input ports of a CSTR. *J. Chem. Phys.*, **91**, 6131–41.

Hourai, M., Kotake, Y., and Kuwata, K. (1985). Bifurcation structure of the Belousov–Zhabotinskii reaction in a stirred flow reactor. *J. Phys. Chem.*, **89**, 1760–4.

Hudson, J. L. and Mankin, J. C. (1981). Chaos in the Belousov–Zhabotinskii reaction. *J. Chem. Phys.*, **74**, 6171–7.

Hudson, J. L., Hart, M., and Marinko, D. (1979). An experimental study of multiple peak periodic and nonperiodic oscillations in the Belousov–Zhabotinskii reaction. *J. Chem. Phys.*, **71**, 1601–6.

Hudson, J. L., Laba, P., and Mankin, J. C. (1986). Experiments on low-amplitude forcing of a chemical oscillator. *J. Phys. Chem.*, **90**, 3430–4.

Hudson, J. L., Mankin, J., McCullough, J., and Lamba, P. (1981). Experiments on chaos in a continuous stirred tank reactor. In *Nonlinear phenomena in chemical dynamics*, (ed. C. Vidal and A. Pacault), pp. 44–8. Springer, Berlin.

Ibison, P. and Scott, S. K. (1990). Complex oscillations and chaos in the Belousov–Zhabotinskii reaction. *J. Chem. Phys. Faraday Trans.*, **86**, 3695–700.

Janz, R. D., Vanecek, D. J., and Field, R. J. (1980). Composite double oscillation in a modified version of the Oregonator model of the Belousov–Zhabotinskii reaction. *J. Chem. Phys.*, **73**, 3132–8.

Lamba, P. and Hudson, J. L. (1987). Experiments on bifurcations to chaos in a forced chemical reactor. *Chem. Eng. Sci.*, **42**, 1–8.

Luo, Y. and Epstein, I. R. (1989). Alternative feedback pathway in the mixed Landolt chemical oscillator. *J. Phys. Chem.*, **93**, 1398–401.

Luo, Y., Orban, M., Kustin, K., and Epstein, I. R. (1989). Mechanistic study of oscillations and bistability in the Cu(II)-catalyzed reaction between H_2O_2 and KSCN. *J. Am. Chem. Soc.*, **111**, 4541–52.

Marek, M. (1979). Dissipative structures in chemical systems—theory and experiment. In *Synergetics: far from equiibrium*, , (ed. A. Pacault and C. Vidal), p. 12–17. Springer, Berlin.

Marek, M. (1985). Periodic and aperiodic regimes in forced chemical oscillations. In *Temporal order* (ed. L. Rensing and N. I. Jaeger), pp. 105–15. Springer, Berlin.

Marek, M. and Stuchl, I. (1975). Synchronization in two interacting oscillatory systems. *Biophys. Chem.*, **3**, 241–8.

Marek, M. and Svobodova, E. (1975). Nonlinear phenomena in oscillatory systems of homogeneous reactions—experimental observations. *Biophys. Chem.*, **3**, 263–73.

Marek, M., Dolnik, M., and Schrekber, I. (1986). Dynamic patterns in interaction chemical cells and effects of external periodic forcing. In *Selforganization by nonlinear irreversible processes*, (ed. W. Ebeling and H. Ulbright), pp. 133–6. Springer, Berlin.

Maselko, J. (1980a). Experimental studies of complicated oscillations. The system Mn^{2+}–malonic acid–potassium bromate–sulphuric acid. *Chem. Phys.*, **51**, 473–80.

Maselko, J. (1980b). Experimental studies of chaos-type reactions. The system Mn^{2+}–oxalacetic acid–H_2SO_4–$KBrO_3$. *Chem. Phys. Lett.*, **73**, 194–9.

Maselko, J. and Epstein, I. R. (1984). Chemical chaos in the chlorite–thiosulfate reaction. *J. Chem. Phys.*, **80**, 3175–8.

Maselko, J. and Swinney, H. L. (1986). Complex periodic oscillations and Farey arithmetic in the Belousov–Zhabotinskii reaction. *J. Chem. Phys.*, **85**, 6430–41.

Nagashima, H. (1982). Experiments on chaotic responses of a forced Belousov–Zhabotinskii reaction. *J. Phys. Soc. Jpn.*, **51**, 21–2.

Nagy, A. and Treindl, L. (1986). **320**, 344–5.

Nakajima, K. and Sawada, Y. (1980). Experimental studies on the weak coupling of oscillatory chemical reaction systems. *J. Chem. Phys.*, **72**, 2231–4.

Noyes, R. M. (1986). Comparison of the Field–Körös–Noyes and Field–Försterling parametrizations of the bromate–cerous reaction. *J. Phys. Chem.*, **90**, 5407–9.

Noyes, R. M., Field, R. J. Försterling, H.-D., Körös, E., and Ruoff, P. (1989). Controversial interprerations of Ag^+ perturbations of the Belousov–Zhabotinskii reaction. *J. Phys. Chem.*, **93**, 270–4.

Orban, M. (1986). Oscillations and bistability in the Cu(II)-catalyzed reaction between H_2O_2 and KSCN. *J. Am. Chem. Soc.*, **106**, 6893–8.

Orban, M. and Epstein, I. R. (1982). Complex periodic and aperiodic oscillation in the chlorite–thiosulfate reaction. *J. Phys. Chem.*, **86**, 3907–10.

Orban, M. and Epstein, I. R. (1985). A new halogen-free chemical oscillator: the reaction between sulfide ion and hydrogen peroxide in a CSTR. *J. Am. Chem. Soc.*, **107**, 2302–5.

Orban, M. and Epstein, I. R. (1987). Chemical oscillators in group VIA: the Cu(II)-catalyzed reaction between hydrogen peroxide and thiosulfate ion. *J. Am. Chem. Soc.*, **109**, 101–6.

Orban, M. and Körös, E. (1978). Chemical oscillations during the uncatalyzed reaction of organic compounds with bromate 1. Search for chemical oscillators. *J. Phys. Chem.*, **82**, 1672–4.

Orban, M., De Keper, P., and Epstein, I. R. (1982a). Systematic design of chemical oscillators, part 10. Minimal bromate oscillators: bromate–bromide– catalyst. *J. Am. Chem. Soc.*, **104**, 2675–8.

Orban, M., Dateo, C., De Kepper, P., and Epsrein, I. R. (1982b). Chlorite oscillators: new experimental examples, tristability and preliminary classification. *J. Am. Chem. Soc.*, **104**, 5911–18.

Orban, M., De Kepper, P., and Epstein, I. R. (1982c). An iodine-free chlorite-based oscillator. The shlorite–thiosulfate reaction in a continuous flow stirrer tank reactor. *J. Phys. Chem.*, **86**, 431–3.

Ouyang, Q., Boissonade, J., Roux, J. C., and De Kepper, P. (1989). Sustained reaction–diffusion structures in an open reactor. *Phys. Lett.*, A **134**, 282–6.

Peng, B. and Showalter, K. (1990). In preparation.

Pikovsky, A. S. (1981). A dynamic model for periodic and chaotic oscillations in the Belousov–Zhabotinskii reaction. *Physs. Lett.*, A **85**, 13–16.

Rabai, G., Kustin, K., and Epsein, I. R. (1989). A systematically designed pH oscillator: the hydrogen periodixe–sulfite–ferrocyanide reaction in a continuous-flow stirred tank reactor. *J. Am. Chem. Soc.*, **111**, 3870–4.

Reckley, J. S. and Showalter, K. (1981). Kinetic stability in the permanganate oxidation of oxalate. *J. Am. Chem. Soc.*, **103**, 7012–13.

Richetti, P. and Arneodo, A. (1985). The periodic–chaotic sequences in chemical reactions: a scenario close to homoclinic conditions? *Phys. Lett.*, A **109**, 359–66.

Richetti, P., De Kepper. P. Roux, J. C., and Swinney, H. L. (1987a). A crisis in the Belousov–Zhabotinskii reaction: experiment and simulation. *J. Stat. Phys.*, **48**, 977–90.

Richetti, P., Roux, J. C., Argoul, F., and Arneodo, A. (1987b). From quasiperiodicity to chaos in the Belousov–Zhabotinskii reaction. II. Modelling and theory. *J. Chem. Phys.*, **86**, 3339–56.

Ringland, J. and Turner, J. S. (1984). One dimensional behaviour in a model of the Belousov–Zhabotinskii reaction. *Phys. Lett.*, A **105**, 96–6.

Rinzel, J. and Schwartz, I. B. (1984). One variable map predictions of Belousov–Zhabotinskii mixed mode oscillations. *J. Chem. Phys.*, **80**, 5610–15.

Rinzel, J. and Troy, W. C. (1982). Bursting phenomena in a simplified Oregonator flow system model. *J. Chem. Phys.*, **76**, 1775–89.

Rössler, O. E. (1981). Chaos and chemistry. In *Nonlinear phenomena in chemical dynamics*, (ed. C. Vidal and A. Pacault), pp. 79–87. Springer, Berlin.

Rössler, O. E. and Wegmann, K. (1978). Chaos in the Zhabotinskii reaction. *Nature*, **271**, 89–90.

Roux, J. C. (1983). Experimental studies of bifurcations leading to chaos in the Belousov–Zhabotinskii reaction. *Physica*, **7D**, 57–68.

Roux, J. C. and Rossi, A. (1984). Quasiperiodicity in chemical dynamics. In *Non-equilibrium dynamics in chemical systems*, (ed. C. Vidal and A. Pacault), pp. 141–5. Springer, Berlin.

Roux, J. C. and Swinney, H. L. (1981). Topology of chaos in a chemical reaction. In *Nonlinear phenomena in chemical dynamics*, (ed. C. Vidal and A. Pacault), pp. 38–43. Springer, Berlin.

Roux, J. C., Rossi, A. Bachelart, S., and Vidal, C. (1980). Representation of a strange attractor from an experimental study of chemical turbulence. *Phys. Lett.*, A **77**, 391–3.

Roux, J. C., Rossi, A. Bachelart, S., and Vidal, C. (1981). Experimental observations of complex dynamical behaviours during a chemical reaction. *Physica*, **2D**, 395–403.

Roux, J. C., Simoyi, R. H., and Swinney, H. L., (1983). Observations of a strange attractor. *Physica*, **8D**, 257–66.

Roux, J. C., Turner, J. S., McCormick, W. D., and Swinney, H. L. (1982). Experimental observations of complex dynamics in a chemical reaction. In *Nonlinear problems: present and future*, (ed. A. R. Bishop, D. K. Campbell, and B. Nicolaenko). North-Holland, Amsterdam.

Ruoff, P. and Noyes, R. M. (1986). An amplified Oregonator model simulating alternative excitabilities, transitions in types of oscillation, and temporary bistability in a closed system. *J. Chem. Phys.*, **84**, 1413–23.

Ruoff, P., Varga, M., and Körös, E. (1988). How bromate oscillators are controlled. *Acc. Chem. Res.*, **21**, 326–32.

Sasaki, Y. (1988). Bistability and oscillations in the bromate–bromide–cerium(III) system in continuous-flow stirred tank reactor. A simulation using a new set of rate constants of the FKN scheme. *Bull. Chem. Soc. Jpn.*, **61**, 4071–5.

Schmitz, R. A., Graziani, K. R., and Hudson, J. L. (1977). Experimental evidence of chaotic states in the Belousov–Zhabotinskii reaction. *J. Chem. Phys.*, **67**, 3040–4.

Schneider, F. W. (1985). Periodic perturbations of chemical oscillators: experiments. *Ann. Rev. Phys. Chem.*, **36**, 347–78.

Schwartz, I. B. (1984). Random mixed modes due to fluctuations in the Belousov–Zhabotinskii reaction. *Phys. Lett.*, A **102**, 25–31.

Sheppard, J. G. (1990). Some computational effects in the simulation of various model chemical reactions. In *Spatial inhomogeneities and transient behaviour in chemical kinetics*, (ed. P. Gray, G. Nicolis, F. Baras, P. Borckmans, and S. K. Scott), p. 672. Manchester University Press.

Showalter, K., Noyes, R. M., and Bar-Eli, K. (1978). A modified Oregonator model exhibiting complicated limit cycle behavior in a flow system. *J. Chem. Phys.*, **69**, 2514–24.

Simoyi, R. H., Wolf, A., and Swinney, H. L. (1982). One-dimensional dynamics in a multicomponent chemical reaction. *Phys. Rev. Lett.*, **49**, 245–8.

Sørensen, P. G. (1974). Comment. *Faraday Symp. Chem. Soc.*, (Physical chemistry of oscillatory phenomena), **9**, 88.

Sørensen, P. G. (1979). Experimental investigation of behaviour and stability properties of attractors corresponding to burst phenomena in the open Belousov–Zhabotinskii reaction. *Ann. NY Acad. Sci.*, **316**, 667–75.

Stuchl, I. and Marek, M. (1982). Dissipative structures in coupled cells: experiments. *J. Chem. Phys.*, **77**, 2956–63.

Swinney, H. L. and Maselko, J. (1985). Comment on 'Renormalization, unstable manifolds and the fractal structure of mode locking'. *Phys. Rev. Lett.*, **55**, 2366.

Swinney, H. L. and Roux, J. C. (1984). Chemical chaos. In *Non-equilibrium dynamics in chemical systems*, (ed. C. Vidal and A. Pacaylt), pp. 124–40. Springer, Berlin.

Tam, W. T., Vastano, J. A., Swinney, H. L., and Horsthemke, W. (1988). Regular and chaotic spatiotemporal patterns. *Phys. Rev. Lett.*, **61**, 2163–6.

Turner, J. S. (1984). Complex periodic and nonperiodic behavior in the Belousov–Zhabotinskii reaction. *Adv. Chem. Phys.*, **LV**, 205–17.

Turner, J. S., Roux, J. C., McCormick, W. D., and Swinney, H. L. (1981). Alternating periodic and chaotic régimes in a chemical reaction—experiment and theory. *Phys. Lett.*, A **85**, 9–12.

Tyson, J. J. (1979). Oscillations, bistability and echo waves in models of the Belousov–Zhabotinskii reaction. *Ann. NY Acad. Sci.*, **316**, 279–95.

Tyson, J. J. (1982). Scaling and reducing the Field–Körös–Noyes mechanism of the Belousov–Zhabotinskii reaction. *J. Phys. Chem.*, **86**, 3006–12.

Tyson, J. J. (1984). Relaxation oscillations in the revised Oregonator. *J. Chem. Phys.*, **80**, 6079–82.

Tyson, J. J. (1985). A quantitative account of oscillations, bistability and traveling waves in the Belousov–Zhabotinskii reaction. In *Oscillations and traveling waves in chemical systems*, (ed. R. J. Field and M. Bjrger), Ch. 3, pp. 93–144. Wiley, New York.

Varga, M., Györgyi, L., and Körös, E. (1985). Bromate oscillators: elucidation of the source of bromide ion and modification of the chemical mechanism. *J. Am. Chem. Soc.*, **107**, 4780–1.

Vastano, J. A. (1988). *Bifurcations in spatiotemporal systems. Ph.D. Dissertation.* University of Gexas, Austin.

Vavilin, V. Z., Zhabotinskii, A. M., and Zaikin, A. N. (1968). Effect of ultraviolet radiation on the self-oscillatory oxidation of malonic acid derivatives. *Russ. J. Phys. Chem.*, **42**, 1649–51.

Vidal, C., Roux, J. C., Rossi, A. and Bachelart, S. (1979). Etude de la transition vers la turbulence chemique dans la reaction de Belousov–Zhabotinskii. *CR Acad. Sci.*, **C289**, 73–6.

Vidal, C., Bachelart, S., and Rossi, A. (1982). Bifurcations en cascade conduisant à la turbulence dans la reaction de Belousov–Zhabotinskii. *J. Phys.*, **43**, 7–14.

Wegmann, K. and Rössler, O. E. (1978). Different kinds of chaotic oscillations in the Belousov–Zhabotinskii reaction. *Z. Naturforsch.*, **33a**, 1179–83.

Zaikin, A. N. and Zhabotinskii, A. M. (1973). A study of a self-oscillatory chemical reaction II. influence of periodic external force. In *Biological and biochemical oscillators*, (ed. B. Chance, E. K. Pye, A. K. Ghosh, and B. Hess), pp. 81–8. Academic Press, New York.

Zhabotinskii, A. M. (1964a). Periodic processes of the oxidation of malonic acid in solution (study of the kinetics of Belousov's reaction). *Biofizika*, **9**, 306.

Zhabotinskii, A. M. (1964b). Periodic liquid-phase oxidation reactions. *Dokl. Akad. Nauk. SSSR*, **157**, 392–5.

Zhabotinskii, A. M. (1985). The early period of systematic studies of oscillations and waves in chemical systems. In *Oscillations and traveling waves in chemical systems*, (ed. R. J. Field and M. Burger), pp. 1–5. Wiley, New York.

Zhabotinskii, A. M. and Rovinsky, A. B. (1987). Mechanism and nonlinear dynamics of an oscillating chemical reaction. *J. Stat. Phys.*, **48**, 959–76.

Zhabotinskii, A. M., Zaikin, A. N., and Rovinsky, A. B. (1982). Oscillating oxidation of cerium(III) ions by bromate in a flow system with a controlled inlet of bromide ion. *React. Kinet. Catal. Lett.*, **20**, 29–31.

9

Gas-phase reactions

The existence of oscillations and the potential for chaotic behaviour in gas-phase reactions have received a much less widespread coverage than their solution-phase counterparts. An introduction to this topic can be found in Gray and Scott (1985), Griffiths (1985a), and in the companion to this book (Gray and Scott 1990). Recent reviews have also been given by Griffiths (1985b, 1986) and Griffiths and Scott (1987).

Perhaps the simplest non-linear behaviour of interest here shown by gas-phase systems is that of thermal explosion. In many ways this is the equivalent of the solution-phase clock reaction. The reactants are mixed and a relatively slow evolution begins, but after some (usually short) time there is a rapid acceleration in rate leading to virtually complete conversion. This process relies on the exothermic nature of many chemical processes and their sensitivity to temperature through the reaction rate constant. Gas-phase systems are much more prone to significant departures from isothermal operation, and temperature plays the role of 'autocatalyst'. There can also be genuine chemical autocatalysis in gas-phase reactions, particularly oxidation processes involving oxygen. Such branched-chain reactions can also provide spectacularly non-linear responses. If self-heating and chemistry conspire together, almost anything seems possible.

9.1. Transient oscillations from self-heating: an experimental realization of the Salnikov model

Di-tertiary butyl peroxide (DTBP) is an organic species that undergoes a simple first-order decomposition reaction

(1)
$$(CH_3)_3COOC(CH_3)_3 \rightarrow 2CH_3OCH_3 + C_2H_6$$

$$-d[DTBP]/dt = k[DTBP] \tag{9.1}$$

with a significant rate at temperatures above about 400 K. The change in number of moles gives an increasing pressure under constant volume conditions and this allows the extent of reaction to be monitored relatively easily as a function of time. At low temperatures within this range and especially if the reactant is heavily diluted with an inert gas such as N_2, the reaction is very well behaved and is widely used to demonstrate first-order kinetics in undergraduate courses.

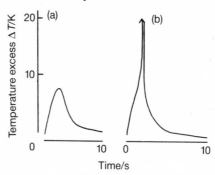

FIG. 9.1. Typical temperature–time histories for the exothermic decomposition of DTPB showing the development of self-heating ΔT of the reacting gas: (a) subcritical behaviour with $p = 3$ Torr, $T_a = 475$ K showing about 8 K self-heating; (b) supercritical thermal runaway for $p = 3$ Torr, $T_a = 480$ K.

The detailed kinetic mechanism is not of great importance here, but is believed to involve the unimolecular breaking of the O–O bond, followed by elimination of methyl radicals which may then combine:

$$(2) \qquad (CH_3)_3COOC(CH_3)_3 \rightarrow 2(CH_3)_3CO$$

$$(3) \qquad (CH_3)_3CO \rightarrow (CH_3)_2CO + CH_3$$

$$(4) \qquad 2CH_3 \rightarrow C_2H_6$$

leading to the stoichiometry (1). Step (2) is usually rate determining, so the overall kinetics are insensitive to any changes brought about in steps (3) and (4).

The overall process (1) is exothermic: $\Delta U^0 = -180$ kJ mol^{-1} and the overall reaction rate constant k has been found to have an activation energy of 152 kJ mol^{-1}. If O_2 is added to the reacting vapour, the effect is to increase the exothermicity (by changing the final products through steps that compete with (4) above) but this does not affect the kinetics which remain first order with the same Arrhenius parameters.

At sufficiently high reaction temperatures, the exothermic heat release accompanying reaction can lead to self-heating. The primary requirement for self-heating is that the reaction rate and, hence, the heat release rate should at some stage become sufficiently high so as to exceed the natural heat transfer rates (conduction, Newtonian cooling, etc.). The evolution of self-heating in an experiment can be followed using a coated fine-wire thermocouple positioned at some point within the reacting gas (usually close to the centre). With a reference junction on the outside wall of the vessel, the thermocouple records directly the temperature difference or extent of self-heating ΔT. Figure 9.1(a) shows a typical trace appropriate to a relatively low partial pressure of the reactant or a low ambient temperature.

The reacting gas temperature (i.e. that inside the vessel) here increases above that of the surroundings, but the temperature excess remains small—the maximum ΔT in this particular example is about 8 K. The corresponding pressure and mass spectrometer records show a steady consumption of the reactant. Because of the self-heating, the reaction rate initially increases with increasing extent of reaction—the indication of an acceleratory process, here through thermal feedback.

The maximum temperature excess observed increases if either the ambient temperature or the initial partial pressure of the DTBP are increased (both of these increase the rate of this first-order process with a positive activation energy). For small changes, the response changes smoothly, but if the increase in T_a or p_{DTBP} is to large, there is a qualitative change in the reaction evolution. Figure 9.1(b) shows a typical response for a 'thermal explosion'. There is an induction period in which the temperature excess increases to a value not much different from that observed for the previous subcritical behaviour, but this is now followed by a rapid acceleration during which the gas temperature increases swiftly to a high transient excess. There is virtually complete removal of DTBP during this ignition pulse and the temperature then falls back to ambient.

The conditions separating these two forms of response (subcritical and supercritical) give rise to a sharp p–T_a ignition boundary as shown in Fig. 9.2. The exact location of this limit depends on many factors, such as the vessel size and geometry and the presence or absence of O_2 and inert diluents (Egieban *et al.* 1982; Griffiths and Singh 1982; Griffiths and Mullins 1984), but can be relatively successfully predicted from classical thermal explosion theory (Boddington *et al.* 1977, 1982*a, b*). The length of the induction period

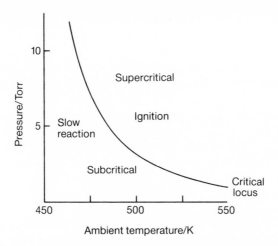

FIG. 9.2. A typical p–T_a explosion limit separating subcritical and supercritical experimental conditions for the exothermic decomposition of DTBP.

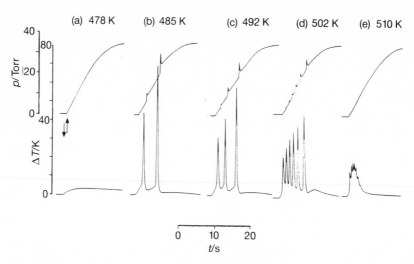

Fig. 9.3. A sequence of 'repetitive thermal explosions' accompanying the slow admission of DTBP to a reactor via a capillary showing self-heating and pressure records for five different ambient temperatures: (a) $T_a = 478$ K, for which the rapid admission experiment shows subcritical evolution; (b)–(e) are supercritical in the rapid admission experiment.

can also be estimated, and expressed in terms of the 'degree of supercriticality'. This latter can be represented as the difference between the actual operating pressure and the corresponding 'critical' pressure (i.e. the pressure on the limit) for the operating ambient temperature. Then

$$t_{ind} \propto 1/(p - p_{cr})^{1/2} \qquad (9.2)$$

i.e. the period gets shorter as the system moves further into the ignition region above the limit, but can become (arbitrarily) long close to the limit.

The application of such theories and interpretations implicitly requires that the reactants should be admitted to the vessel in as short a time as possible (preferably instantaneously) and the above experiments come close to that.

Of more interest in the present context, however, is an experimental situation in which the admission process is not fast, but becomes comparable or even slower than the reaction timescale. Griffiths *et al.* (1988; Gray and Griffiths 1989) employed a reactor into which the reactants were admitted via a capillary tube. In this way it was possible to obtain a series of 'repetitive thermal explosions'. Example time series are shown in Fig. 9.3.

A relatively simple explanation for the oscillatory waveforms observed under these conditions of gradual admission can be given. As the gas enters the vessel at some fixed ambient temperature, the partial pressure of the reactant increases and so we move up a vertical line in the p_{DTBP}–T_a

parameter plane (the ignition diagram, Fig. 9.2). Once the pressure increases beyond the critical pressure for that ambient temperature there will be a thermal ignition (with perhaps some induction period). This ignition decreases the partial pressure of the reactant below the critical pressure (and probably close to zero), although the total pressure will actually increase. Continued inflow of fresh reactant on the slower timescale through the capillary increases p_{DTBP} again, and a second ignition can occur if the critical pressure is reached again. This process may repeat a number of times until the pressures inside the vessel and in the reservoir (i.e. at each end of the capillary) become equal.

This simple interpretation is of some value, but does not deal too easily with all the experimental observations. For instance there are both upper and lower limits to the temperature for which multiple ignitions (oscillations) are found for any given final pressure and inflow characteristics. Equivalently, there are upper and lower final pressure limits for oscillations at any fixed ambient temperature. The lower limits correspond simply to conditions for which the system does not reach the ignition limit. The upper limits are less intuitive, but relate to the matching of inflow and the supercritical reaction rate, so the ignition becomes sustained and steady rather than quenching due to reaction consumption after each excursion.

Fortunately there is a simple quantitative model for this system—one we have already seen. The experiments described constitute a realization of the Salnikov scheme of Section 3.5, with some minor modification. In particular, the role of the first step $P \rightarrow A$, which was taken to be a chemical reaction with zero exothermicity and no activation energy, is now played by the continuous slow inflow. The DTBP decomposition reaction then provides the exothermic first-order step $A \rightarrow B$ + heat. The train of oscillations must now be finite, as the inflow only continues for a finite time and is decreasing in magnitude as the pressure difference decreases. Numerically computed traces using the known thermokinetic parameters are shown in Fig. 9.4 and successfully reproduce the experiments.

This design of transient oscillations can clearly be extended to many other exothermic processes. Coppersthwaite and Griffiths (Coppersthwaite 1990) have produced similar behaviour in the $H_2 + Cl_2$ system. As in the above case, the oscillatory excursions can lead to the attainment of dramatically higher temperatures than those that occur in neighbouring pseudo-steady-state responses and so there may be considerable hazards associated with the onset of this mode of reaction.

9.2. Clock reactions and the *pic d'arrêt*

The idea of a thermal explosion as a fast-timescale clock reaction may seem

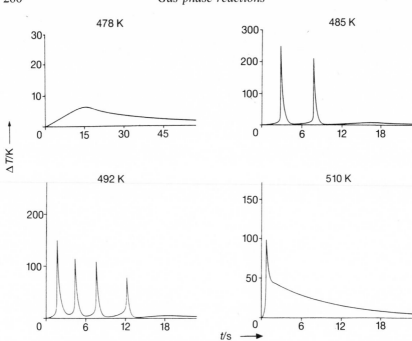

FIG. 9.4. Modelling of the slow admission experiments from Fig. 9.3 based on a modified Salnikov scheme (Section 3.5).

to push the unified view of solution-phase and gas-phase non-linear behaviour to an extreme. There is, however, a second situation in which this analogy is clearly justified to the full. The oxidation of hydrocarbons will feature strongly in later sections of this chapter, but here we can consider one aspect—the evolution of reaction for relatively fuel-rich mixtures. Snee and Griffiths (1989) studied the oxidation of cyclohexane under such conditions, at atmospheric pressure in a wide range of vessel sizes. Their investigation was primarily concerned with investigating 'minimum ignition temperatures'— the lowest temperature of a preheated flask for which ignition of a $1 \, cm^3$ volume of cyclohexane would occur after injection of the fuel. More standard kinetic studies showed that the reaction could exhibit 'quadratic auto-catalysis' in the absence of any self-heating, i.e. the dependence of the reaction rate on the extent of reaction (measured in terms of the limited oxygen concentration) has a simple parabolic form as shown in Fig. 9.5. In the presence of self-heating which inevitably accompanied reaction at higher temperatures, this parabolic dependence became skewed, as is also shown in the figure.

Griffiths and Phillips (1989) found similar autocatalysis in the oxidation

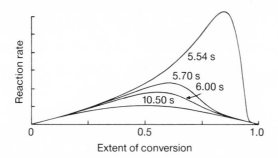

FIG. 9.5. Dependence of reaction rate on extent of conversion for the oxidation of fuel-rich n-butane + oxygen mixtures: under isothermal conditions ($t_{ch} = 10.50$ s) a simple parabolic dependence indicative of 'quadratic autocatalysis' is observed; in the presence of self-heating ($t_{ch} < 6$ s) the dependence becomes skewed to higher extents. The 'chemical timescale' t_{ch} is related to the ambient temperature, $t_{ch} = 1/k(T_a)$, where k is the autocatalytic reaction rate constant. (Reproduced with permission from Griffiths and Phillips (1989), © Royal Society of Chemistry.)

of n-butane, again under fuel-rich conditions (n-butane:O_2:N_2 = 5:1:4) at reduced pressure (500 Torr) over the range 589–600 K. The fuel–air mixture is admitted to a heated vessel and the reaction can be monitored by a mass spectrometer ($m/e = 32$ for O_2), photomultiplier (chemiluminescence from electronically excited formaldehyde CH_2O^*) and a fine-wire thermocouple as described above. Figure 9.6 shows four different O_2 concentration time series for temperatures within this range. With $T_a = 589$ K, no self-heating is observed at any stage during the reaction. The time trace shows a classic autocatalytic or clock reaction form, with a long induction period followed by an accelerating rate and then slowing due to reactant consumption. A plot of rate (i.e. slope of the time trace) against the extent of conversion is again of simple parabolic form appropriate to quadratic autocatalysis. At higher temperatures, the induction periods shorten, and the increase in magnitude of the rate at the end of this period becomes more marked. Significant transient temperature rises (self-heating) accompany the maximum reaction rate, up to 15 K for $T_a = 600$ K. An additional feature here is that the reaction produces a chemiluminescent intermediate CH_2O^* at appreciable concentrations during the period of high rate and this emission can be observed just before the reaction stops due to (virtually) complete O_2 consumption. This phenomenon has been termed the *pic d'arrêt* (Déchaux and Lucquin 1968, 1971; Déchaux *et al.* 1968).

Griffiths and Phillips were able to match the form of their experimental traces using a simple model based on quadratic autocatalysis and self-heating (Kay *et al.* 1989). They included a modification that allowed a surface-controlled termination step, reflecting the sensitivity of their results to

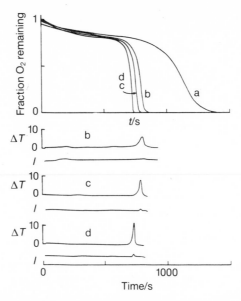

Fig. 9.6. Experimental measurements of O_2 concentration, self-heating ΔT, and light emission intensity I for the oxidation of fuel-rich n-butane mixtures: a, $T_a = 589$ K showing isothermal reaction with $\Delta T = I = 0$ for all time; b–d, reaction develops increasing transient non-isothermal behaviour with increasing extents of self-heating and eventually an accompanying light emission; b, $T_a = 594.5$ K; c, $T_a = 598$ K; d, $T_a = 600$ K. (Reproduced with permission from Griffiths and Phillips (1989), © Royal Society of Chemistry.)

pretreatment and ageing of their vessel. The model, then, has the form

$$A \rightarrow B$$

$$A + B \rightarrow 2B$$

$$B + S \rightarrow \text{inert}$$

along with the appropriate heat-balance equation.

9.3. The $H_2 + O_2$ reaction

The 'classic' gas-phase combustion reaction must be that between hydrogen and oxygen to form water, with stoichiometry

$$2H_2 + O_2 \rightarrow 2H_2O.$$

Although this is an exothermic process, much of the 'interesting' behaviour in this system occurs for relatively low reactant temperatures and pressures and self-heating does not make much of a contribution to determining the

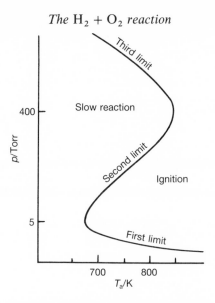

FIG. 9.7. A typical $p-T_a$ explotion limit diagram for the spontaneous oxidation of hydrogen showing the first, second, and third limits separating explosion from slow reaction.

qualitative evolution. The primary feedback mechanisms are chemical, involved 'branched-chain' kinetics as detailed below.

In a closed vessel, the behaviour of a mixture of H$_2$ and O$_2$, perhaps in the presence of other diluents, typically takes one of two distinct forms: slow reaction or ignition. Figure 9.7 shows a schematic $p-T_a$ ignition diagram. The locus of points separating slow reaction and ignition makes up the three 'ignition limits', numbered in terms of increasing pressure. Thus at temperatures to the left of any of the three limit branches, the oxidation of hydrogen proceeds relatively slowly and smoothly. If such an experiment is repeated at a higher temperature or with the pressure varied so as to have crossed one of the limit branches, then the process is one of a rapid 'explosion', the reactants are converted to H$_2$O in milliseconds, and there is a considerable transient self-heating.

The exact locations of the limits are determined by factors such as the vessel size, packing and any coating of the inner surface (these affect the first limit most), and the initial gas composition (mainly the second limit). Kinetic mechanisms of varying detail can be found in standard kinetics (Benson 1960; Mulcahy 1973) or combustion texts (Semenov 1958; Barnard and Bradley 1985; Lewis and von Elbe 1987) and more specialized reviews (Baldwin and Walker 1972; Dixon-Lewis and Williams 1977).

There have been isolated reports of oscillations ('repetitive ignitions') in closed-vessel studies (Warren 1957; Sahetchian *et al.* 1986), but the most well-established observations of such behaviour come from modern studies in open systems, CSTRs in particular, which we now discuss.

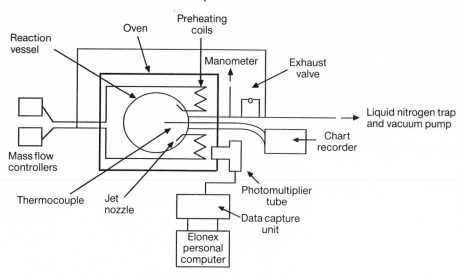

FIG. 9.8. Schematic representation of typical experimental set-up for the gas-phase reactions such as $H_2 + O_2$ in a CSTR.

9.4. The $H_2 + O_2$ reaction in a CSTR

The $H_2 + O_2$ reaction can be readily performed in a well-stirred continuous-flow reactor. A suitable experimental system is shown in Fig. 9.8. The reactants are preheated separately before entering the reactor (a 0.5 dm³ sphere). Mixing may either be achieved mechanically (Gray *et al.* 1984*a*, *b*) with a magnetically driven stirrer, or by 'jet mixing' (Gray *et al.* 1987*a*; Baulch *et al.* 1988*a*, *b*; Griffiths *et al.* 1990*a*, *b*). In the latter case, the reactants enter through nozzles at sonic velocities. Various tests have been performed to establish that these methods provide acceptable stirring for the operating pressures, temperatures, and volumetric gas flow rates of interest.

The pressure within the reactor is controlled by a needle valve at the exit: a particular choice of operating pressure, total volumetric flow rate, and ambient temperature determines the mean residence time spent by molecules within the CSTR.

9.4.1. *Equimolar mixtures*

The spontaneous reaction of 1:1 $H_2:O_2$ mixtures in a CSTR leads to a similar ignition limit structure to that shown above for a closed vessel. The location of the second limit for a typical residence time ($t_{res} = 5.2 \pm 0.7$ s) is shown in Fig. 9.9 along with some other boundaries that have no equivalent in closed-vessel studies.

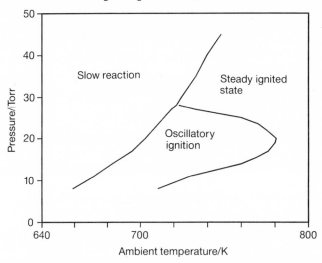

FIG. 9.9. The p–T_a second ignition limit and the boundary between oscillatory ignition and the steady ignited state for an equimolar mixture of H$_2$ + O$_2$ in a CSTR.

At the lowest ambient temperatures in the p–T_a plane, the reaction is slow, similar to that found to the left of the second explosion limit in closed vessels. As the limit is approached, and particularly at the higher ambient temperatures and pressures within this region, stationary-state extents of self-heating can be detected (up to about 10 K). Similar, but transient, temperature excesses are also observed in closed systems in the slow reaction zone (Griffiths *et al.* 1981; Kordylewski and Scott 1984) with corresponding consumption of the reactants, but there is no light emission.

Ignition occurs as the ambient temperature is raised so as to cross the ignition limit (ignition will also accompany a decrease in pressure across the limit but it is difficult to maintain the residence time constant throughout such a procedure, so ambient temperature is normally the parameter varied during a given experiment). At relatively low pressures ($p \lesssim 30$ Torr in Fig. 9.9) the ignited state is oscillatory, i.e. we observe repetitive ignitions. Typical time series (thermocouple, photomultiplier, and mass spectrometer responses) are shown in Fig. 9.10. The transient temperature rise occurs too quickly with this 'relaxation ignition' waveform for the thermocouple and, in particular, a traditional chart recorder to follow accurately: it certainly exceeds the few hundred kelvin indicated in the figure, but the record gives a quantitative measure that can be used in various ways discussed below. The light emission from the ignition pulse provides another useful marker, despite a rather unpromising origin. It is believed that this emission is caused by excited sodium ions taken from the surface of the vessel by the reaction. Thus the intensity of emission cannot be usefully assigned to any obvious radical concentration or a simple state of the system, and hence this signal

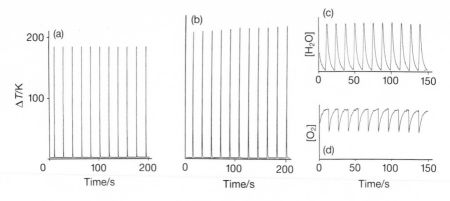

FIG. 9.10. Time series for (a) self-heating, (b) light emission, (c) H_2O concentration, and (d) O_2 concentration accompanying oscillatory ignition of an equimolar mixture of $H_2 + O_2$ in a CSTR (the experimental conditions for (c) and (d) differ slightly from those for (a) and (b)). (Figures 9.8–9.10 from Griffiths *et al.* (1990*b*).)

is used uncalibrated. The mass spectrometer records indicate rapid consumption of 50 per cent of the O_2 present, with correspondingly complete H_2 consumption, and the formation of the product H_2O.

The repetitive nature of the ignition pulses can be rationalized qualitatively if we note that the product is an inhibitor of the reaction. A quantitative account will be given later. Following the ignition the reactor contains H_2O and the excess O_2. Now fresh fuel (and oxidant) enters because of the constant inflow, and the product concentration decreases. The incoming H_2 does not react immediately because of the high concentration of the inhibitor H_2O. Only once the water concentration has fallen to some generally much lower value can it cease to inhibit, at which point the reactor has been significantly recharged with H_2 and O_2 so a second ignition can occur. This process then repeats as long as the flows are maintained and provided the operating conditions (such as the ambient temperature) are not changed so that the inhibiting effect becomes unable to compete with the branched-chain processes.

As the ambient temperature is varied across the region of oscillatory ignition, the relaxation character remains basically unchanged for most conditions, but its amplitude and period vary. Figure 9.11 shows a typical variation in period with the ambient temperature for a system with a mean residence time of 4 s (Baulch *et al.* 1988*a*). The period decreases smoothly from a high value at ambient temperature just above the ignition limit (where the rate of change of period is greatest). At higher T_a the oscillations cease, but the period does not fall to zero. We will consider the extinction of oscillations at both the low and high ambient temperature ends of the range in more detail below.

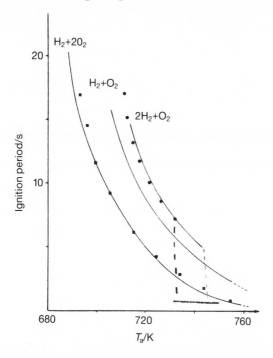

FIG. 9.11. The variation of the period between successive oscillatory ignitions with ambient temperature for three different compositions ($p = 16$ Torr, $t_{res} = 4$ s): with the fuel-lean H$_2$ + 2O$_2$ and H$_2$ + O$_2$ mixtures the period decreases monotonically as T_a is increased from a large (possibly infinite) value at the explosion limit to a small but non-zero value at the upper Hopf bifurcation; the stoichiometric mixture 2H$_2$ + O$_2$ shows a more complex multi-valued response. (Reproduced with permission from Baulch *et al.* (1988a), © The Combustion Institute.)

The final region in Fig. 9.9 is that of a 'steady flame': a stable stationary-state response with high extents of fuel consumption. In the present non-adiabatic system and with the relatively low inflow rates used, the stationary-state temperature excess associated with this state is, perhaps surprisingly, relatively low—typically less than 100 K. The stationary state is a stable focus as any perturbation decays back via a series of dampled overshoots and undershoots.

At high operating pressures in the range of interest ($p > 30$ Torr) the system jumps directly from the slow reaction state to the stable ignited state. This appears to have the classic characteristics of a saddle–node bifurcation in terms of Section 3.2. There is hysteresis here; the extinction of the ignited state occurs at a lower ambient temperature than that for the 'jump up'.

9.4.2. *Extinction of oscillatory ignition*

In this section we seek to categorize the two modes through which oscillatory ignition is extinguished, either by reducing the ambient temperature across the ignition limit or by increasing T_a across the second boundary in Fig. 9.9 into the region of steady flame. The second of these is slightly simpler and so we begin with this.

Figure 9.12 shows a sequence of time series (the thermocouple record) for increasing ambient temperatures with a fixed operating pressure of 18 Torr. Well below the upper boundary, the oscillation still has a relaxation form. The corresponding power spectrum shows many contributing overtones as well as the fundamental frequency peak. As the ambient temperature is raised, so the waveform becomes 'smoother', smaller in amplitude, and slightly shorter in period. The power spectra show increasingly fewer harmonics. With remarkable control over the ambient temperature, Griffiths *et al.* (1990*b*) have obtained an almost purely sinusoidal trace (a single peak in the power spectrum) with very small amplitude. The sustained oscillatory character vanishes if T_a is increased by 0.1 K giving the stable flame state (a stable focus). These same waveforms are reproduced exactly on reducing T_a again, with no hysteresis.

The above scenario has all the qualitative features of a classic supercritical Hopf bifurcation. This is born out further by following the death of the limit cycle reconstructed from the time series, which are shown for four examples in Fig. 9.13. Clearly the limit cycle decreases in size, finally shrinking to a stable point as the ambient temperature is increased. There is also reasonable agreement with this interpretation from the quantitative tests that can be made. The period grows linearly with the distance from the bifurcation point, i.e. with $(T_{a,H} - T_a)$ where $T_{a,H}$ is the ambient temperature of the Hopf bifurcation point (800.6 K in the present example). The amplitude is expected to scale as the square root of this distance $(T_{a,H} - T_a)^{1/2}$: the appropriate plot does show a linear relationship over some range, but the linear portion does not appear to extrapolate to zero as required. It is likely that this discrepancy has more to do with the difficulty of maintaining constant ambient temperatures and flow rates rather than with any high-order degeneracy about the behaviour of the system.

At low temperatures the extinction of oscillatory ignition occurs as we cross the ignition limit. No hysteresis has been observed in this transition. We have seen above that the ignition limit has the characteristics of a saddle–node bifurcation (a turning point in the stationary-state locus). We can thus imagine that for $p < 30$ Torr for the present system, the stationary state on the upper branch to which we might move is now unstable and surrounded by a stable limit cycle. The loss of stability and the birth of the stable limit cycle is simply the supercritical Hopf bifurcation just described.

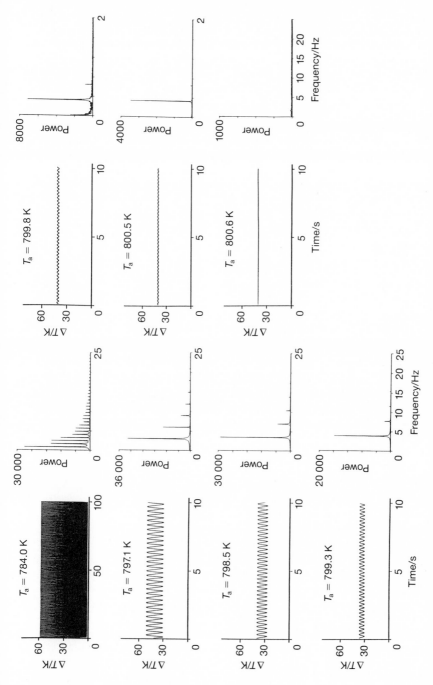

Fig. 9.12. The quenching of oscillatory ignition at high ambient temperatures showing time series and Fourier power spectra typical of a classical supercritical Hopf bifurcation.

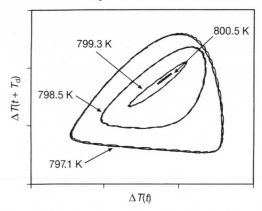

FIG. 9.13. Reconstructed limit cycles corresponding to the loss of oscillatory ignition via a supercritical Hopf bifurcation from Fig. 9.12. (Figures 9.12–9.13 from Griffiths *et al.* (1990*b*).)

We also, thus, expect that the extinction at this low-temperature end will involve the formation of a homoclinic orbit. The characteristics of such a scenario are the growth in oscillatory period and a 'hard' extinction from a large-amplitude oscillation as the critical ambient temperature is approached. There are two possible cases to be considered as sketched in Fig. 9.14(a) and (b) which show how the stationary-state and oscillatory solution loci might vary with T_a. In the first case, the limit cycle born at the Hopf point forms a homoclinic orbit on the middle (saddle point) branch. In such a case we might expect to see hysteresis between the onset of oscillations at the ignition

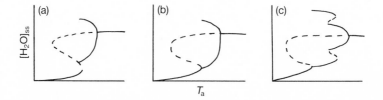

FIG. 9.14. Schematic bifurcation diagrams showing possible dependences of the $H_2 + O_2$ reaction steady state on ambient temperature: (a) oscillatory ignition is entered on increasing T_a from a saddle–node bifurcation (hard excitation) of the lowest and middle branches, terminates at high T_a at a supercritical Hopf bifurcation and at low T_a at a normal homoclinic orbit on the saddle branch thus allowing some hysteresis between oscillation and the low reaction steady state; (b) similar to (a) except that the homoclinic orbit is formed at the saddle–node bifurcation, thus giving no hysteresis and a different lengthening; (c) as (b) except that the limit cycle locus has developed two turning points and a region of two stable oscillatory forms (birhythmicity) now exists.

limit (the turning point) on increasing T_a and the homoclinic extinction as T_a is decreased. In fact, this could be of such small extent that it would be impossible to observe experimentally. There is, however, another, more quantitative test (Kaas-Petersen and Scott 1988). For a homoclinicity of the form in Fig. 9.14(a), the period is expected to lengthen logarithmically as the extinction point is approached

$$\text{period} \propto \ln(T_a - T_{a,cr}). \tag{9.3}$$

With the second form, shown in Fig. 9.14(b), the homoclinicity occurs exactly on the nose of the locus, i.e. on the saddle–node turning point. In this case, there is expected to be a different lengthening law, with the period growing as the inverse square root

$$\text{period} \propto 1/(T_a - T_{a,cr})^{1/2}. \tag{9.4}$$

Griffiths *et al.* (1990*a*) observed period lengthening, obtaining periods that were up to 80 residence times (160 s for the 2 s residence time case examined in detail) as shown in Fig. 9.15. Their plot of log(period) against log($T_a - T_{a,cr}$) agreed very well within experimental error with the second of the two forms above, having a slope close to $-\frac{1}{2}$. Thus we can conclude that the extinction of oscillatory ignition at the ignition limit is an example of saddle–node infinite period (or SNIPER) bifurcation.

9.4.3. *Complex oscillations for stoichiometric mixtures*

A similar p–T_a ignition diagram is found for stoichiometric mixtures (H$_2$:O$_2$ = 2:1), as shown in Fig. 9.16. There is, however, now an additional

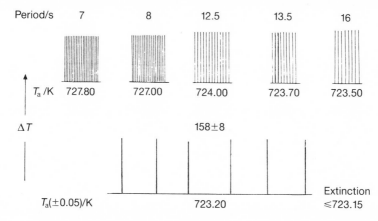

FIG. 9.15. The lengthening of the oscillatory ignition period for the hydrogen–oxygen reaction in a CSTR as the ambient temperature is decreased to the explosion limit ($T_{a,cr} = 723.15$ K) for a stoichiometric mixture with $p = 16$ Torr and $t_{res} = 2$ s. (Reproduced with permission from Griffiths *et al.* (1990*a*), © Manchester University Press.)

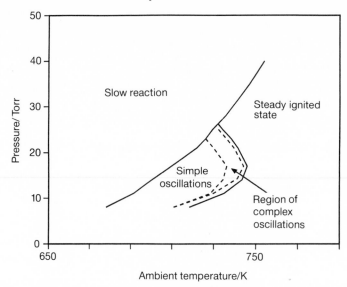

FIG. 9.16. .Oscillatory ignition and steady ignition limits for stoichiometric $2H_2 + O_2$ mixture in a CSTR showing region of complex oscillations.

region for the system with $t_{res} = 2\,(\pm 0.2)$ s which has been labelled 'complex ignitions'. For pressures in the approximate range $15 < p/\text{Torr} < 25$, this region occupies a non-zero width within the region of relaxation (period-1) ignitions. On entering this complex ignition region by increasing the ambient temperature, the ignition pulse develops a second maximum which interrupts the simple cooling (Fig. 9.17(a)). Clearly there is a second reaction phase entering into the period. The two-dimensional projection of the attracting limit cycle shows a small loop. The second maximum may disappear at slightly higher ambient temperatures, but it remains as a shoulder on the ignition pulse (Fig. 9.17(b)) and the loop remains in the attractor. Additional maxima and loops appear as the region is traversed.

Thus the $H_2 + O_2$ reaction can support complex mixed-mode oscillations of a generally similar form to those reported for the Belousov–Zhabotinskii system in the previous chapter. To date, all the observations in the present system appear to have been periodic.

At lower operating pressures, the region of complex ignitions narrows abruptly and is replaced by a region of birhythmicity. Figure 9.18 shows the variation of oscillatory period with the ambient temperature typical of such behaviour. The large-amplitude oscillations born on crossing the ignition limit decrease in period as T_a is increased, but at some point there is an abrupt change in waveform and period to small-amplitude high-frequency oscillations. If the ambient temperature is now decreased, the second form survives with increasing period until a second, discontinuous jump occurs

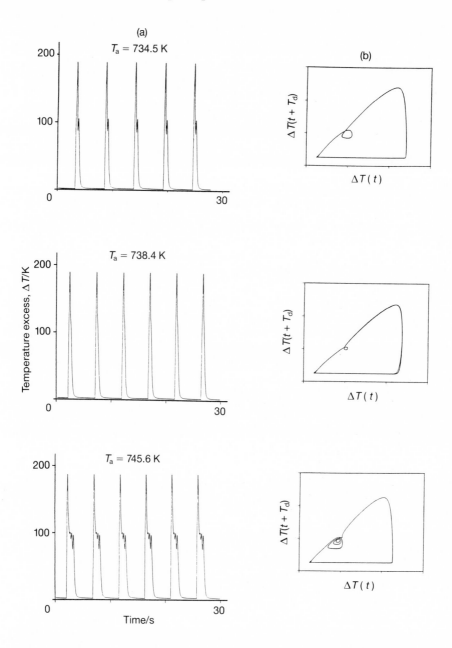

FIG. 9.17. Example complex oscillations for 2H$_2$ + O$_2$ mixtures corresponding to Fig. 9.16 showing time series and reconstructed attractors.

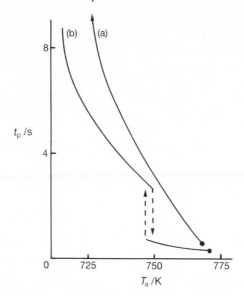

FIG. 9.18. The unfolding of a birhythmic dependence of oscillatory ignition period on ambient temperature by varying the residence for a stoichiometric hydrogen–oxygen reaction with $p = 14$ Torr: a, $t_{res} = 4$ s; b, $t_{res} = 2$ s. (Adapted and reproduced with permission from Baulch *et al.* (1988*a*), © The Combustion Institute.)

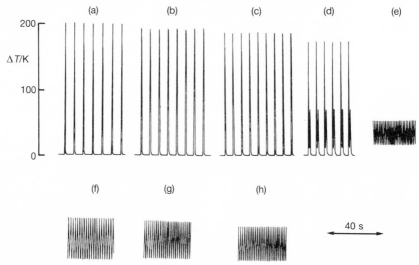

FIG. 9.19. Birhythmicity for a stoichiometric $2H_2 + O_2$ system showing two coexisting waveforms with $p = 20$ Torr and $t_{res} = 4$ s: (a) and (f) $T_a = 751$ K; (b) and (g) $T_a = 753$ K; (c) and (h) $T_a = 756$ K; (d) evidence of a complex ignition trace for $T_a = 758$ K; (e) $T_a = 761$ K. Traces (a)–(e) are found as the ambient temperature is increased, (e)–(h) as T_a is decreased. (Reproduced with permission from Baulch *et al.* (1988*a*), © The Combustion Institute.)

and the system moves back to the large-amplitude relaxation ignitions. There is hysteresis in these transitions, so for a narrow range of ambient temperature either oscillatory form may be realized depending on the previous history. Typical time series are shown in Fig. 9.19.

9.5. Modelling studies of the $H_2 + O_2$ reaction

Three distinct classes of modelling approaches can be identified: those using big models, small models, and medium-sized schemes respectively.

State-of-the-art kinetic schemes for modelling, say, the details of a hydrogen–oxygen flame may typically involve 44–62 reactions amongst eight or more species (Dougherty and Rabitz 1980; Dixon-Lewis 1984; Warnatz 1984). Full-scale numerical studies for the present range of experimental conditions might employ 32–33 steps based on the Baldwin–Walker model (Table 9.1) (Baldwin *et al.* 1965, 1974; Baulch *et al.* 1988*b*), whilst moderate schemes such as that of Foo and Yang (1971) may use 15–17 (Baldwin *et al.* 1967*a,b*; Kordylewski and Scott 1984; Chinnick *et al.* 1986*b*).

At the other extreme are the simplest models that set out simply to account for the location of, say, the second explosion limit and aim for analytical formulae. For the present system we then need only consider the following steps:

(H0)	initiation	$H_2 + O_2 \rightarrow$ radicals
(H1)	propagation	$OH + H_2 \rightarrow H_2O + H$
(H2)	branching	$H + O_2 \rightarrow OH + O$
(H3)	branching	$O + H_2 \rightarrow OH + H$
(H4)	termination	$\begin{cases} H + O_2 + M \rightarrow HO_2 + M \\ \quad\quad HO_2 \rightarrow \text{inert.} \end{cases}$
(H5)		

The initiation step produces radicals (perhaps $HO_2 + H$) which then either multiply through the branching cycle (H1)–(H3) or are terminated via formation of the relatively unreactive species HO_2 in steps (H4) and (H5). In a CSTR, branching must also compete with the net outflow of radicals. The termination step (H4) is a 'three-body' process involving the species M. The latter is any species present capable of removing energy from, and hence stabilizing, the HO_2 adduct. As different species are more or less efficient at this stabilization, the effective rate constant for this step will vary with the mixture composition, both in terms of the possible presence of diluents added to the inflow and to the production of species during the reaction. Thus H_2O which is a very efficient stabilizer of the HO_2 in step (H4) acts as an inhibitor of the system by enhancing the termination process relative to the branching cycle.

Table 9.1

The Baldwin–Walker mechanism for the hydrogen–oxygen reaction

---

$$H_2 + O_2 \rightarrow 2OH$$

$$H_2 + O_2 \rightarrow HO_2 + H$$

$$H_2 + OH \rightarrow H_2O + H$$

$$H + O_2 \rightarrow OH + O$$

$$O + H_2 \rightarrow OH + H$$

$$H + O_2 + M \rightarrow HO_2 + M$$

$$H \rightarrow \text{inert at wall}$$

$$O \rightarrow \text{inert at wall}$$

$$OH \rightarrow \text{inert at wall}$$

$$HO_2 + HO_2 \rightarrow H_2O_2 + O_2$$

$$HO_2 + H \rightarrow 2OH$$

$$HO_2 + H \rightarrow H_2O + O$$

$$HO_2 + H \rightarrow H_2 + O_2$$

$$HO_2 + H_2 \rightarrow H_2O_2 + H$$

$$H_2O_2 + M \rightarrow 2OH + M$$

$$H_2O_2 + H \rightarrow H_2 + HO_2$$

$$H_2O_2 + H \rightarrow H_2O + OH$$

$$H_2O_2 + OH \rightarrow H_2O + HO_2$$

$$H_2O_2 + O \rightarrow OH + HO_2$$

$$H_2O + O \rightarrow 2OH$$

$$H_2O + H \rightarrow H_2 + OH$$

$$OH + O \rightarrow O_2 + H$$

$$OH + H \rightarrow H_2 + O$$

$$OH + OH \rightarrow H_2O + O$$

$$H + OH + M \rightarrow H_2O + M$$

$$H + H + M \rightarrow H_2 + M$$

$$OH + OH + M \rightarrow H_2O_2 + M$$

$$O + O + M \rightarrow O_2 + M$$

$$HO_2 + OH \rightarrow H_2O + O_2$$

$$H_2O + M \rightarrow H + OH + M$$

$$H_2 + M \rightarrow 2H + M$$

$$O_2 + M \rightarrow 2O + M$$

Quantitative accounts of the conditions for the ignition limit based on such a scheme can be found in Gray *et al.* (1984*a, b*), Chinnick *et al.* (1986*a, b*), Baulch *et al.* (1988*b*), and in Gray and Scott (1990). By assuming that the overall third-body efficiency of the reacting gas mixture changes with the instantaneous composition, and that periods between successive ignition pulses are determined primarily by the rate of outflow of the inhibitor H$_2$O to be replaced by less efficient participants in step (H4), analytical expressions for the oscillatory ignition period can be obtained (Gray *et al.* 1985*b*, 1987*a*; Chinnick *et al.* 1986*a*). These work well for conditions that are not very close to the second limit, but fail as the extinction is approached when the inverse square-root form becomes important as described above.

Chinnick *et al.* (1986*b*) also considered extensions of the above model that would show isothermal oscillations. They found that by adding only the extra step

(H6) $$H + HO_2 \rightarrow H_2O + O$$

sufficient extra non-linearity was introduced for oscillatory responses to be observed computationally. The justification for this single addition was purely *ad hoc*, in the spirit of identifying 'the simplest oscillator' for the H$_2$ + O$_2$ system—much like the initial derivation of the Oregonator scheme for the BZ reaction. More recent investigations of this model and its bifurcation structure (Tomlin 1990) has shown that the development of oscillatory amplitude and other features are really not in keeping with the observed experimental behaviour, and the current interest is, as always, to 'improve' the qualitative match again at the minimum possible expense in complexity.

Better qualitative, and to some extent quantitative, matching of experiment and model can be achieved through the medium- and larger-scale models mentioned earlier, in particular that in Table 9.1. Complex oscillations appear to require that the reaction exothermicity and departures from isothermal operation be included, with the extra feedback mechanism through self-heating.

9.6. The CO + O$_2$ reaction

The other stoichiometrically simple oxidation reaction is that of carbon monoxide. Because the oxidation of carbon monoxide is important as the final stage of the oxidation of almost all hydrocarbon species, the earliest kinetic studies began as long ago as the beginning of the present century (Kühl 1903; Bodenstein and Ohlmer 1905). Sagulin (1928) and Semenov (1929) discovered an ignition limit peninsula similar to that of the H$_2$ + O$_2$ reaction discussed above, whilst Prettre and Laffitte (1929) reported the existence of a long-lived, steady chemiluminescent glow. One of the most frustrating features of the $CO + O_2$ system, and that which has been the primary cause of the slow development not just of our mechanistic interpretation but even of our understanding of the basic experimental facts, is the extreme sensitivity to small traces of hydrogen or hydrogenous species such as water and methane (Kopp *et al.* 1930; Topley 1930).

Ashmore and Norrish (1951) were apparently the first to observe and report the oscillatory behaviour of the $CO + O_2$ reaction. They employed a closed vessel, but could sustain over 100 pulses of chemiluminescent glow separated by periods of no emission, over a period of several minutes. Each pulse of emission was accompanied by a small stepwise decrease of the pressure within the reaction cell. For other experimental conditions, a smaller number of larger-amplitude excursions were found. Figure 9.20 shows these

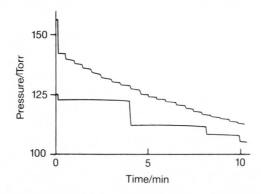

FIG. 9.20. The 'lighthouse effect' for $CO + O_2$ in a chloropicrin-treated closed vessel showing two sequences of stepwise pressure decreases that are accompanied by oscillatory chemiluminescence or 'glow'. Upper trace, $T_a = 1050$ K, 20 mm diameter vessel; lower trace, $T_a = 1025$ K, 33 mm vessel diameter. (Reproduced with permission from Ashmore and Norrish (1951), © Macmillan Journals.)

first observations of the phenomena christened 'the lighthouse effect' by its discoverers because of the regular luminescence.

The vessel employed by Ashmore and Norrish had previously been used to study the inhibiting effects of the compound chloropicrin (CCl_3NO_2). Cleaning the vessel surface removed the oscillatory response, which could be recovered by exposing the surface to chloropicrin again (the inhibitor would then be pumped out of the reactor before the reactants were admitted). We must, therefore, add a sensitivity to the nature of the surface to the sensitivity of the reaction to hydrogen. The latter effect was further examined, in terms of the influence of water vapour on the position of the various p–T_a limits, by Hoare and Walsh, by Gordon (1952; Gordon and Knipe 1955), and particularly by Linnett and various coworkers in an extended systematic survey (Dickens *et al.* 1964; Linnett *et al.* 1968). These latter authors also observed oscillatory glow (which they termed 'multiple explosions') in essentially 'clean' reaction vessels (coated with Al_2O_3 but no inhibitor treatment). The oscillations are somewhat ephemeral in such studies, coming and going with no obvious pattern in apparently identical experiments.

McCaffery and Berlad (1976) obtained very long oscillatory trains which they monitored spectroscopically (CO_2^* and OH) whilst other observations are due to Aleksandrov and Azatyan (1977). Many of the apparent enigmas of the closed-vessel experiments have been resolved to a great extent through the work of Bond *et al.* (1981, 1982). These authors showed that the 'ignition limits' identified by earlier workers in extensively dried mixtures were more likely the limits for the onset of the chemiluminescent glow. It seems that branched-chain ignition is not a feature of pure $CO + O_2$ and hence many of the kinetic mechanisms proposed are inappropriate. (A recent mechanistic summary has been given by von Elbe and Lewis (1986).) Steady glow and oscillatory glow appear to be alternatives, and exist at the same pressure and ambient temperature conditions. Surface conditioning has the strongest influence in determining whether the glow will be steady or oscillatory in a particular experiment in such systems: a sequence of identical experiments in an initially 'new' vessel will typically show steady glow for the first few runs, oscillatory glow for subsequent repetitions, but then steady glow again in its old age. Because we are dealing with closed systems, the glow is only a transient response, whether steady or oscillatory. In fact, oscillatory glow often dies into a steady glow and both forms of emission typically cease well before complete consumption of the reactants is approached (and can often be revived by increasing the vessel temperature).

Table 9.2 lists some of the important experimental studies of $CO + O_2$ in closed systems whilst Figs 9.21 and 9.22 show examples of oscillatory trains and the conditions for their existence. In more recent years, the tendency to move to open systems mentioned earlier in connection with the solution-phase systems has also provided perhaps the greatest steps forward in this reaction, as we discuss in the next section.

9.7. CO + O₂ reaction in a CSTR

Of the studies mentioned above, only Kopp *et al.* (1930) employed a flow reactor at any stage. Gray *et al.* (1984*b*, 1985*a*, *b*) began a series of studies of this reaction in a CSTR, for which reproducible behaviour seems easier to achieve. The reaction exhibits the four primary responses identified above: slow dark reaction, steady glow, oscillatory glow, and ignition. In a flow system the first two responses become true stationary states, the oscillatory glow is indefinitely sustained, and ignition can be an oscillatory process, as with H_2 oxidation. The sensitivity to hydrogenous species remains, and this was countered in the above studies by the deliberate, continuous addition

TABLE 9.2
Experimental studies of ignition and oscillations in carbon monoxide oxidation

Studies of spontaneous reaction and self-ignition limits

1903	Kühl	Preliminary experiments
1905	Bodenstein and Ohlmer	
1928	Sagulin	Discoveries of explosion peninsula
1929	Semenov	
1929	Prettre and Laffitte	Discovery of glow peninsula
1930	Kopp, Kowalskii, Sagulin, and Semenov	Effect of mixture composition, flow reactor
1930	Topley	Effect of H_2O and drying
1930	Cosslett and Garner	Effect of drying on ignition limits
1932	Hadman, Thompson, and Hinshelwood	Dried gases over P_2O_5
1938	Buckler and Norrish	Sensitivity to H_2 demonstrated
1953	Dixon-Lewis and Linnett	Studied complete range of fuel composition from H_2 to CO
1954	Hoare and Walsh	Extensive drying, mapping separation of glow, and ignition peninsulas
1954	Heslop	Theses
1956	Dickens, Dove	
1955	Gordon and Knipe	Freeze drying of reactants, displaced limits to 960 K
1955	von Elbe, Lewis, and Roth	Addition of inert gases
1965	Baldwin, Jackson, Walker, and Webster	Elementary rate constants for $CO + OH \rightarrow CO_2 + H$ and $CO + HO_2 \rightarrow CO_2 + OH$
1966	Brokaw	Influence of H_2 on detonation velocity of $CO + O_2$ mixtures
1980	Nalbandyan, Arustaniyan, Shakhnazaryan, and Philipossyan	Slow oxidation studies: numerical kinetic modelling

Studies of oscillatory reaction

1939–51	Ashmore and Norrish	Oscillatory glow first reported (discovered 1939)
1964	Dickens, Dove, and Linnett	Extensive drying and oscillations
1968	Linnett, Reuben, and Wheatley	Al_2O_3 coating, oscillations
1972–5	Bond, Gray, and Griffiths	Systematic study of steady and oscillatory glow regions over wide H_2 range: closed vessel
1976	McCaffrey and Berlad	Long oscillatory trains
1977	Aleksandrov and Azatyan	Oscillations near first limit
1979–82	Gray, Griffiths, and Scott	Systematic study of steady and oscillatory glow regions over wide H_2 range: well-stirred flow reactor

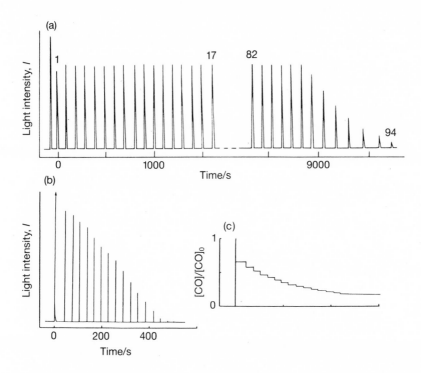

FIG. 9.21. Examples of long and short oscillatory trains of oscillatory glow for CO + O₂ mixtures in closed vessels: (a) chloropicrin-treated reactor with $p = 33$ Torr, $T_a = 841$ K giving a total of 94 emissions over 3 h; (b) and (c), light emission and CO consumption for 'clean reactor' with $p = 22$ Torr, $T_a = 900$ K.

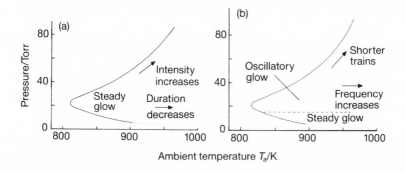

FIG. 9.22. The coexistence of (a) steady and (b) oscillatory glow in successive series of experiments for 'dry' CO + O₂ reaction in a clean, closed vessel.

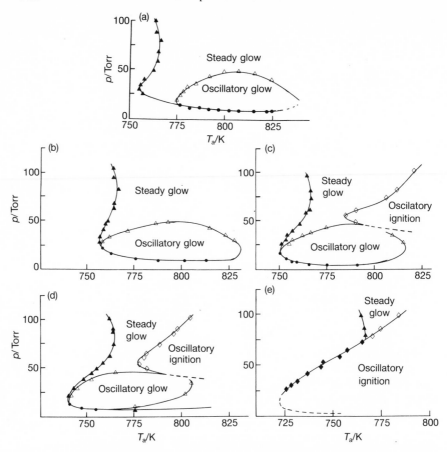

FIG. 9.23. The p–T_a limits for slow dark reaction, steady glow, oscillatory glow and oscillatory ignition, and the influence of added hydrogen on the CO + O$_2$ reaction in a CSTR with t_{res} = 8 s: (a) no added H$_2$; (b) with 150 p.p.m. H$_2$; (c) with 1500 p.p.m. H$_2$; (d) with 7500 p.p.m. H$_2$; (e) with 10 per cent H$_2$.

of known quantities of H$_2$ to the carbon monoxide inflow generally to swamp any unknown impurity level.

Oscillatory and steady glow now occupy separate p–T_a regions, and the changes in the relative positions of the regions for these various responses and the limits between them as the H$_2$ content is varied are shown in Fig. 9.23. With no added hydrogen, ignition is not found within the temperature range available with pyrex glass vessels. The oscillatory glow occurs with no measurable self-heating, so is an isothermal phenomenon. As the concentration of H$_2$ increases, so small transient temperature excursions accompany the chemiluminescence, and ignition becomes possible. In the latter form, there are considerable temperature excursions and much more intense

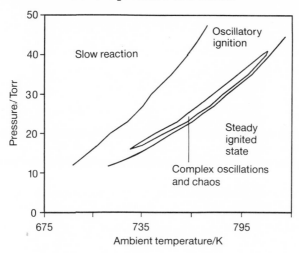

FIG. 9.24. Details of the p–T_a regions for simple and complex oscillatory ignition for $CO + O_2$ with 1 per cent H_2 in a CSTR with $t_{res} = 16$ s.

emissions (with large contributions at wavelengths different from that corresponding to CO_2^*, the emitter in the glow region).

The behaviour of the $CO + O_2$ system in a CSTR can also become more complicated. The experiments of Gray *et al.* could not resolve clearly the changing behaviour observed close to the boundary between oscillatory glow and ignition, nor did they probe into the ignition region in the way described above for the $H_2 + O_2$ system. Recent experiments (Johnson and Scott 1990) have paid more attention to these latter aspects, for systems with about 0.5 per cent added H_2 in particular. Figure 9.24 shows a section of the p–T_a diagram at temperatures above the limit for oscillatory ignition (similar to the second limit for the $H_2 + O_2$ system). There is again a region of 'complex ignitions', but the scenario accompanying a traverse through this region differs from that reported above.

The typical oscillatory ignition trace observed once the ignition limit has been crossed is that of a relaxation oscillation. As T_a is increased, we enter the 'complex' region and the time trace from the thermocouple or photo-multiplier reveals a classic period doubling, as shown in Fig. 9.25. For many pressures and residence times, the period-2 solution survives across the whole range of complex behaviour, and we return to period-1 relaxation ignitions via a period halving.

For some conditions, however, a second period doubling arises giving a period-4 solution. As we might then expect from our experiences with simple models, this is frequently accompanied by a whole (subharmonic) cascade leading through hints of period-8 etc. into an aperiodic or chaotic response. Various periodic windows have been observed and sustained in this remarkable

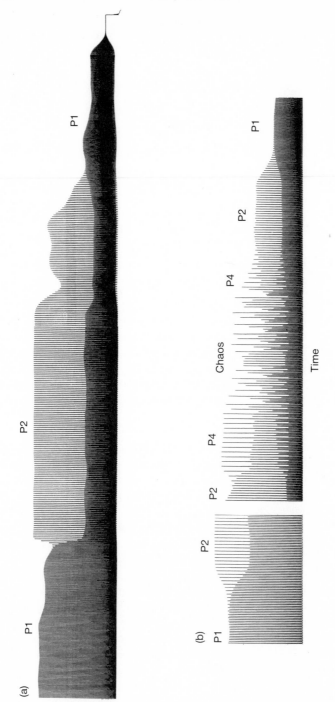

FIG. 9.25. Bifurcations to complex oscillatory ignition on increasing T_a for $CO + O_2$ with 1 per cent H_2 in a CSTR with $t_{res} = 16$ s: (a) period doubling and period halving; (b) a full period-doubling cascade to chaos followed by the reverse sequence.

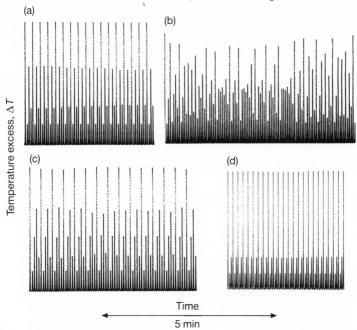

FIG. 9.26. Example time series for sustained waveform patterns observed for the system from Fig. 9.25(b): (a) period-4; (b) aperiodic sequence; (c) period-5; (d) period-3. (Figures 9.24–9.26 from Johnson and Scott (1990).)

work, as shown in Fig. 9.26. There is then a return to simple periodicities, period-2, and finally period-1 before a supercritical Hopf bifurcation at the highest T_a.

The basic structure above the oscillatory ignition limit for these H$_2$-doped CO + O$_2$ mixtures is thus very similar to that shown by the hydrogen + oxygen system itself. However, rather than showing mixed-mode waveforms, we have a wonderful example of the period-doubling route to chaos (and back again).

9.8. Mechanistic interpretation of the CO + O$_2$ reaction

Various early interpretations of ignition, and to some extent oscillation, in the CO + O$_2$ reaction tried to identify possible branching processes based on, say, electronically excited CO$_2$ (energy branching) or various exotic higher oxides such as C$_2$O$_3$. The most successful approaches, and those that seem most firmly based on the experimental facts as they are now viewed, stress the important interrelationship between the CO + O$_2$ and the H$_2$ + O$_2$ systems.

Full-scale numerical attacks frequently employ the Baldwin and Walker scheme of Table 9.1, augmented with only a few extra steps for the carbon monoxide intervention: typical of such studies are Yang (1974a, b), Yang and Berlad (1974), and Babushok *et al.* (1982). The most significant extra steps for the present range of pressures and temperatures are

(C1) $CO + OH \rightarrow CO_2 + H$

(C2) $CO + HO_2 \rightarrow CO_2 + OH$

(C3) $CO + O + M \rightarrow CO_2^* + M$.

Reaction (C1) is the primary route to formation of CO_2 but otherwise is simply a propagation process, as is the relatively slow step (C2) although this may be significant as it converts HO_2 to the more active radical OH. Step (C3) leads to the formation of electronically excited CO_2 (a triplet state) that is the source of chemiluminescence in this system. It is also a termination process and becomes significant at very low hydrogen concentrations. Other possible contributors to the mechanism involve the species HCO formed in the step

(C4) $H + CO + M \rightarrow HCO + M$.

The possible fates for this are redissociation, reaction with O_2, or a series of radical–radical reactions unlikely to be important except during the peak of the ignition phase:

(C5) $HCO + M \rightarrow H + CO + M$

(C6) $HCO + O_2 \rightarrow HO_2 + CO$

(C7) $H + HCO \rightarrow H_2 + CO$

(C8) $O + HCO \rightarrow OH + CO$

(C9) $OH + HCO \rightarrow H_2O + CO$.

Reaction (C6) is similar to the third-body formation of HO_2 in the $H_2 + O_2$ mechanism (in which carbon monoxide can also participate as the species M). The Arrhenius parameters for most of these steps are known with reasonable precision across the relevant temperature and pressure range, and critical evaluations and recommended values are listed in Baulch *et al.* (1972a, b, 1976). Additionally, there may be some reactions that take place or involve the vessel surface. These are generally not always well characterized and some will be discussed below.

A qualitative picture of the underlying kinetics appears to be emerging. The reaction is clearly dominated by hydrogen–oxygen dynamics, even when the concentration of hydrogenous species is relatively low. A more logical sequence can be presented, however, if we begin 'at the other end' and consider the effect of replacing H_2 by CO in a hydrogen–oxygen system.

Initially, the carbon monoxide plays little role, other than as a third body in the termination step (H4). Carbon monoxide is less efficient than H_2 as a third body and so this early replacement weakens the termination process, favouring ignition and thus shifting the oscillatory ignition limit to lower ambient temperatures. The greater the extent of this replacement, the greater the shift in the limit. Similar enhancement of the ignition limit is achieved by replacing the H_2 by the chemically inert diluent N_2 which has a similar third body efficiency to carbon monoxide (Baulch *et al.* 1988*b*).

Only when the replacement procedure has reduced the H_2 mole fraction to less than a few per cent, does carbon monoxide begin to exert a significant chemical role. The extra termination step (C3) competes successfully with the branching reaction (H3) for O atoms. This then begins to shift the ignition limit back to higher ambient temperatures. With very low mole fractions of H_2 we have a heavily inhibited hydrogen–oxygen system.

The distinction between oscillatory ignition and oscillatory glow can also be cast in terms of this interpretation. In each case we can imagine a 'normal' $H_2 + O_2$ branched-chain explosion. For an ignition, experiments and modelling show that there is not only H_2 consumption but also full oxidation of the carbon monoxide—the initial $H_2 + O_2$ criticality somehow is sufficient to 'trigger' a complete combustion process. In the oscillatory glow, however, the carbon monoxide appears to remain basically impervious to the advances being made by the $H_2 + O_2$ ignition. The latter causes relatively little carbon monoxide consumption (but sufficient to register a measurable emission on a sensitive photomultiplier detection system) and no measurable temperature excursion. Again, it is believed that the deciding influence between full fuel ignition or just H_2 consumption is the inhibiting effect of termination step (C3), which becomes relatively more effective as proportionately more H_2 is consumed in the early stages of the reaction, combined with the formation of H_2O.

The whole process can easily be oscillatory in a CSTR as there is a replacement inflow of fresh H_2 to fuel the next 'ignition' pulse in either case. For closed systems, however, the resetting of the clock requires a source of H_2 (or at least of available H atoms). Some suggestions have been made to this end (Babushok *et al.* 1982; Gray *et al.* 1987*b*), such as a possible role for the water–gas shift process

$$CO + H_2O \rightleftharpoons CO_2 + H_2.$$

This might perhaps occur on a suitably prepared surface—hence the sensitivity of closed-vessel studies to conditioning of the reactor and the apparent relative insensitivity of the oscillatory glow in a CSTR—and account for the rather different experimental conditions required for oscillation in the two types of system. Chinnick and Griffiths (1986) have based a numerical simulation of the restarting of oscillations in a closed vessel that can occur on increasing the ambient temperature, as described above, on this interpretation.

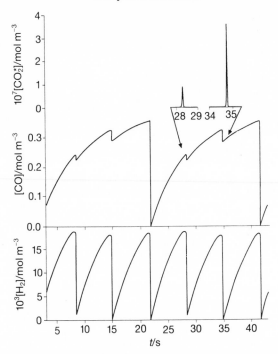

FIG. 9.27. Complex oscillatory ignition waveforms (of the mixed-mode form) pre-dicted numerically for $CO + O_2 + 0.4$ per cent H_2 with $T_a = 765$ K and $p = 45$ Torr showing concentrations of excited (emitting) species CO_2^* and reactants CO and H_2: there is full H_2 consumption at each excursion but full CO consumption occurs only every third pulse. (Reproduced with permission from Griffiths and Sykes (1989b), © Royal Society of Chemistry.)

The complex ignition waveforms in a CSTR have been modelled by Griffiths and Sykes (1989b) using a 38 reaction scheme. Example time series are shown in Fig. 9.27 and a computed p–T_a diagram is compared with the experimental observation in Fig. 9.28.

At the other extreme from the large mechanisms come the simple model schemes. Early amongst these is an 'inhibited branched-chain' model proposed by Gray (1970) and extended by Yang (1974a, b). This two-variable scheme can be written in the form

branching	$X + \cdots \rightarrow 2X$	rate $= k_b x$
termination	$X + \cdots \rightarrow$ inert	rate $= k_t x$
propagation	$X + \cdots \rightarrow X + Y$	rate $= k_p x$
inhibition	$X + Y \rightarrow$ inert	rate $= k_i xy$
removal	$Y + \cdots \rightarrow$ inert	rate $= k_r y$.

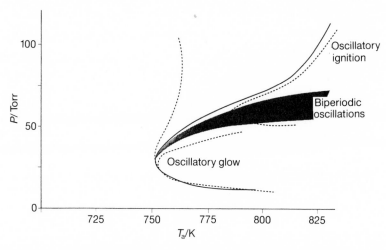

FIG. 9.28. Comparison of experimental (–––) and numerically predicted (——) p–T_a ignition limits for CO + O₂ containing 0.15 per cent H₂ also showing region of 'biperiodic' (complex) oscillations. (Reproduced with permission from Griffiths and Sykes (1989*b*), © Royal Society of Chemistry.)

The species Y produced in the third step is an inhibitor: the fourth step provides a quadratic non-linearity. This is not sufficient to produce sustained oscillations in this scheme—neither in a closed vessel with reactant consumption ignored nor in a CSTR. Yang's modification was to allow some 'saturating' effects in the termination step that introduce a second non-linearity. Thus he wrote the rate law for the second step in a modified form: rate = $k_t x/(1 + rx)$ where r is some extra parameter.

The particular 'chemical' interpretations placed on this scheme by its two inventors need not be dwelt on here. A slightly simplified form with the same saturating termination step coupled to quadratic autocatalysis (chain branching) has been studied in some mathematical detail (Merkin *et al.* 1985*a,b*; Scott 1985; Brindley *et al.* 1988) and discussed in the context of carbon monoxide oxidation in a CSTR (Gray *et al.* 1987*b*). The saturating term, with its rational form, is used widely in surface reaction modelling as the Langmuir–Hinshelwood rate law (and an equivalent form arises in the Michaelis–Menten treatment for enzyme kinetics) and simply requires a finite rather than infinite number of surface sites on which the termination process can occur. Even with this extra non-linearity, the model remains two variable and hence cannot predict complex responses or chaos. Learning from the H₂ + O₂ experiences, these latter may require the inclusion of self-heating effects.

9.9. Exothermic reactions in a CSTR

The final sentence notwithstanding, the previous few sections have discussed systems for which the driving mechanisms for complex behaviour have been based mainly on chemical feedback. Elsewhere in this book the seminal role of the simple FONI model of a single, exothermic first-order reaction in a non-adiabatic CSTR has been stressed, and naturally there have been attempts made to realize such a system experimentally. Two of the most successful of such studies have involved gas-phase reactions, although inevitably there are various practical considerations that have made this a tougher nut to crack *in vitro* than *in numero* or *in calculo*.

9.9.1. *The decomposition of DTBP*

The decomposition of di-tertiary butyl peroxide (DTBP) has been mentioned earlier (Section 9.1) as an exemplary first-order exothermic process either diluted in N_2 or even for much of the time in the presence of O_2 despite the possibility of extra downstream reactions in this latter case. Gray *et al.* (1984c) have attempted to use this system to test at least some of the various predictions outlined in Chapter 3 and the papers referred to therein. There are considerable experimental hurdles that must be overcome: a stable and accurately controllable supply of the reactant vapour in N_2 or O_2 must be obtained and must enter the vessel with as little prereaction as possible so that the inflow characteristics are well defined. The first of these can be achieved using a vapour saturator: the carrier gas passes through the DTBP liquid at a slightly elevated temperature. The vapour/carrier gas is then cooled so excess vapour condenses leaving a saturated mixture at this second temperature. A knowledge of the vapour-pressure dependence on temperature then allows the apparent inflow composition to be calculated.

In order to prevent reaction before entry to the vessel the inlet port needs to be cooled as efficiently as possible. This, however, slightly complicates the issue—the simplest formulation of the equations arises when the inflow and ambient (oven) temperatures T_0 and T_a are equal, but that is not possible in the experiment. Some apparent simplification can be achieved by defining a more complex reference temperature that is a weighted mean of T_0 and T_a involving also the residence and Newtonian cooling times. Another difference between experimental and analytical procedures is that the vessel temperature is the most convenient parameter in the former, whereas the residence time seems to be universally chosen for theoretical approaches, allowing the more interesting responses such as isolas and mushrooms.

In a non-adiabatic CSTR, the existence of multiple stationary states is governed by the exothermicity per unit mass, the mean residence time, and the Newtonian cooling time. Despite limitations on the range of these

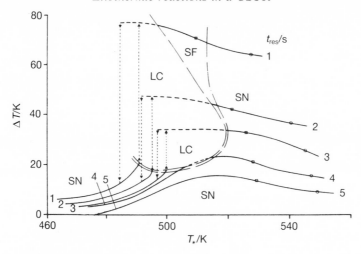

FIG. 9.29. Extent of self-heating as a function of ambient temperature for DTBP decomposition in a CSTR for mean residence times of 1–5 s as indicated showing multiplicity and oscillations for $t_{res} < 4$ s: the local stability of the stationary state is indicated as stable node (SN), stable focus (SF), or limit cycle (LC). (Reproduced with permission from Gray *et al.* (1984*c*), © Institute of Chemical Engineers.)

parameters that are available in the DTBP system, Gray *et al.* observed hysteresis loops (bistability) and their unfolding, and the onset and extinction of oscillations. Figure 9.29 shows the variation of the stationary-state temperature rise above the weighted mean reference T_* as a function of the latter temperature for five different mean residence times. As the residence time increases, so multistability and oscillatory behaviour are lost.

The oscillations in this system typically arise at Hopf bifurcation points and can be extinguished at either Hopf points or via homoclinic orbit formation (Fig. 9.30). Griffiths *et al.* (1985) successfully reproduced many of these features with a relatively detailed kinetic model involving 14 elementary steps—12 of which cope with the extra reactions arising from the oxidation of methyl radicals that occurs in the presence of added O_2—and, of course, the effects of self-heating. Both hard and soft excitations are shown in Fig. 9.31. Sheintuch and Luss (1988) have interpreted these various data in terms of the bifurcation diagram, types of attractor, and 'organizing centres' described in Chapter 3.

9.9.2. The oxidation of ethane

At its simplest, the reaction between ethane and oxygen in a CSTR shows classic FONI behaviour. Gray *et al.* (1984*d*) examined this system experimentally for various fuel-rich mixtures at atmospheric pressure over the

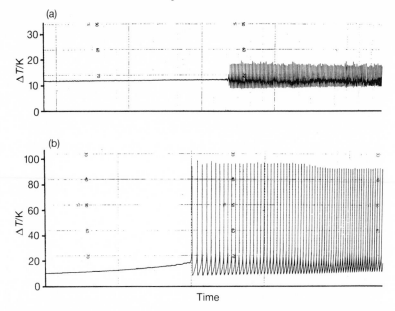

FIG. 9.30. Experimental traces showing (a) soft excitation at the onset of sustained oscillations on increasing T_a for DTBP decomposition and (b) hard excitation for DTPB oxidation in a CSTR. (Reproduced with permission from Griffiths *et al.* (1985), © Royal Society of Chemistry.)

temperature range 500–800 K: in particular for the cases $C_2H_6:O_2 = 8:1$, 6:1, 4:1, 2:1, and 1:1 which can be compared with the stoichiometric composition of 0.28:1. An important quantity for the FONI systems is the adiabatic temperature excess ΔT_{ad}—the temperature rise that would accompany complete reaction in a reactor with no heat transfer through the walls (its dimensionless equivalent is the group B in eqn (3.61)). Multiple stationary states are only possible in an adiabatic CSTR if this is sufficiently high (so the dimensionless form has a value larger than four). For a non-adiabatic system, the adiabatic temperature rise becomes scaled by a factor involving the residence time t_{res} and the Newtonian cooling time characterizing heat transfer t_N: we can use the maximum non-adiabatic temperature excess ΔT_{na} related to the adiabatic value through

$$\Delta T_{na} = \frac{\Delta T_{ad}}{1 + (t_{res}/t_N)}. \tag{9.5}$$

Again, multiplicity requires that this temperature rise should be sufficiently large (its dimensionless equivalent is usually denoted B^* and for multiplicity we need $B^* > 4$). Gray *et al.* estimated $t_N = 0.6$ s for their reactor. Again there are slight technical problems in that the inflow and oven temperatures were not quite equal requiring the use of a weighted mean reference

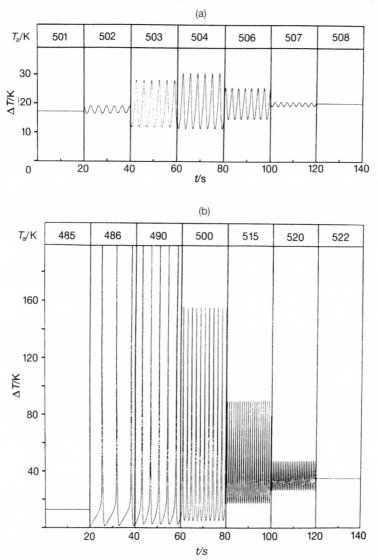

FIG. 9.31. Numerically computed traces corresponding to experimental records in Fig. 9.30. (Reproduced with permission from Griffiths *et al.* (1985), © Royal Society of Chemistry.)

temperature T_* for quantitative comparisons: in terms of T_0 and T_a the appropriate mean would be given by

$$T_* = \frac{(T_0/t_{res}) + (T_a/t_N)}{(1/t_{res}) + (1/t_N)}. \qquad (9.6)$$

The dimensionless forms are then obtained by multiplying ΔT_{ad} and ΔT_{na} by

E/RT_*^2 where E is the activation energy of the reaction. As an approximation, however, we can take $T_0 \approx T_a$ in which case the right-hand side of eqn (9.6) simplifies to give $T_* = T_a$: this will be a better estimate the longer the residence time, but even for $t_{res} = 15$ s, the shortest reported in their study, Gray *et al.* calculated less than 10 K difference (in 700 K) between T_* and T_a.

Increasing the fuel:oxygen ratio decreases ΔT_{ad} and hence, for a fixed residence time, decreases ΔT_{na} and the dimensionless forms. This reduction comes partly from the diluting effect of the extra fuel above that required for complete combustion with available oxygen, but also because of the changing overall stoichiometry and hence the effective reaction exothermicity amongst the actual, partial oxidation products. Increasing the residence time will, from eqn (9.5), decrease ΔT_{na}.

With the 2:1 mixture, the above comments are born out in full. For residence times of 60 s or less, the reaction shows an S-shaped hysteresis loop (Fig. 9.32) as the ambient (oven) temperature is varied. For T_a less than

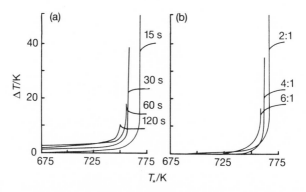

FIG. 9.32. Variation of stationary-state temperature excess with ambient temperature in ethane oxidation for (a) various residence times with $C_2H_6:O_2 = 2:1$, showing ignition for $t_{res} \leqslant 60$ s but a smoother dependence for $t_{res} = 15$ s showing trend towards loss of critical phenomena on increasing fuel:oxygen ratio. (Reproduced with permission from Gray *et al.* (1984d), © The Royal Society.)

about 750 K, the system exhibits a low reaction stationary state, with $\Delta T_{ss} < 10$ K. This state is a stable node. The primary products arising from the low extents of conversion prevailing here are ethene, with minor quantities of formaldehyde, ethylene oxide, and acetaldehyde. As the oven temperature is increased, so there is an abrupt jump in the response. Typically the system shows a transient overshoot before settling to a high reaction state, with a significantly larger stationary-state temperature excess. (The overshoot helps locate the ignition point.) The ambient temperature at which ignition occurs increases slightly as the residence time decreases or as the fuel:oxygen ratio decreases. The upper state has virtually complete O_2

consumption and up to about 70 per cent ethane conversion (to partially oxygenated products—mainly ethene, but also significant amounts of H_2O and measurable quantities of CO_2, C_2H_4O, CH_2O, and CH_3CHO). An additional feature of this upper stationary state is the accompanying chemiluminescence (from excited formaldehyde CH_2O^*).

If the ambient temperature is now reduced, the system stays in the upper state through the region of hysteresis, until reaching the extinction point at some lower T_a. The upper state can have stable focal character, with small perturbations decaying via damped oscillations.

At longer residence times, the sharpness of the transitions is lost and the response may have unfolded into a monotonic dependence of temperature excess on oven temperature. This is not completely resolved for the 2:1 system, but clearly becomes the case for the richer mixtures as suggested by the arguments above.

The 1:1 composition, on the other hand, shows additional behaviour, particularly with the shorter residence times. The system can now also show dynamic bifurcations into sustained oscillatory states. These typically are associated with the upper branch of high reactivity. On decreasing the oven temperature along this branch, the reaction undergoes an apparently supercritical Hopf bifurcation to yield small-amplitude 'cool-flame' oscillations (Fig. 9.33(a)) in both the reactant temperature and the intensity of the chemiluminescence. The amplitude grows quite rapidly as T_a is reduced

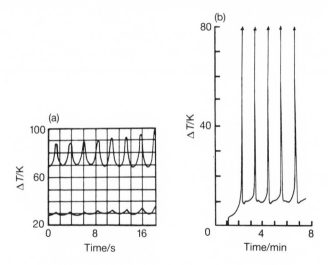

FIG. 9.33. The oxidation of equimolar ethane mixtures in a CSTR: (a) sustained oscillations in temperature excess and light emission intensity for 'cool flame' behaviour with $T_a = 763$ K and $t_{res} = 15$ s; (b) oscillatory ignition with $T_a = 795$ K and $t_{res} = 15$ s. (Reproduced with permission from Gray *et al.* (1984*d*), © The Royal Society.)

further, and there are features suggesting that the oscillations are extinguished via homoclinic orbit formation. There are some operational problems in confirming this latter point unequivocally as the large-amplitude excursion produces not only high transient temperatures but also dangerously large pressure pulses. For the same practical reason, the transition from the lower stationary state into the oscillations has not been examined, although we may expect the ignition to be from a saddle–node bifurcation as above with the system moving on to the large-amplitude limit cycle surrounding the now locally unstable upper state and that has grown from the Hopf bifurcation point at higher T_a.

At the highest ambient temperatures a second bifurcation to large-amplitude oscillations was observed (Fig. 9.33(b)), but again not investigated in any detail to save the apparatus. There are also some extra features that go beyond those characteristic of a simple first-order deceleratory reaction and that reflect the rather intricate chemistry that can accompany hydrocarbon oxidations. Much of this may arise through further reactions of the partially oxygenated intermediates in this somewhat complex chemical soup. To make life a bit easier, it makes sense to look at the oxidation of these species on their own (as much as is possible). To that end, we move on to consider the oxidation of acetaldehyde (ethanal CH_3CHO), one of the components identified above. This will introduce many of the typical non-linear kinetic forms of response characteristic of 'thermokinetic' systems where both thermal and chemical feedback processes occur and combine.

9.10. Thermokinetic systems: the oxidation of acetaldehyde

Another reason for the interest in the oxidation of small fuel molecules is that the combustion of typical larger fuels—octane etc.—almost inevitably proceeds in its early stages by breaking carbon–carbon bonds to produce C_2 and C_3 species, some partially oxygenated. Acetaldehyde is, in some senses, an archetypal compound for combustion—exhibiting most of the 'typical' forms of behaviour that can be related to technological problems such as engine knock and premature ignition. (There are, however, some special features of acetaldehyde combustion that will be commented on with regard to more general oxidation mechanisms below.)

Again, we will concentrate on CSTR studies: early observations from closed vessels, and especially in unstirred reactors, have many complicating aspects not directly due to chemistry. Well-stirred flow reactors have become commonly used for hydrocarbon combustion since the early 1970s (Felton and Gray 1974; Felton *et al.* 1976; Caprio *et al.* 1977; Gray *et al.* 1981*a,b*).

The ranges of (ambient) temperatures and pressures of interest are typically $5 \leqslant (p/kN\ m^{-2}) \leqslant 25$ (i.e. 40–200 Torr) and $450 \leqslant (T/K) \leqslant 650$ for a range of fuel:oxygen ratios and mean residence times. Figure 9.34 shows a

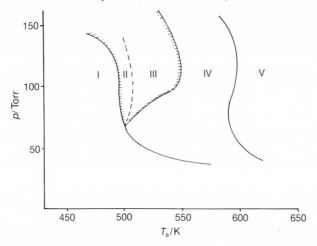

FIG. 9.34. Experimental p–T_a diagram for $CH_3CHO + O_2$ in a CSTR with $t_{res} = 3$ s showing the relative locations of regions I–V. (Reproduced with permission from Gray *et al.* (1981*a*), © The Royal Society.)

p–T_a ignition diagram for $t_{res} = 3$ s and an equimolar mixture $CH_3CHO:O_2 = 1:1$. Five different regions are identified, separated by boundaries representing changes in qualitative behaviour (i.e. loci of bifurcations). The responses in these regions are termed as follows: I, steady dark reaction; II, oscillatory (two-stage) ignition; III, complex oscillatory ignition; IV, oscillatory cool flames; V, steady glow.

9.10.1. Steady dark reaction

This response is found at the lowest ambient temperatures. There is no chemiluminescence, but reaction can be followed in terms of measurable stationary-state temperature excesses and by the mass spectrometric determination of the stationary-state concentration of various reactant and product species. The maximum extents of self-heating are typically about 40 K. The temperature rise, ΔT_{ss}, increases with the ambient temperature, attaining its maximum value at the boundaries between region I and regions II or IV: ΔT_{ss} is also a measure of the stationary-state heat release rate R which thus also increases with T_a across this region (R has a positive temperature coefficient).

Typical products of the incomplete combustion in this mode are peracetic acid CH_3CO_3H, methanol CH_3OH, formaldehyde CH_2O, and methane CH_4 as well as CO_2 and H_2O, but no CO is formed. The stationary state here has stable nodal character: if the inflow is perturbed momentarily and then restored, the return to the stationary state is monotonic (no overshoot).

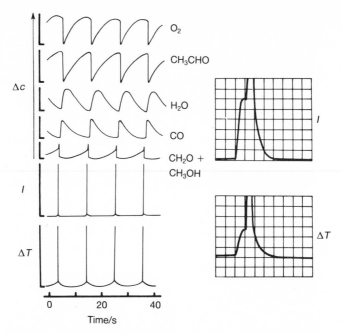

FIG. 9.35. Details of the concentration, light emission, and temperature excess time series for oscillatory two-stage ignition during acetaldehyde oxidation (region II) in a CSTR. (Reproduced with permission from Gray *et al.* (1981*a*), © The Royal Society.)

9.10.2. Oscillatory (two-stage) ignition

The $CH_3CHO + O_2$ reaction shows a simple oscillatory ignition in a CSTR. The excursions are typically ignition events, with rapid, sharp decreases in the concentrations of the reactants and increases in the products (primarily CO and H_2O). There is an intense light emission and a large temperature excursion.

The ignition waveform actually shows some important fine structure. The two-stage nature of the ignition in this system is shown in Fig. 9.35: similar detail is observed in the one-off ignition in a closed reactor. The first stage leads to an increase in the gas temperature of up to about 250 K, during which time partial oxidation products such as CH_2O and CH_3OH build up. This initial stage leads on to a rapid acceleration in the reaction and the development of the true ignition stage: the partial oxidation species disappear as CO and H_2O are formed.

Region II is typically entered from region I by raising the ambient temperature. As T_a is increased across the boundary, there is hard excitation to the ignition pattern, i.e. the system 'jumps' into fully developed large-amplitude ignitions immediately. At short residence times the boundary between regions I and II may show hysteresis: if T_a is reduced again the 'hard extinction' transition from oscillatory ignition to steady dark reaction occurs at a lower value than the upwards transition. At longer residence times, however, this hysteresis disappears and the abrupt change between the two patterns occurs at the same ambient temperature in both directions.

9.10.3. Oscillatory cool flames

It is easier if we discuss region IV before region III. The system displays regular oscillatory pulses of chemiluminescent reaction. Some examples of these cool flames are shown in Fig. 9.36. The temperature excursion is always less than 200 K and may become as small as only 10 K at higher ambient temperatures. (In fact, we may suspect that sufficiently fine control over the ambient temperature would allow us to observe the amplitude decreasing smoothly to zero at the boundary with region V.)

The chemical species formed during oscillatory cool-flame oxidation are CO, H_2O, CH_2O, CH_3OH, and CH_4, with lower yields of CH_3CO_3H, ethane C_2H_6, and hydrogen peroxide H_2O_2. The emission which accompanies these excursions originates from electronically excited formaldehyde CH_2O^*: the emission typically does not fall to zero between the maxima. Pugh *et al.* (1987) measured the concentrations of OH radicals and of acetaldehyde during cool-flame oscillations by laser-induced fluorescence and ultraviolet absorption spectroscopy respectively. They could distinguish no phase difference between these and the thermocouple and pressure records, suggesting that within experimental resolution (0.1 s) these quantities are all in phase.

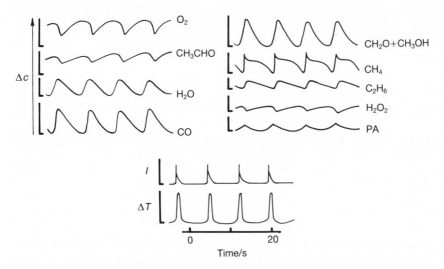

FIG. 9.36. Details of the concentration, light emission, and temperature excess time series for oscillatory cool flames during acetaldehyde oxidation (region IV) in a CSTR. (Reproduced with permission from Gray *et al.* (1981*a*), © The Royal Society.)

9.10.4. *Steady glow*

Region V lies at the highest temperatures. As it is entered from region IV, by raising T_a, the cool flames give way to stable steady-state glow. The stationary state is a stable focus: small perturbations decay via a damped oscillatory return. The chemiluminescence is still from excited CH_2O, and other partially oxidized products such as CH_3OH are formed as well as CO_2, CO, and H_2O.

The stationary-state temperature rise can be up to 80 K close to the boundary with region IV. In contrast to the steady dark reaction mode, however, in region V the temperature excess ΔT_{ss} typically decreases with increasing T_a. Thus the heat release rate R has a 'negative temperature coefficient' (n.t.c.) as shown in Fig. 9.37. (It is important to distinguish

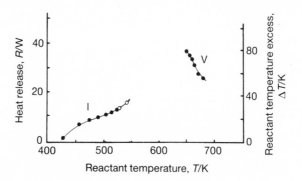

Fig. 9.37. The stationary-state heat release rate as a function of reacting gas temperature for regions I and V, showing the 'negative temperature coefficient' for the steady glow in region V. (Reproduced with permission from Gray *et al.* (1981*a*), © The Royal Society.)

between heat release rate and reaction rate here. The n.t.c. arises primarily because the distribution of partially oxidized products, and with it the exothermicity of the reaction, changes with the ambient temperature. The exothermicity decreases as T_a increases across this region.)

9.10.5. *Additional features of oscillatory cool flames*

The results presented by Gray *et al.* (1981*a,b*) show a monotonically decreasing oscillatory amplitude and increasing frequency with increasing ambient temperature across region IV. If the amplitude does decrease to zero at the boundary with region V, we can thus interpret this as a locus of supercritical Hopf bifurcation points, with stable limit cycle emerging around an unstable state as T_a is decreased and the stationary state becoming a focus for higher ambient temperatures.

There have, however, also been observations of more complex responses in region IV and at its upper boundary (Jones and Gray 1983; Gray and Jones 1984). These authors report the occurrence of birhythmicity, similar to that described earlier for the $H_2 + O_2$ system. The amplitude of the cool flames does not always vary monotonically or smoothly with the ambient temperature across the region. Gray and Jones observed distinct jumps from large- to small-amplitude waveforms on increasing T_a as shown in Fig. 9.38.

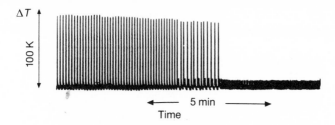

FIG. 9.38. Abrupt transition between the large- and small-amplitude cool flames that give rise to birhythmicity in acetaldehyde oxidation on increasing the ambient temperature by 1 K. (Reproduced with permission from Gray and Jones (1984), © The Combustion Institute.)

The jump back from small to large amplitude occurs at a lower ambient temperature, so for a range of conditions the two oscillatory solutions coexist.

As a further twist, the discontinuous jump from the large-amplitude oscillations can take the system into a stable stationary state corresponding to the steady glow of region V. If T_a is then decreased, the system stays in the stationary state (there is a coexistence of steady and oscillatory states) until it undergoes a supercritical Hopf bifurcation into small-amplitude limit cycles. At some lower ambient temperature, there is a jump back to the larger-amplitude waveform. Suggested bifurcation diagrams showing the folding of a hysteresis loop in the limit cycle locus appropriate to these observations have been given and are shown in Fig. 9.39.

Gray and Jones also found that they could observe quite complex and almost apparently chaotic behaviour close to the conditions for transition between the different limit cycle solutions. These were always transient responses and eventually settled to one of the periodic waveforms and may well have their origin in experimental control rather than in thermokinetics. Harding *et al.* (1988) also report apparently aperiodic traces close to the Hopf bifurcation point, as shown in Fig. 9.40. Again, however, they conclude that this does not arise from the kinetics alone but from the interaction of experimental noise with the nearby Hopf point.

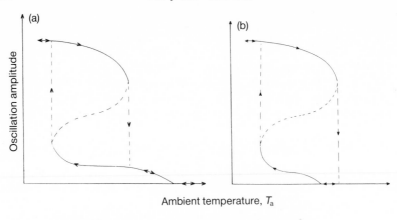

Fig. 9.39. Suggested bifurcation diagrams plotting oscillatory amplitude as a function of ambient temperature for acetaldehyde oxidation reaction showing birhythmicity. In each case a small-amplitude limit cycle emerges off the axis from a supercritical Hopf bifurcation as T_a is reduced and coexists over some of its range with an unstable cycle (– – –) and another stable cycle (——): (a) the region of birhythmicity lies completely below the Hopf bifurcation point so the large-amplitude cool flames give way to small-amplitude oscillations on increasing T_a; (b) the upper limit cycle branch 'overhangs' the Hopf bifurcation, so the larger-amplitude cool flames are extinguished into a steady state. (Reproduced with permission from Gray and Jones (1984), © The Combustion Institute.)

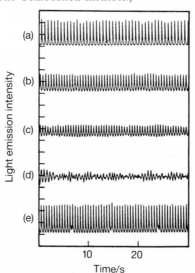

Fig. 9.40. Periodic and apparently aperiodic time series of light emission during cool flame oxidation of acetaldehyde in a CSTR. (Reproduced with permission from Harding *et al.* (1988), © American Institute of Physics.)

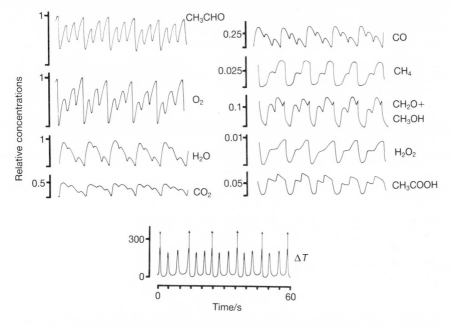

FIG. 9.41. Details of the concentration and temperature excess time series for complex ignition during acetaldehyde oxidation (region III) in a CSTR. (Reproduced with permission from Gray *et al.* (1981*b*), © The Combustion Institute.)

9.10.6. Complex ignition

Region III lies between the oscillatory cool flames (IV) and the simple (two-stage) ignitions (II). Perhaps not surprisingly then, the waveform here now contains both ignition and cool-flame features. Some examples are shown in Fig. 9.41. A complete period consists of a number, n, of cool-flame excursions and a two-stage ignition pulse. Thus we can talk of an $(n + 2)$-stage ignition. (In region II we could say that $n = 0$ whilst for region IV we have $n \to \infty$.)

Returning to region III, then, close to the boundary with region II the ignition excursions are interspersed by just one cool flame ($n = 1$ or three-stage ignition). As T_a is increased more cool flames appear between the ignitions. At present the experimental record is a seven-stage ignition ($n = 5$) observed close to the boundary with region IV.

9.11. Forcing of acetaldehyde cool flames

Pugh *et al.* (1986*a, b*) report a series of experiments that constitute one of the few attempts at periodically forcing a gas-phase oscillatory system. In

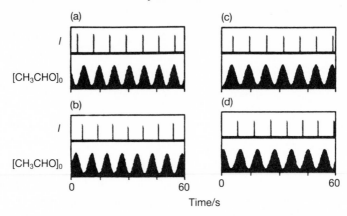

FIG. 9.42. Quasiperiodicity and entrainment on passing through the 1:1 Arnol'd tongue for acetaldehyde oxidation in a CSTR with a 20 per cent forcing amplitude: upper trace shows system response to forced fuel inflow (lower trace); (a) and (d) lie outside and (c) and (d) inside the entrainment band. (Reproduced with permission from Pugh *et al.* (1986*a*), © American Institute of Physics.)

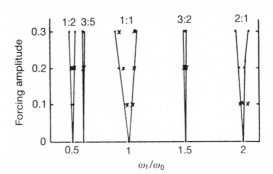

FIG. 9.43. Entrainment bands (Arnol'd tongues) for forced acetaldehyde oxidation in a CSTR. (Reproduced with permission from Pugh *et al.* (1986*a*), © American Institute of Physics.)

their first sequence, they examined the response of a cool-flame oscillation (with natural period $t_p \approx 8.5$ s) to sinusoidal variations in either the acetaldehyde or the oxygen inflow rate. These perturbations were achieved by regulating the voltage applied to the mass flow control valves: generally there is a linear relationship between applied voltage and mass flow rate for a wide range with these controllers.

Figure 9.42 shows the response of the system to four different forcing periods of the fuel inflow. The forcing amplitude is such that there is a ± 20 per cent variation in $[CH_3CHO]_0$. For cases (a) and (d) the ratio of the forcing frequency ω_f to that of the autonomous system ω_0 is 1.06 and 0.96 respectively. These lie outside the region of 1:1 entrainment (Fig. 9.43) and so

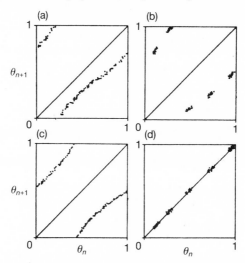

FIG. 9.44. Onset and loss of the 3:5 entrainment in forced acetaldehyde oxidation shown by stroboscopic (next-phase) maps: (a) $\omega_f/\omega_0 = 0.555$, showing a quasi-periodic response; (b) $\omega_f/\omega_0 = 0.610$, showing five discrete points indicating an entrained response; (c) $\omega_f/\omega_0 = 0.775$, showing quasiperiodicity; (d) the five mapping points from (b) are brought on to the identity line in a fifth-phase map θ_{n+5}–θ_n. (Reproduced with permission from Pugh *et al.* (1986a), © American Institute of Physics.)

the response is quasiperiodic. For (b) and (c), with frequency ratios of 1.056 and 0.965, the response has become fully entrained with with forcing. Pugh *et al.* report entrainment limits for five different frequency ratios and three amplitudes, giving the resonance horns shown in Fig. 9.43.

The data from these experiments can also be displayed in a more succinct form in terms of 'next-phase maps'. These are constructed from successive measurements of the 'phase difference' θ_n—here defined as the phase separation between a maximum in the light emission and the previous maximum in the fuel inflow concentration. The map then plots θ_{n+1} versus θ_n. Figure 9.44 shows a sequence of next-phase maps for a traverse through the 3:5 entrainment band: (a) and (c) have $\omega_f/\omega_0 = 0.555$ and 0.775 respectively, for which the response is quasiperiodic giving a full mapping curve. For $\omega_f/\omega_0 = 0.610$, case (b), the system lies within the entrainment band, and the map has five discrete points (within experimental precision). The same data for the entrained state can be plotted as the fifth next-phase map (θ_{n+5} versus θ_n, Fig. 9.44(d)), which brings the five points on to the identity line.

Experiments in which both inflow feeds were varied periodically together have also been performed (Pugh *et al.* 1986b). Various games can be played here. If the two inflows are perturbed with the same frequency, the phase

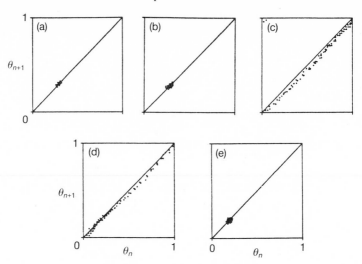

FIG. 9.45. Acetaldehyde oxidation forced through both fuel and oxygen inflows, with same frequency but different phases: O_2 perturbation lags fuel forcing by (a) 0°, 1:1 entrainment; (b) 90°, 1:1 entrainment; (c) 180°, quasiperiodic response; (d) 90°, still showing quasiperiodicity and hence revealing hysteresis effects with respect to (b); 0°, 1:1 entrainment. (Reproduced with permission from Pugh *et al.* (1986b), © American Institute of Physics.)

relationship between them can be varied. If the phase difference is initially zero, the system may become fully entrained if the forcing frequency is sufficiently close to the natural frequency, as described above. This entrainment will then remain as the phase difference is increased slightly. Pugh *et al.* were able to maintain entrainment with a 90° phase difference (the O_2 maximum trailing the fuel), but with an out-of-phase (180° difference) situation, the system responds quasiperiodically as indicated by the next-phase map (Fig. 9.45(c)). When the phase difference of the inflows was reduced back to 90°, the quasiperiodic response remained where there had been entrainment before. There thus appear to be coexisting stable responses with hysteresis. Full entrainment returns by the time the phase lag has been reduced to zero.

In another sequence, the system was set up entrained to the perturbations in one of the reactants (CH_3CHO) and the frequency of the second then varied. This allows the possibilities of an overall quasiperiodic forcing as discussed in Chapter 6. Typically the motion corresponds to evolution on some three-dimensional surface of a three-frequency torus. If the system response stays entrained with the acetaldehyde perturbations we see a vertical line in the 'two-phase' plot that maps the nth phase difference between the light emission maximum and the O_2 maximum, $\theta_{n,O}$, as a function of the nth phase difference with relation to the acetaldehyde

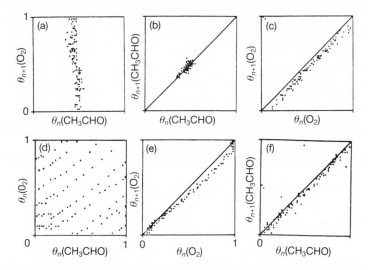

FIG. 9.46. Acetaldehyde oxidation forced through both fuel and oxygen inflows with different frequencies: (a)–(c) reveal entrainment to acetaldehyde inflow but quasi-periodicity with respect to $[O_2]_0$; (a) two-phase plot showing single value for phase with respect to fuel but full range with respect to O_2; (b) next-phase map with respect to fuel, giving single entrainment on identity line; (c) next-phase map with respect to O_2 showing quasiperiodicity: (d)–(f) response is quasiperiodic with respect to both inflows (non-chaotic strange attractor). (Reproduced with permission from Pugh *et al.* (1986*b*), © American Institute of Physics.)

perturbation $\theta_{n,A}$, as shown in Fig. 9.46(a). The corresponding (single) next-phase maps for the two reactants individually also reveal this entrainment and quasiperiodicity with regard to the fuel and O_2 (Fig. 9.46(b) and (c) respectively).

For other combinations of forcing frequencies, the response may become unentrained to either frequency. The two-phase plot then shows a series of scattered points (Fig. 9.46(d)) and the next-phase maps for both reactants have quasiperiodic forms (Fig. 9.46(e) and (f)).

9.12. Kinetic mechanisms for acetaldehyde oxidation

The most successful numerical computations for the acetaldehyde + oxygen system involve 60 different elementary steps and allow for non-isothermal operation of the CSTR. Figure 9.47 compares the predicted and computed ignition diagrams from such a study (Griffiths and Sykes 1989*a*)—a severe test of the proposed mechanism. Also shown are some computed complex ignitions with $n = 6$ and $n = 8$ correspondiing to eight- and 10-stage ignition respectively. (It is interesting that the nine-stage waveform does appears to

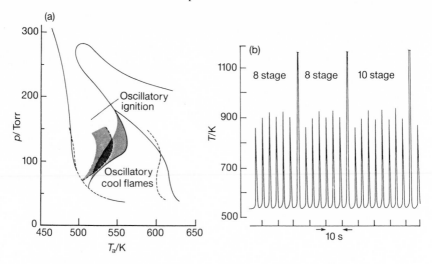

Fig. 9.47. Numerical predictions for acetaldehyde oxidation based on 60 reaction model: (a) comparison of computed (——) and experimental (– – –) p–T_a limits and (shaded) regions of hysteresis; (b) eight- and 10-stage complex ignitions for $T_a =$ 536.892 K and 536.893 K respectively. (Reproduced with permission from Griffiths and Sykes (1989a), © The Royal Society.)

be missing—we will return to this point later when discussing the results of Wang and Mou in Section 9.13.)

Many of the steps in these large schemes are included to give a good quantitative match during the full ignition process. If we restrict our ambitions to a qualitative interpretation of the cool-flame behaviour and the onset of ignition, a smaller scheme can be abstracted (Gibson *et al.* 1984). Table 9.3 lists 25 steps for this purpose.

The first four reactions provide an initiation process. In particular, it leads to the production of methyl radicals CH_3. The important steps for the cool-flame clockwork are (5) to (15), and of these the heart is the pair of reactions (6) and (7). The latter in fact constitute the forward and reverse steps of the addition process

(6, 7) $$CH_3 + O_2 \rightleftharpoons CH_3O_2.$$

It is this equilibrium, and its temperature dependence, which holds the key to the observed behaviour. At low reactant temperatures, the equilibrium lies to the right, in favour of CH_3O_2. This methylperoxy species is then further oxidized through reaction (11):

(11) $$CH_3O_2 + CH_3CHO \rightarrow CH_3O_2H + CH_3CO.$$

This produces a CH_3CO to continue the chain, through step (5), and the species CH_3O_2H. The latter decomposes in a subsequent step to produce

TABLE 9.3

Reduced kinetic model for acetaldehyde oxidation. Rate constant data are given in Gibson *et al.* (1984)

(1)	$CH_3CHO + O_2 \rightarrow CH_3CO + HO_2$
(2)	$CH_3CO + O_2 \rightarrow CH_3CO_3$
(3)	$CH_3CO_3 + CH_3CHO \rightarrow CH_3CO_3H + CH_3CO$
(4)	$CH_3CO_3H \rightarrow CH_3 + CO_2 + OH$
(5)	$CH_3CO + M \rightarrow CH_3 + CO + M$
(6)	$CH_3 + O_2 \rightarrow CH_3O_2$
(7)	$CH_3O_2 \rightarrow CH_3 + O_2$
(8)	$CH_3 + CH_3 \rightarrow C_2H_6$
(9)	$CH_3O_2 + CH_3CO \rightarrow 2CH_3O + O_2$
(10)	$CH_3O + CH_3O \rightarrow CH_3OH + CH_2O$
(11)	$CH_3O_2 + CH_3CHO \rightarrow CH_3O_2H + CH_3CO$
(12)	$CH_3O_2H \rightarrow CH_3O + OH$
(13)	$OH + CH_3CHO \rightarrow CH_3CO + H_2O$
(14)	$CH_3O + O_2 \rightarrow CH_2O + HO_2$
(15)	$OH + CH_2O \rightarrow HCO + H_2O$
(16)	$HCO + M \rightarrow H + CO + M$
(17)	$HCO + O_2 \rightarrow CO + HO_2$
(18)	$OH + CO \rightarrow CO_2 + H$
(19)	$H + O_2 \rightarrow OH + O$
(20)	$H + O_2 + M \rightarrow HO_2 + M$
(21)	$O + CH_2O \rightarrow HCO + OH$
(22)	$HO_2 + CH_2O \rightarrow HCO + H_2O_2$
(23)	$HO_2 + HO_2 \rightarrow H_2O_2 + O_2$
(24)	$H_2O_2 + M \rightarrow 2OH + M$
(25)	$OH + H_2O_2 \rightarrow HO_2 + H_2O$

two more radicals:

(12) $CH_3O_2H \rightarrow CH_3O + OH.$

This process is known as 'degenerate branching', to distinguish it from a direct branching step such as $H + O_2 \rightarrow OH + O$. Because of the branching, the reaction rate increases. The methoxy radical reacts with O_2 via step (14), but the formaldehyde molecule produced is in an excited electronic state, allowing for the chemiluminescent emission observed experimentally. The overall reaction is exothermic, so the accelerating reaction leads to self-heating and higher reactant temperatures.

At higher temperatures, however, the equilibrium formed by steps (6) and (7) shifts to the left, in favour of the dissociated methyl and O_2. At high T, the reaction pathway involves simple methyl radical chemistry, principally the termination via the recombination reaction (8). The total radical concentration falls and, hence, so does the reaction rate. Finally, in the cycle, the temperature of the gas will fall back towards T_a either by Newtonian heat loss or by the outflow of heated gas: the equilibrium then moves back to the right, and the next cool flame can be initiated.

It is important that the above arguments involve both the chemical non-linearity of degenerate branching *and* the thermal feedback effects on the equilibrium steps: neither a purely kinetic nor a purely thermal mechanism can explain the full complexity of acetaldehyde cool flames.

The final group of reactions (16) to (25), become important as the system moves towards a full ignition. A linking role is played by an increasing formaldehyde concentration, through step (15) which produces the formyl radical HCO. This, in turn, leads to CO and then H atoms. Once hydrogen atoms have been produced, the key branching reaction (19) can occur. This scenario requires the ignition to be at least a two-stage event, with control in the first stage with reactions (5) to (14) and in the ignition phase with steps (16) to (25). If the cool-flame stage does not produce a sufficiently high CH_2O concentration, the ignition steps do not come into play: one or more additional cool-flame stages may be needed before the ignition stage is initiated, or indeed the latter may not occur at all.

Slightly smaller mechanisms have been employed by other workers. Wang and Mou (1985) list 20 steps to help justify their very simplified model scheme (see below), but do not report on the direct computation from this scheme. Their main source is the mechanism developed initially by Halstead *et al.* (1971, 1973, 1975; Kirsch and Quinn 1976). This 'Thornton' scheme has also been modified more recently by Harrison and Cairnie (Table 9.4) and applied to the modelling of ignition and cool-flame phenomena developing in the vicinity of heated surfaces or on hot pipes (Harrison *et al.* 1988). The difference between this and the mechanism in Table 9.3 lies mainly in the final set of 'ignition' reactions involving the smaller, more active radicals.

The various experiments described above by Ross and coworkers (Pugh *et al.* 1986*a*, *b*, 1987; Harding and Ross 1988; Harding *et al.* 1988) have been compared with a 12 reaction model with five species. These are based on an earlier suggestion that a degenerate branching cycle might be built around the species peracetic acid CH_3CO_3H, and do not include the crucial

TABLE 9.4

The Thornton scheme for acetaldehyde oxidation. Rate constant data are given in Harrison and Cairnie (1988)

$$CH_3CHO + O_2 \rightarrow CH_3CO + HO_2$$

$$CH_3CO \rightarrow CH_3 + CO$$

$$CH_3CO + O_2 \rightarrow CH_3CO_3$$

$$CH_3 + O_2 \rightarrow CH_3O_2$$

$$CH_3O_2 \rightarrow CH_3 + O_2$$

$$CH_3O_2 + CH_3CHO \rightarrow CH_3O_2H + CH_3CO$$

$$CH_3CO_3 + CH_3CHO \rightarrow CH_3CO_3H + CH_3CO$$

$$HO_2 + CH_3CHO \rightarrow H_2O_2 + CH_3CO$$

$$CH_3 + CH_3CHO \rightarrow CH_4 + CH_3CO$$

$$CH_3O + CH_3CHO \rightarrow CH_3OH + CH_3CO$$

$$CH_3O + O_2 \rightarrow HCHO + HO_2$$

$$CH_3O_2 + CH_3O_2 \rightarrow 2CH_3O + O_2$$

$$CH_3CO_3 + CH_3CO_3 \rightarrow 2CH_3 + 2CO_2 + O_2$$

$$CH_3CO_3H \rightarrow CH_3 + CO_2 + OH$$

$$CH_3O_2H \rightarrow CH_3O + OH$$

$$H_2O_2 + M \rightarrow 2OH + M$$

$$CH_3 + CH_3 \rightarrow C_2H_6$$

$$CH_3O_2 + CH_3O_2 \rightarrow CH_3OH + HCHO + O_2$$

$$HO_2 + HO_2 + M \rightarrow H_2O_2 + O_2 + M$$

$$CH_3O_2 + HO_2 \rightarrow CH_3O_2H + O_2$$

$$OH + CH_3CHO \rightarrow CH_3CO + H_2O$$

equilibrium $CH_3 + O_2 \rightleftharpoons CH_3O_2$. The almost vital role of the latter and of the effects on it of the dynamic temperature variations bear repetition again at this stage.

9.13. Model schemes for acetaldehyde oxidation

The simplest model that still accounts for many of the qualitative features described above is that proposed by Yang and Gray (1969a, b; Gray 1969; Yang 1969). This thermokinetic scheme involves chain branching and thermal feedback, and can be written in terms of the following steps:

(GY1)	initiation	$A \rightarrow X$	rate $= k_i a$
(GY2)	branching	$X \rightarrow 2X$	rate $= k_b x$
(GY3)	termination	$X \rightarrow S_1$	rate $= k_{t1} x$
(GY4)	termination	$X \rightarrow S_2$	rate $= k_{t2} x.$

The two variables here are the concentration of X and the reacting gas temperature. The model has been most frequently analysed in a 'closed-vessel' form with the concentration of the pool chemical reactant A assumed constant. Reactions (GY2) and (GY4) are taken to be exothermic. Thermal feedback acts via the temperature dependence of the reaction rate constants for steps (GY1), (GY2), and (GY4) which have an Arrhenius form $k(T) = A \exp(-E/RT)$, whilst step (GY3) is assumed to be virtually temperature independent ($E_{t1} = 0$). Two important additional specifications must be added: the activation energies must satisfy the inequalities $E_{t2} > E_b > E_{t1}$ whilst the exothermicities need to be ordered in the same way, $q_{t2} > q_b > q_{t1}$ (where $q = -\Delta H$).

The mass- and energy-balance equations for this scheme appropriate to a closed vessel can be written as

$$\frac{\mathrm{d}x}{\mathrm{d}t} = k_i a + \phi x \tag{9.7}$$

$$C\frac{\mathrm{d}T}{\mathrm{d}t} = q_i k_i a + \Theta x - \chi(T - T_a). \tag{9.8}$$

Here ϕ is the net branching factor and Θ a quantity related to the exothermicities q of the three steps (GY2)–(GY4):

$$\phi = k_b - k_{t1} - k_{t2} \qquad \Theta = q_b k_b + q_{t1} k_{t1} + q_{t2} k_{t2}.$$

These are both functions of temperature, through the rate constants. The quantities C and χ are the effective heat capacity and heat transfer coefficient respectively.

The stationary-state radical concentration and temperature are given by solutions of the equations

$$x_{ss} = -k_i a/\phi(T_{ss}) \tag{9.9}$$

$$q_i k_i a + \Theta(T_{ss})x_{ss} - \chi(T_{ss} - T_a) = 0. \tag{9.10}$$

A convenient way of representing the behaviour of this model is to use a thermal diagram. Eliminating the radical concentration x_{ss} between eqns (9.9) and (9.10), the stationary-state condition can be written in

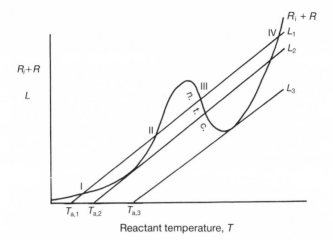

FIG. 9.48. Typical dependence of the total heat release rate $R + R_i$ on reactant temperature T for the Yang and Gray model showing a region of negative temperature coefficient and possible multiple intersections and tangencies with the heat loss L. (Reproduced with permission from Yang and Gray (1969a), © American Chemical Society.)

the form

$$q_i k_i a - \frac{k_i a \Theta(T_{ss})}{\phi(T_{ss})} = \chi(T_{ss} - T_a) \tag{9.11}$$

$$\underbrace{\hphantom{q_i k_i a}}_{R_i} + \underbrace{\hphantom{\frac{k_i a \Theta(T_{ss})}{\phi(T_{ss})}}}_{R} \quad \underbrace{\hphantom{\chi(T_{ss} - T_a)}}_{L}$$

The left-hand side of this equation $(R_i + R)$ is shown as a function of the reacting gas temperature T in Fig. 9.48. This sum represents the total rate of heat release at the stationary state. There is an initial exponential rise in $R_i + R$ as T increases, but eventually the locus shows a maximum. There is then a region of negative temperature coefficient in which the rate of heat release decreases with increasing temperature. This arises because the high activation energy, termination step (4), becomes dominant over branching at these temperatures. There is, however, an ultimate increase in R, even without net branching the reaction is exothermic and thermal runaway can occur. The right-hand side of eqn (9.11) describes a straight line with a gradient given by the heat transfer coefficient χ and an intercept corresponding to the ambient temperature T_a. Stationary-state solutions are located by the intersections of R and L.

If the ambient temperature is low or if the heat transfer conditions are such that L is steep, the first intersection of the heat generation and loss lines corresponds to only small extents of self-heating and to a low radical concentration. Such a situation holds for $T_{a,1}$ in Fig. 9.48: the intersection marked I is typical of a slow reaction state. The second intersection, marked II, is a saddle point. Other intersections between R and L also exist.

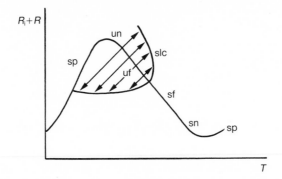

Fig. 9.49. Typical variation of local stability along the n.t.c. branch for the Yang and Gray model: sn, stable node; sf, stable focus; uf, unstable focus; un, unstable node; sp, saddle point; slc, stable limit cycle.

If the ambient temperature is increased to $T_{a,2}$, the heat release and heat loss rate R and L become tangential: the lowest two intersections merge at a saddle–node bifurcation. Beyond this, the system may jump to a higher intersection such as III. This lies on, or close to, the branch of R corresponding to the n.t.c. There is much more self-heating and a higher radical concentration in this state. A typical variation in local stability of the stationary-state intersections along this third branch is indicated in Fig. 9.49. The solutions lowest along this (decreasing) branch, i.e. those with highest temperatures, are generally stable: either nodes or foci. As we move up this branch, to lower gas temperatures, however, there is typically a (supercritical) Hopf bifurcation from which a stable limit cycle grows. The limit cycle terminates by forming a homoclinic orbit. If this is formed with the saddle point on branch II, the system will move back to the low reaction state. On the other hand, if the homoclinic orbit is formed with the saddle point on branch IV, there is the opportunity of escape to high temperatures, typical of the onset of ignition.

The present scheme will not model ignition processes for which reactant consumption must become important, but we can see here many experimental features. There is a jump from a low reaction state into either a steady higher reaction state (steady glow perhaps) or into oscillations (oscillatory cool flames). A transition from steady glow to cool flames can also occur. The relative positions of the point to which the system jumps on branch III and the Hopf point are sensitive to the total pressure assumed in the model parameters. The pressure also determines whether the system can indeed jump to this branch with negative temperature coefficient. Figure 9.50 shows the thermal diagram for a higher pressure indicating that the middle branch is not accessible and the system jumps from slow reaction to ignition.

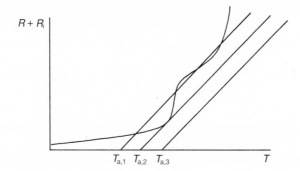

$R + R_i$

$T_{a,1}$ $T_{a,2}$ $T_{a,3}$ T

FIG. 9.50. Typical relative positions of heat release and heat loss lines for higher reactant pressures in the Yang and Gray model such that the system does not jump to the middle branch after the tangency for $T_{a,2}$ but gives rise to ignition. The 'glow' branch might still be accessible for $T_{a,1}$ given a suitable perturbation of the system.

The two types of homoclinicity described above indicate that on decreasing the ambient temperature we can expect cool-flame oscillations to give way either to a return to state I or to ignition, depending on the total pressure and other parameters. Again this is in qualitative agreement with experimental observation.

9.14. Complex oscillations and ignition

The simple Yang and Gray scheme does not account for the ignition process, because reactant consumption is ignored. It also cannot account for complex ignitions: the model as discussed so far has two variables, so only simple period-1 limit cycles can occur. Wang and Mou (1985) have gone some way towards improving the model in both these respects. They have explicitly included fuel consumption (but not that of O_2). This raises the model to a three-variable scheme. Also, they have introduced an additional high-temperature branching step to the mechanism

(GY5) $A + X \rightarrow 2X$ rate $= k_{b2}ax$.

This step involves a direct branching between the fuel and the radical: it is exothermic and has a higher activation energy than the other steps.

The reaction rate and energy-balance equations given by Wang and Mou can be written as

$$\frac{da}{dt} = k_f(a_0 - a) - k_i a - k_{b2}ax \tag{9.12}$$

$$\frac{dx}{dt} = -k_f x + k_i a + \phi x + k_{b2}ax \tag{9.13}$$

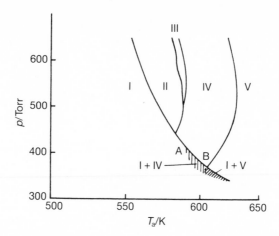

FIG. 9.51. Computed p–T_a diagram with the Wang–Mou modification of the Yang and Gray model. (Reproduced with permission from Wang and Mou (1985), © American Institute of Physics.)

$$C \frac{\mathrm{d}T}{\mathrm{d}t} = q_i k_i a + \Theta X + q_{b2} k_{b2} ax - \chi(T - T_a). \qquad (9.14)$$

(Apparently these equations have no heat transfer term associated with the outflow: if the flow rate is not going to be varied during a given experiment, however, this extra effect can be included by redefining χ.)

With this scheme, a p–T_a ignition diagram of the correct qualitative form can be predicted: Wang and Mou obtained an improvement in the fit with experiment by varying the rate constants slightly from those used by Yang and Gray—and it should be noted that the Yang and Gray set are not derived from known rate data. Most significantly, this model predicts complex oscillatory waveforms, of the general form 1^n, with successive large excursions separated by a number, n, of small, cool flames. Some examples of these computations are shown in Fig. 9.52. These responses exist over a relatively narrow range of parameter values, e.g. 586–590.1 K for a total pressure of 560 mm Hg. The complex ignition region is bounded at low ambient temperature by simply oscillatory ignition and by the cool-flame region at high T_a (Fig. 9.52) as observed experimentally.

A closer examination of the behaviour in the region of complex oscillations revealed a wealth of fine structure (Wang and Gaspard 1990), summarized in Fig. 9.53. If we traverse the region by increasing the ambient temperature, so entering from the region of simple (two-stage) ignition, then the waveform develops through the sequence $1^1, 1^3, 1^5, \ldots$, i.e. 1^n with n odd and increasing. After a region of apparent irregularity, a different sequence with $1^8, 1^6, 1^4$, and 1^2 (n even and decreasing as T_a increases) emerges. Finally, there is a small range of T_a for which the oscillation has a high periodicity $\sim 1^{12}$, found

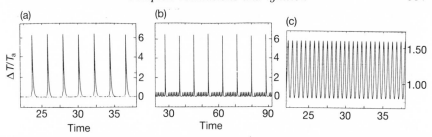

FIG. 9.52. Representative time series for (a) two-stage ignition, (b) nine-stage ignition, and (c) cool flame oxidation in the Wang–Mou scheme for acetaldehyde oxidation, corresponding to ambient temperatures of 585, 589.455, and 620 K. (Reproduced with permission from Wang and Mou (1985), © American Institute of Physics.)

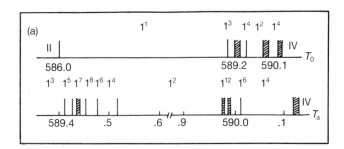

FIG. 9.53. Bifurcation sequence across region III (complex ignitions) from Wang and Mou model with $p = 560$ Torr, with 1^n indicating n small excursions between each two-stage ignition, i.e. an $(n + 2)$-stage ignition process: (a) the range $586.0 \leqslant T_a/K \leqslant 590.1$; (b) details of the range $589.4 \leqslant T_a/K \leqslant 590.1$. (Reproduced with permission from Wang and Mou (1985), © American Institute of Physics.)

close to the boundary with the cool-flame region IV. Wang (1989) also found period-doubling cascades within the cool-flame region for this model.

These latter details have not been confirmed experimentally: Gray *et al.* observed simply that n increases through the sequence 1, 2, 3, 4 as T_a is increased but the computations of Griffiths and Sykes shown earlier do have this feature. Gaspard and Wang (1987) have produced an extraordinarily detailed analysis of the origin and development of the 'mixed-mode' oscillations, i.e. complex ignition, in this model.

The scenario described for a particular pressure (553 Torr) is as follows. For ambient temperatures above 623 K, the stationary state is a stable focus corresponding to steady glow. On decreasing T_a through this value, there is a supercritical Hopf bifurcation yielding the cool-flame limit cycle. The latter is stable until $T_a = 590.242\,73$, when it loses stability at the beginning of a period-doubling cascade that leads to chaotic cool flames. Further changes in the waveform arise. With $T_a < 590.1295$, the mixed-mode waveform of the

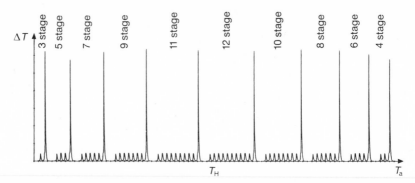

FIG. 9.54. Sequence of complex ignitions for different ambient temperatures on either side of the homoclinicity at $T_H = 589.459\,801\,925$ for the Wang–Mou scheme: three-stage, $T_a = 589.1$ K; five-stage, $T_a = 589.42$ K; seven-stage, $T_a = 589.45$ K; nine-stage, $T_a = 589.4575$ K; 11-stage, $T_a = 589.4595$ K; 12-stage, $T_a = 589.46$ K; 10-stage, $T_a = 589.4613$ K; eight-stage, $T_a = 589.47$ K; six-stage, $T_a = 589.485$ K; four-stage, $T_a = 589.80$ K. See also Table 9.5. (Reproduced with permission from Gaspard and Wang (1987), © Plenum Publishing.)

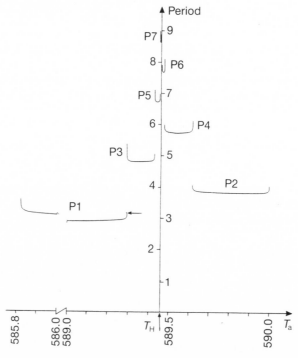

FIG. 9.55. Variation of oscillatory period for the various complex ignition states found on either side of the homoclinicity for the Wang–Mou scheme. (Reproduced with permission from Gaspard and Wang (1987), © Plenum Publishing.)

complex ignitions occurs. The origin of these mixed-mode oscillations is a homoclinic tangency of the inset and outset of the original, and now unstable, cool-flame limit cycle. This tangency occurs at $T_a = 589.459\,801\,925 \pm 15 \times 10^{-9}$ (a remarkable feat of location by Gaspard and Wang). This homoclinicity does not throw off a stable chaotic solution, rather it leads to these complex periodic solutions. The strong contraction of the flow associated with 'chemical' models seems to favour periodicity. Figure 9.54 shows some of the mixed-mode forms for ambient temperatures just above and just below this homoclinicity. The feature mentioned above becomes evident. For the higher T_a the complex ignitions have an even number of small excursions before the two-stage ignition, the number decreasing as we move further from the tangency condition. For lower T_a, there are odd numbers of small peaks. Figure 9.55 shows how the period of a complete repeating unit varies with T_a across its region of existence. Notice that the period lengthens as each bifurcation from one mixed-mode form to the next is approached.

The mixed-mode waveform at lowest T_a has just one small peak, and gives way to simple two-stage ignition for $T_a < 585.8$. Similarly the sequence on increasing T_a leads to a mixed-mode form with two small peaks. At slightly high ambient temperatures, there is then a very complicated sequence of different periodicities as listed in Table 9.5.

The extremely narrow ranges of ambient temperature over which the different responses exist clearly mean that there is virtually no chance of observing these bifurcation sequences experimentally for acetaldehyde oxidation. However, Gaspard and Wang suggest that these are 'typical' scenarios and might well be more amenable in other systems, and the results described above show just how powerful modern computational techniques have become in non-linear science.

9.15. Other hydrocarbon oxidations

Cool-flame oscillations are characteristic of the spontaneous oxidation reactions of a great many hydrocarbon fuels (Lignola and Reverchon 1986, 1987; Westbrook and Dryer 1984). One particularly important effect of this mode of reaction is that of engine knock which occurs during the heating of petrol–air mixtures in the compression stage of the internal combustion cycle. If we denote a general hydrocarbon as RH, we may then hope to prepare a generalized version of Table 9.3.

An initiation process involving $RH + O_2$ can produce alkyl radicals of the form R, and again there is an important equilibrium of the form

$$R + O_2 \rightleftharpoons RO_2.$$

Gas-phase reactions

TABLE 9.5
Number N of small-amplitude oscillations between each two-stage ignition for the Mou–Wang model as a function of ambient temperature in the region of complex ignitions (region III). The broken line indicates the homoclinicity $T_a = 589.459\,801\,925$. (Adapted and reproduced with permission from Gaspard and Wang (1987), © Plenum Publishing.)

Ambient temperature T_0		N
564	−585.8	0 (region II)
585.821 1093	−589.291 9558	1
589.291 9577	−589.428 1664	3
589.428 1674	−589.428 1675	7
589.428 1678	−589.454 4529	5
589.454 4534	−589.458 9649	7
589.458 9657	−589.459 6739	9
589.459 68	−589.459 78	11
589.459 785	−589.459 795	13
589.4598	−589.459 801	15
589.459 8017	−589.459 8018	17
589.459 8016	−589.459 801 91	19
589.459 801 94		20
589.459 802		18
589.459 8022	−589.459 803	16
589.459 807		14
589.459 81		12
589.459 8520	−589.460 1312	10
589.460 1319	−589.462 0057	8
589.462 0066	−589.475 4037	6
589.475 4054	−589.618 178	4
589.618 2911	−589.996 5818	2
589.996 85		22
589.997		12
589.997 49		10
589.997 5		12
589.997 5025	−589.997 504	14
589.997 5045	−589.997 5048	16
589.997 5049		20
589.997 505		18
589.997 505 15		20
589.997 505 25		22
589.997 505 28		24
589.997 5053	−589.997 505 31	26

(*continued on next page*)

TABLE 9.5 (*continued*)

589.997 505 33	37
589.997 505 34	29
589.997 505 35	27
589.997 505 355 –589.997 505 365	25
589.997 505 37 –589.997 505 39	23
589.997 5054	21
589.997 505 45 –589.997 5055	19
589.997 508 –589.9977	8
589.999 –590	6
590.01 –590.1226	4
590.122 7	21
590.123	46
590.125	12
590.126	14
590.14	∞ (region **IV**)

A degenerate branching sequence can then be proposed on similar principles to reactions (11) and (12) above, i.e.

$$RO_2 + RH \rightarrow RO_2H + R$$

$$RO_2H \rightarrow RO + OH.$$

In general, however, this is not a satisfactory explanation (Griffiths and Scott 1987). With acetaldehyde, the –CHO group presents a particularly labile hydrogen atom. The energy barrier for the intermolecular H abstraction in step (11) is relatively low, about 44 kJ mol^{-1}. For other hydrocarbons, however, the relevant bond may be a primary, secondary, or tertiary C–H: the H abstraction then has a significantly higher activation energy, $E > 60$ kJ mol^{-1}, and so is too slow to account for observed cool-flame behaviour.

As a more general route, intramolecular H atom abstraction appears to offer branching potential. In particular, abstraction from a β C–H bond permits the reaction to proceed through a favoured six-atom ring. Thus an alkyl radical may go through a sequence of the form

$$-CH-CH_2-CH_2- + O_2 \rightarrow \overset{\displaystyle O-O}{\underset{|}{-CH}}-CH_2-CH_2-$$

$$\overset{\displaystyle O-O \quad\; H}{\underset{|\qquad\;\; |}{-CH-CH_2-CH}}- \rightarrow \overset{\displaystyle O_2H}{\underset{|}{-CH}}-CH_2-CH-$$

$$\begin{array}{c}\overset{\displaystyle O_2H}{\underset{\displaystyle |}{}} \\ -CH-CH_2-CH- \; + \; O_2 \end{array} \quad \rightarrow \quad \begin{array}{c}\overset{\displaystyle O_2H}{\underset{\displaystyle |}{}} \\ -CH-CH_2-CH- \\ | \\ O_2 \end{array}$$

$$\begin{array}{c}O_2H \\ | \\ -C-CH_2-CH- \\ | \qquad | \\ H \quad\; O-O \end{array} \quad \rightarrow \quad \begin{array}{c}O_2H \\ | \\ -C-CH_2-CH- \\ | \\ O_2H \end{array}$$

$$\begin{array}{c}O_2H \\ | \\ -C-CH_2-CH- \\ | \\ O_2H \end{array} \quad \rightarrow \quad -CO + CH_2\text{-CHCH} + 2OH.$$

This sequence of O_2 additions and intramolecular H atom abstractions finally leads to fragmentation of the fuel into smaller molecular species as well as providing degenerate branching. The detailed structure of the original hydrocarbon is also important as the rates of different steps will depend crucially on whether the particular H atom being abstracted is primary, secondary, or tertiary: thus there will be different kinetic implications for branched- or straight-chain hydrocarbons.

The alternative non-branching fate for the R radical is the formation of the conjugate alkene, P say, via

$$R + O_2 \; \rightarrow \; P + HO_2.$$

9.16. Miscellaneous combustion systems

A number of other interesting systems may be mentioned briefly here. Soltzberg *et al.* (1987) have described a 'flashback' oscillator comprising a gas burner (e.g. a glass-blowing torch) with a pilot wire (platinum) suspended 5 cm above the burner rim (Fig. 9.56). The heating of this pilot wire allows the flame to relight spontaneously after extinguishing through a flashback. By varying the flow rate and the fuel:oxygen ratio various patterns from simple period-1 to bursting waveform can be obtained, as well as the possibility of coexisting oscillatory and stable flame states (Fig. 9.57). Mostly these phenomena require a fuel-rich mixture. Whilst the full mechanism of this oscillatory process is not established, some gross features have been modelled (Soltzberg *et al.* 1990) using a piecewise two-variable model that allows for the rather unusual dependence of the gas ignition temperature on the composition, which has a general cuspoidal shape (Fig. 9.58). With a

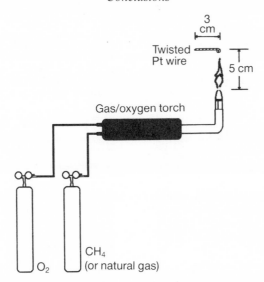

Fig. 9.56. Schematic representation of the apparatus for the 'flashback oscillator' comprising a 3 cm Pt pilot wire above a methane–O_2 flame. (Reproduced with permission from Soltzberg *et al.* (1987), © American Chemical Society.)

slightly larger model, incorporating the pilot wire temperature and the two-reactant concentration, the basic trend towards increasing simplicity as the mixture gets leaner can be reproduced. Some computed temperature time series are shown in Fig. 9.59.

Combustion waves also occur in solid-phase systems, in particular the so-called 'gasless' pyrotechnics (e.g. $Mg + K_2Cr_2O_7$ mixtures). Typically these powders will be mixed and then laid as a 'fuse' that is ignited at one end. The mixture can be judged such that there is exactly sufficient oxygen in the solid components as required by the stoichiometry of the reaction. In theory, then, no gas-phase oxygen need participate. The progress of the combustion front along the fuse can be monitored in various ways, typically by embedding fine-wire thermocouples in the powder.

Whilst the achievement of a steady burning velocity is not unusual, more complex responses are also known. So-called 'spinning waves' have been observed, and more recently a model has been proposed to account for the suggestions that the burning velocity may oscillate about some mean as the wave progresses—and even become aperiodic (Bayliss and Matkowsky 1987, 1990; Bayliss *et al.* 1989).

9.17. Conclusions

A number of features may have become clear in the previous discussions. Gas-phase reactions can support a great range of non-linear behaviour and

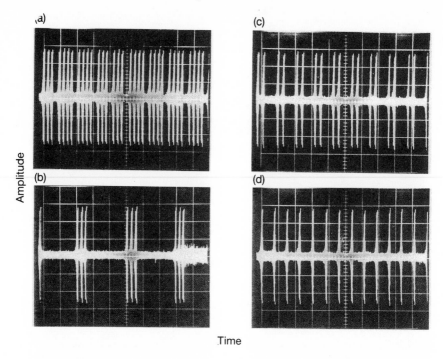

FIG. 9.57. Oscilloscope traces from sound recordings from the flashback oscillator showing periodic and complex bursting patterns for different experimental conditions: (a) natural gas flow 1485 ml min^{-1}, O_2 1350 ml min^{-1}; (b) 1150 and 570 ml min^{-1}; (c) 1300 and 570 ml min^{-1}; (d) 1485 and 740 ml min^{-1}. (Reproduced with permission from Soltzberg *et al.* (1987), © American Chemical Society.)

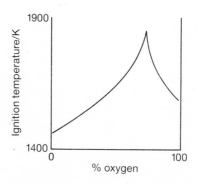

FIG. 9.58. Schematic representation of a typical dependence on gas composition of ignition temperature for CH_4–O_2 mixtures that provides the non-linearity for the flashback oscillator. (Courtesy of L. Soltzberg, 1990.)

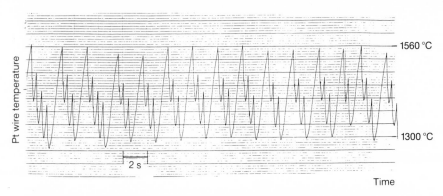

Fig. 9.59. Computed complex oscillation for the flashback oscillator based on the dependence in Fig. 9.58. (Courtesy of L. Soltzberg, 1990.)

often in reactions that also have great technical importance. They have been, however, less widely exploited than solution-phase systems. Partly, the reduced pressures and higher ambient temperatures at which reaction occurs cause more operational problems (at least at first sight) than the simple pumping of liquids at room temperature. The current chapter has reviewed in detail some of the 'simpler' responses—multistability and oscillations rather than concentrating on complex oscillations and chaos. The latter seem to exist, but are awaiting deeper study which they surely will reward.

References

Aleksandrov, E. N. and Azatyan, V. (1977). Intermittent combustion of carbon monoxide under static conditions. *Combust. Explos. Shock Waves*, **12**, 407–11.

Ashmore, P. G. and Norrish, R. G. W. (1951). A phenomenon of successive ignitions. *Nature*, **167**, 390–3.

Babushok, V. I., Novikov, E. A., and Babkin, V. S. (1982). Oscillatory regime of the oxidation of carbon monoxide in the gas phase. *Dokl. Akad. Nauk SSSR*, **271**, 878–81.

Baldwin, R. R. and Walker, R. W. (1972). Branching-chain reactions: the hydrogen–oxygen reaction. In *Essays in chemistry*, (ed. J. A. Barnard, R. D. Gillard, and R. F. Hudson), Vol. 3, pp. 1–37. Academic Press, London.

Baldwin, R. R., Jackson, D., Walker, R. W., and Webster, S. J. (1965). Use of the $H_2 + O_2$ reaction in evaluating velocity constants. *10th Int. Symp. on Combustion*, pp. 423–33. The Combustion Institute, Pittsburgh, PA.

Baldwin, R. R., Jackson, D., Walker, R. W., and Webster, S. J. (1967a). Interpretation of induction periods in the hydrogen + oxygen reaction in aged boric-acid-coated vessels. *Trans. Faraday Soc.*, **63**, 1665–75.

Baldwin, R. R., Jackson, D., Walker, R. W., and Webster, S. J. (1967b). Interpretation of the slow reaction and second limit of hydrogen + oxygen mixtures by computer methods. *Trans. Faraday Soc.*, **63**, 1676–86.

Baldwin, R. R., Fuller, M. E., Hillman, J. S., Jackson, D., and Walker, R. W. (1974). Second limit of hydrogen + oxygen mixtures: the reaction $H + HO_2$. *J. Chem. Soc. Faraday Trans. 1*, **70**, 635–41.

Barnard, J. A. and Bradley, J. N. (1985). *Flame and combustion*. Chapman and Hall, London.

Baulch, D. L., Drysdale, D. D., Horne, D. G., and Lloyd, A. C. (1972a). *Evaluated kinetic data for high temperature reactions*, Vol. 1. Butterworths, London.

Baulch, D. L., Drysdale, D. D., and Horne, D. G. (1972b). *Evaluated kinetic data for high temperature reactions*, Vol. 2. Butterworths, London.

Baulch, D. L., Drysdale, D. D., Duxbury, J., and Grant, S. (1976). *Evaluated kinetic data for high temperature reactions*, Vol. 3. Butterworths, London.

Baulch, D. L., Griffiths, J. F., Pappin, A. J., and Sykes, A. F. (1988a). Stationary-state and oscillatory combustion of hydrogen in a well-stirred flow reactor. *Combust. Flame*, **73**, 163–85.

Baulch, D. L., Griffiths, J. F., Pappin, A. J., and Sykes, A. F. (1988b). Third-body interactions in the oscillatory oxidation of hydrogen in a well stirred flow reactor. *J. Chem. Soc. Faraday Trans. 1*, **84**, 1575–86.

Bayliss, A. and Matkowsky, B. J. (1987). Fronts, relaxation oscillations and period doubling in solid fuel combustion. *J. Comput. Phys.*, **71**, 147–68.

Bayliss, A. and Matkowsky, B. J. (1990). Two routes to chaos in condensed phase combustion. *SIAM J. Appl. Math.*, **50**, 437–59.

Bayliss, A., Matkowsky, B. J., and Minkoff, M. (1989). Period doubling gained, period doubling lost. *SIAM J. Appl. Math.*, **49**, 1047–63.

Benson, S. W. (1960). *Foundations of chemical kinetics*. McGraw-Hill, New York.

Boddington, T., Gray, P., and Wake, G. C. (1977). Criteria for thermal explosions with and without reactant consumption. *Proc. R. Soc.*, **A357**, 403–22.

Boddington, T., Gray, P., and Scott, S. K. (1982a). Temperature distributions, critical conditions and scaling for exothermic materials under different boundary conditions. Part 1. *J. Chem. Soc. Faraday Trans. 2*, **77**, 801–12.

Boddington, T., Gray, P., and Scott, S. K. (1982b). Temperature distributions, critical conditions and scaling for exothermic materials under different boundary conditions. Part 2. *J. Chem. Soc. Faraday Trans. 2*, **77**, 813–23.

Bodenstein, M. and Ohlmer, F. (1905). Heterogene katalytische Reaktionen III: Katalyse des Kohlenoxydnallgases durch Kieselsäure. *Z. Phys. Chem.*, **53**, 166–76.

Bond, J. R., Gray, P., and Griffiths, J. F. (1981). Oscillations, glow and ignition in carbon monoxide oxidation. I. Glow and ignition in a closed reaction vessel and the effects of added hydrogen. *Proc. R. Soc.*, **A375**, 43–64.

Bond, J. R., Gray, P., Griffiths, J. F., and Scott, S. K. (1982). Oscillations, glow and ignition in carbon monoxide oxidation. II. Oscillations in the gas-phase reaction in a closed system. *Proc. R. Soc.*, **A381**, 293–314.

Brindley, J., Kaas-Petersen, C., Merkin, J. H., and Scott, S. K. (1988). A simple model for sustained oscillations in isothermal branched-chain or autocatalytic reactions in well-stirred, open systems. III. Multiple stationary-states and Hopf bifurcations. *Proc. R. Soc.*, **A417**, 463–96.

Brokaw, R. S. (1966). Ignition kinetics of the carbon monoxide–oxygen reaction. *11th Int. Symp. on Combustion*, pp. 1063–73. The Combustion Institute, Pittsburgh, PA.

Buckler, E. J. and Norrish, R. G. W. (1938). A study of sensitized explosion. *Proc. R. Soc.*, **A167**, 292–342.

Caprio, V., Insola, A., and Lignola, P.-G. (1977). Isobutane cool flames investigation in a continuous stirred tank reactor. *16th Int. Symp. on Combustion*, pp. 1155–63. The Combustion Institute, Pittsburgh, PA.

Chinnick, K. and Griffiths, J. F. (1986). Inhibitory features of the thermal oxidation of carbon monoxide. A kinetic foundation to dynamic instabilities in closed vessels. *J. Chem. Soc. Faraday Trans. 2*, **82**, 881–8.

Chinnick, K., Gibson, C., Griffiths, J. F., and Kordylewski, W. (1986*a*). Isothermal interpretations of oscillatory ignition during hydrogen oxidation in an open system. I. Analytical predictions and experimental measurements of periodicity. *Proc. R. Soc.*, **A405**, 117–28.

Chinnick, K., Gibson, C., and Griffiths, J. F. (1986*b*). Isothermal interpretations of oscillatory ignition during hydrogen oxidation in an open system. II. Numerical analysis. *Proc. R. Soc.*, **A405**, 129–42.

Coppersthwaite, D. P. (1990). *B.Sc. thesis*. University of Leeds.

Cosslet, V. E. and Garner, W. E. (1930). The critical pressures of ignition of dry and wet mixtures of carbon monoxide and oxygen. *Trans. Faraday Soc.*, **26**, 190–5.

Déchaux, J.-C. and Lucquin, M. (1968). Étude de la consommation des réactifs initiaux dans l'oxydation lente du butane. *J. Chim. Phys.*, **65**, 982–91.

Déchaux, J.-C. and Lucquin, M. (1971). Inhibition by nitrogen dioxide of the slow oxidation of butane at low temperatures. *13th Int. Symp. on Combustion*, pp. 205–16. The Combustion Institute, Pittsburgh, PA.

Déchaux, J.-C., Langrand, F., Hermant, G., and Lucquin, M. (1968). Le pic d'arrêt de basse température et la réactivité des mélanges hydrocarbure–oxygène au moment où disparaît l'oxygène. *Bull. Soc. Chim.*, No. 633, 4031–8.

Dickens, P. G. (1956). *B.A. thesis*. University of Oxford.

Dickens, P. G., Dove, J. E., and Linnett, J. W. (1964). Explosion limits of the dry carbon monoxide + oxygen reaction. *Trans. Faraday Soc.*, **60**, 539–52.

Dixon-Lewis, G. (1984). Computer modelling of combustion reactions in flowing systems with transport. In *Combustion chemisty*, (ed. W. C. Gardiner), pp. 21–126. Springer, New York.

Dixon-Lewis, G. and Linnett, J. W. (1953). The oxidation of mixtures of hydrogen and carbon monoxide. *Trans. Faraday Soc.*, **49**, 756–65.

Dixon-Lewis, G. and Williams, D. J. (1977). The oxidation of hydrogen and carbon monoxide. In *Comprehensive chemical kinetics*, (ed. C. H. Bamford and C. F. H. Tipper), Vo. 17, pp. 1–248. Elsevier, Amsterdam.

Dougherty, E. P. and Rabitz, H. (1980). Computational kinetics and sensitivity analysis of hydrogen–oxygen combustion. *J. Chem. Phys.*, **72**, 6571–86.

Dove, J. E. (1956). *Ph.D. thesis*. University of Oxford.

Egieban, O. M., Griffiths, J. F., Mullins, J. R., and Scott, S. K. (1982). Explosion hazards in exothermic materials: critical conditions and scaling rules for masses of different geometry. *19th Int. Symp. on Combustion*, pp. 825–33. The Combustion Institute, Pittsburgh, PA.

Felton, P. G. and Gray, B. F. (1974). Low temperature oxidation in a stirred flow reactor. 1. Propane. *Combust. Flame*, **23**, 295–304.

Felton, P. G., Gray, B. F., and Shank, N. (1976). Low temperature oxidation in a stirred flow reactor, part 2, acetaldehyde (theory). *Combust. Flame*, **27**, 363–76.

Foo, K. K. and Yang, C. H. (1971). On the surface and thermal effects of hydrogen oxidation. *Combust. Flame*, **17**, 223–35.

Gaspard, P. and Wang, X.-J. (1987). Homoclinic orbits and mixed-mode oscillations in far-from-equilibrium systems. *J. Stat. Phys.*, **48**, 151–99.

Gibson, C., Gray, P., Griffiths, J. F., and Hasko, S. M. (1984). Spontaneous ignition of hydrocarbon and related fuels: a fundamental study of thermokinetic interactions. *20th Int. Symp. on Combustion*, pp. 101–9. The Combustion Institute, Pittsburgh, PA.

Gordon, A. S. (1952). The explosive reaction of carbon monoxide and oxygen at the second explosion limit in quartz vessels. *J. Chem. Phys.*, **20**, 340–1.

Gordon, A. S. and Knipe, R. J. (1955). The explosive reaction of carbon monoxide and oxygen at the second explosion limit in quartz vessels. *J. Phys. Chem.*, **59**, 1160–5.

Gray, B. F. (1969). Unified theory of explosions, cool flames and two-stage ignitions, part 1. *Trans. Faraday Soc.*, **65**, 1603–13.

Gray, B. F. (1970). Theory of branching reactions with chain interaction. *Trans. Faraday Soc.*, **66**, 1118–26.

Gray, B. F. and Jones, J. C. (1984). The heat release rates and cool flames of acetaldehyde oxidation in a continuously stirred tank reactor. *Combust. Flame*, **57**, 3–14.

Gray, P. and Griffiths, J. F. (1989). Thermokinetic combustion oscillations as an alternative to thermal explosion. *Combust. Flame*, **78**, 87–98.

Gray, P. and Scott, S. K. (1985). Isothermal oscillations and relaxation ignitions in gas-phase reactions: the oxidations of carbon monoxide and hydrogen. In *Oscillations and traveling waves in chemical systems*, (ed. R. J. Field and M. J. Burger), Ch. 14, pp. 493–528. Wiley, New York.

Gray, P. and Scott, S. K. (1990). *Chemical oscillations and instabilities: non-linear chemical kinetics*. Oxford University Press.

Gray, P., Griffiths, J. F., Hasko, S. M., and Lignola, P.-G. (1981*a*). Oscillatory ignitions and cool flames accompanying the non-isothermal oxidation of acetaldehyde in a well stirred flow reactor. *Proc. R. Soc.*, **A374**, 313–39.

Gray, P., Griffiths, J. F., Hasko, S. M., and Lignola, P.-G. (1981*b*). Novel, multiple-stage ignitions in the spontaneous combustion of acetaldehyde. *Combust. Flame*, **43**, 175–86.

Gray, P., Griffiths, J. F., and Scott, S. K. (1984*a*). Branched-chain reactions in open systems: theory of the oscillatory ignition limit for the hydrogen + oxygen reaction in a continuous-flow stirred-tank reactor. *Proc. R. Soc.*, **394**, 243–58.

Gray, P., Griffiths, J. F., and Scott, S. K. (1984*b*). Oscillatory ignition of the $H_2 + O_2$ and $CO + H_2 + O_2$ reactions in open systems. *20th Int. Symp. on Combustion*, pp. 1809–15. The Combustion Institute, Pittsburgh, PA.

Gray, P., Griffiths, J. F., Hasko, S. M., and Mullins, J. R. (1984*c*). Exotic behaviour in the continuous flow, stirred-tank reactor (CSTR): experimental studies of oscillations and of isola and mushroom patterns in stationary states under gaseous conditions. *8th Int. Symp. on Chemical reaction engineering*, pp. 101–8. Institute of Chemical Engineers, London.

Gray, P., Griffiths, J. F., and Hasko, S. M. (1984*d*). Ignitions, extinctions and thermokinetic oscillations accompanying the oxidation of ethane in an open system (continuously stirred tank reactor). *Proc. R. Soc.*, **A396**, 227–55.

Gray, P., Griffiths, J. F., and Scott, S. K. (1985*a*). Oscillations, glow and ignition in carbon monoxide oxidation in an open system. I. Experimental studies of the ignition diagram and the effects of added hydrogen. *Proc. R. Soc.*, **A397**, 21–44.

Gray, P., Griffiths, J. F., and Scott, S. K. (1985*b*). Oscillations, glow and ignition in carbon monoxide oxidation in an open system. II. Theory of the oscillatory ignition limit in the c.s.t.r. *Proc. R. Soc.*, **A402**, 187–204.

Gray, P., Griffiths, J. F., Pappin, A. J., and Scott, S. K. (1987*a*). The interpretation of oscillatory ignition during oxidation in an open system. In *Complex chemical reaction systems*, (ed. J. Warnatz and W. Jäger), pp. 150–9. Springer, Berlin.

Gray, P., Griffiths, J. F., and Scott, S. K. (1987*b*). Surface effects in the gas-phase oxidation of carbon monoxide. *J. Chim. Phys.*, **84**, 49–53.

Griffiths, J. F. (1985*a*). Thermokinetic oscillations in homogeneous gas-phase oxidations. In *Oscillations and traveling waves in chemical systems*, (ed. R. J. Field and M. J. Burger), Ch. 15, pp. 529–64. Wiley, New York.

Griffiths, J. F. (1985*b*). Thermokinetic interactions in simple gaseous reactions. *Ann. Rev. Phys. Chem.*, **37**, 77–104.

Griffiths, J. F. (1986). The fundamentals of spontaneous ignition of gaseous hydrocarbons and related organic compounds. *Adv. Chem. Phys.*, **64**, 203–303.

Griffiths, J. F. and Mullins, J. R. (1984). Ignition, sealf-heating and the effects of added gases during the thermal decomposition of di-t-butyl peroxide. *Combust. Flame*, **56**, 135–48.

Griffiths, J. F. and Phillips, C. H. (1989). Quadratic autocatalysis and self-heating in hydrocarbon oxidation. *J. Chem. Soc. Faraday Trans. 1*, **85**, 3471–9.

Griffiths, J. F. and Scott, S. K. (1987). Thermokinetic interactions: fundamentals of spontaneous ignition and cool flames. *Prog. Energy Combust. Sci.*, **13**, 161–97.

Griffiths, J. F. and Singh, H. J. (1982). Effects of self-heating during the thermal decomposition of di-t-butyl peroxide. *J. Chem. Soc. Faraday Trans. 1*, **78**, 747–60.

Griffiths, J. F. and Sykes, A. F. (1989*a*). Numerical studies of a thermokinetic model for oscillatory cool flame and complex ignition phenomena in ethanal oxidation under well-stirred flowing conditions. *Proc. R. Soc.*, **A422**, 289–310.

Griffiths, J. F. and Sykes, A. F. (1989*b*). Numerical interpretation of oscillatory glow and ignition during carbon monoxide oxidation in a well-stirred flow reactor. *J. Chem. Soc. Faraday Trans. 1*, **85**, 3059–69.

Griffiths, J. F., Scott, S. K., and Vandamme, R. (1981). Self-heating in the $H_2 + O_2$ reaction in the vicinity of the second explosion limit. *J. Chem. Soc. Faraday Trans. 1*, **77**, 2265–70.

Griffiths, J. F., Hasko, S. M., Shaw, N. K., and Torrez-Mujica, T. (1985). A thermokinetic foundation for oscillatory phenomena in gaseous organic oxidations under well stirred flowing conditions. *J. Chem. Soc. Faraday Trans. 1*, **81**, 343–54.

Griffiths, J. F., Kay, S. R., and Scott, S. K. (1988). Oscillatory combustion in closed vessels: theoretical foundations and their experimental verification. *22nd Int. Symp. on Combustion*, pp. 1597–607. The Combustion Institute, Pittsburgh, PA.

Griffiths, J. F., Kordylewski, W., Pappin, A. J., and Sykes, A. F. (1990*a*). Slowing down of ignition and complex oscillatory phenomena in hydrogen oxidation. In *Spatial inhomogeneities and transient behaviour in chemical kinetics*, (ed. P. Gray, G. Nicolis, F. Baras, P. Borckmans, and S. K. Scott), pp. 237–53. Manchester University Press.

Griffiths, J. F., Johnson, B. R., and Scott, S. K. (1990*b*). Characterisation of oscillations in the $H_2 + O_2$ reaction in a continuous flow reactor. *J. Chem. Soc. Faraday Trans.*, in press.

Hadman, G., Thompson, H. W., and Hinshelwood, C. N. (1932). The oxidation of carbon monoxide. *Proc. R. Soc.*, **A138**, 297–311.

Halstead, M. P. Prothero, A., and Quinn, C. P. (1971). A mathematical tool of the cool flame oxidation of acetaldehyde. *Proc. R. Soc.*, **A322**, 377–403.

Halstead, M. P., Prothero, A., and Quinn, C. P. (1973). Modelling the ignition and cool flame limits of acetaldehyde oxidation. *Combust. Flame*, **20**, 211–21.

Halstead, M. P., Kirsch, L. J., Prothero, A., and Quinn, C. P. (1975). A mathematical model for hydrocarbon auto-ignition at high pressures. *Proc. R. Soc.*, **A346**, 515–38.

Harding, R. H. and Ross, J. (1988). Symptoms of chaos in observed oscillations near a bifurcation with noise. *J. Chem. Phys.*, **89**, 4743–51.

Harding, R. H., Sevcikova, H., and Ross, J. (1988). Complex oscillations in the combustion of acetaldehyde. *J. Chem. Phys.*, **89**, 4737–42.

Harrison, A. J. and Cairnie, L. R. (1988). The development and experimental validation of a mathematical model for predicting hot-surface autoignition hazards using complex chemistry. *Combust. Flame*, **71**, 1–21.

Harrison, A. J., Furzeland, R. M., Summers, R., and Cairnie, L. R. (1988). An experimental and theoretical study of autoignition on a horizontal hot pipe. *Combust. Flame*, **72**, 119–29.

Heslop, W. R. (1954). *D.Phil. dissertation*. University of Oxford.

Hjelmfelt, A., Harding, R. H., Tsujimoto, K. K., and Ross, J. (1990). Theory and experiments on the effects of perturbations on nonlinear chemical systems: generation of multiple attractors and efficiency. *J. Chem. Phys.*, **92**, 3559–68.

Hoare, D. E. and Walsh, A. D. (1954). The oxidation of carbon monoxide. *Trans. Faraday Soc.*, **50**, 37–50.

Johnson, B. R. and Scott, S. K. (1990). Period doubling and chaos during the oscillatory ignition of the $CO + O_2$ reaction. *J. Chem. Soc. Faraday Trans.*, **86**, 3701–5.

Jones, J. C. and Gray, B. F. (1983). Inhibition of acetaldehyde cool flames. *Combust. Flame*, **52**, 211–13.

Kaas-Petersen, C. and Scott, S. K. (1988). Homoclinic orbits in a simple chemical model of an autocatalytic reaction. *Physica D*, **32**, 461–70.

Kay, S. R., Scott, S. K., and Tomlin, A. S. (1989). Quadratic autocatalysis in a non-isothermal CSTR. *Chem. Eng. Sci.*, **44**, 1129–37.

Kirsch, L. J. and Quinn, C. P. (1976). A fundamentally based model of knock in the gasoline engine. *16th Int. Symp. on Combustion*, pp. 233–44. The Combustion Institute, Pittsburgh, PA.

Kopp, D., Kowalskii, A., Sagulin, A. B., and Semenov, N. N. (1930). Entzundungsgrenze des Gemisches $2H_2 + O_2$ und $2CO + O_2$. *Z. Phys. Chem., part B*, **6**, 307.

Kordylewski, W. and Scott, S. K. (1984). The influence of self-heating on the second and third explosion limits in the $O_2 + H_2$ reaction. *Combust. Flame*, **57**, 127–39.

Kühl, H. (1903). Beiträge zur Kinetik des Kohlenoxydknall-Gases. *Z. Phys. Chem.*, **44**, 385–459.

Lewis, B. and von Elbe, G. (1987). *Combustion, flames and explosions of gases*. Academic Press, Orlando, FL.

Lignola, P.-G. and Reverchon, E. (1986). Dynamics of n-heptane and i-octane combustion processes in a jet-stirred flow reaction operated under pressure. *Combust. Flame*, **2**, 177–83.

Lignola, P.-G. and Reverchon, E. (1987). Cool flames. *Prog. Energy Combust. Sci.*, **13**, 75–96.

Linnett, J. W., Reuben, B. G., and Wheatley, T. F. (1968). A photoelectric study of the low pressure explosion limit of the wet $CO + O_2$ reaction. *Combust. Flame*, **12**, 325–32.

McCaffery, B. J. and Berlad, A. L. (1976). Some observations on the oscillatory behaviour of carbon monoxide oxidation. *Combust. Flame*, **26**, 77–83.

Merkin, J. H., Needham, D. J., and Scott, S. K. (1985a). A simple model for sustained oscillations in isothermal branched-chain or autocatalytic reactions in well-stirred, open systems. I. Stationary-states and local stabilities. *Proc. R. Soc.*, **A398**, 81–100.

Merkin, J. H., Needham, D. J., and Scott, S. K. (1985b). A simple model for sustained oscillations in isothermal branched-chain or autocatalytic reactions in well-stirred, open systems. II. Limit cycles and non-stationary states. *Proc. R. Soc.*, **A398**, 101–16.

Mulcahy, M. (1973). *Gas kinetics*. Nelson, London.

Nalbandyan, A. B., Arustamyan, A. M., Shakhnazaryan, I. K., and Philipossyan, A. G. (1980). The kinetics and mechanism of the oxidation of carbon monoxide in the presence of hydrogen. *Int. J. Chem. Kinet.*, **12**, 55–75.

Prettre, M. and Laffitte, P. (1929). Sur l'oxydation de l'oxyde de carbon. *C. R. Hebd. Séances Acad. Sci.*, **189**, 177–9.

Pugh, S. A., Schell, M., and Ross, J. (1986a). Effects of periodic perturbations on the oscillatory combustion of acetaldehyde. *J. Chem. Phys.*, **85**, 868–78.

Pugh, S. A., DeKock, B., and Ross, J. (1986b). Effects of two periodic perturbations on the oscillatory combustion of acetaldehyde. *J. Chem. Phys.*, **85**, 879–86.

Pugh, S. A., Kim, H.-R., and Ross, J. (1987). Measurements of [OH] and [CH_3CHO] oscillations and phase relations in the combustion of CH_3CHO. *J. Chem. Phys.*, **86**, 776–83.

Sagulin, A. B. (1928). Explosion temperatures of gaseous mixtures at different pressures. *Z. Phys. Chem.*, **81**, 275.

Sahetchian, K., Jorand, F., Chamboux, J., and Viossat, V. (1986). Pulsations de pression observées au voisinage de la second limite d'inflammation de l'hydrogène en système clos. *J. Chim. Phys.*, **83**, 685.

Scott, S. K. (1985). Sustained oscillations in simple autocatalytic and branched-chain reactions. *J. Chem. Soc. Faraday Trans. 2*, **81**, 789–801.

Semenov, N. N. (1929). Kinetics of chain reactions. *Chem. Rev.*, **6**, 347–79.

Semenov, N. N. (1958). *Some problems in chemical kinetics and reactivity*. Pergamon, London.

Sheintuch, M. and Luss, D. (1988). Use of observed transitions for classification of dynamics systems—application to cool flames. *Combust. Flame*, **71**, 267–81.

Snee, T. J. and Griffiths, J. F. (1989). Criteria for spontaneous ignition in exothermic, autocatalytic reactions: chain branching and self-heating in the oxidation of cyclohexane in closed vessels. *Combust. Flame*, **75**, 381–95.

Soltzberg, L. J., Boucher, M. M., Crane, D. M., and Pazar, S. S. (1987). Far from equilibrium—the flashback oscillator. *J. Chem. Educ.*, **64**, 1043–6.

Soltzberg, L. J., Griffiths, J. F., and Scott, S. K. (1990). In preparation.

Tomlin, A. S. (1990). *Ph.D. thesis*. University of Leeds.

Topley, B. (1930). The homogeneous isothermal reaction $2CO + O_2 = 2CO_2$ in the presence of water vapour. *Nature*, **125**, 560–1.

von Elbe, G. and Lewis, B. (1986). Free radical reactions in glow and explosion of carbon monoxide–oxygen mixtures. *Combust. Flame*, **63**, 135–50.

von Elbe, G., Lewis, B., and Roth, W. E. (1955). The problem of the second explosion limit in the carbon monoxide–oxygen system. *5th Int. Symp. on Combustion*, pp. 610–16. Reinhold, New York.

Wang, X.-J. (1989). Chaotic oscillations of cool flames. *Combust. Flame*, **75**, 107–9.

Wang, X.-J. and Gaspard, P. (1990). Homoclinicity and multimodal periodic or chaotic oscillations in chemical kinetics. In *Spatial inhomogeneities and transient behaviour in chemical kinetics*, (ed. P. Gray, G. Nicolis, F. Baras, P. Borckmans, and S. K. Scott), pp. 687–90. Manchester University Press.

Wang, X.-J. and Mou, C. Y. (1985). A thermokinetic model of complex oscillations in gaseous hydrocarbon oxidation. *J. Chem. Phys.*, **83**, 4554–61.

Warnatz, J. (1984). Rate coefficients in the C/H/O system. In *Combustion chemistry*, (ed. W. C. Gardiner), pp. 196–360. Springer, New York.

Warren, D. R. (1957). Surface effects in combustion reactions. Part 1. Effects of wall coating on the $H_2 + O_2$ reaction. *Trans. Faraday Soc.*, **53**, 199–205.

Westbrook, C. K. and Dryer, F. L. (1984). Chemical kinetic modelling of hydrocarbon combustion. *Prog. Energy Combust. Sci.*, **10**, 1–57.

Yang, C. H. (1969). Two-stage ignition and self-excited thermokinetic oscillation in hydrocarbon oxidation. *J. Phys. Chem.*, **73**, 3407–13.

Yang, C. H. (1974a). On the explosion, glow and oscillation phenomena in the oxidation of carbon monoxide. *Combust. Flame*, **23**, 97–108.

Yang, C. H. (1974b). Oscillatory and explosive oxidation of carbon monoxide. *Faraday Symp. Chem. Soc.*, **9**, 114–28.

Yang, C. H. and Berlad, A. L. (1974). Kinetics and kinetic oscillations in carbon monoxide oxidation. *J. Chem. Soc. Faraday Trans. 1*, **70**, 1661–75.

Yang, C. H. and Gray, B. F. (1969a). On the slow oxidation of hydrocarbon and cool flames. *J. Phys. Chem.*, **73**, 3395–406.

Yang, C. H. and Gray, B. F. (1969b). Unified theory of explosions, cool flames and two-stage ignitions, part 2. *Trans. Faraday Soc.*, **65**, 1614–22.

10

Heterogeneous catalysis

Heterogeneous catalytic systems have seemed a natural and fertile field for non-linear behaviour and only solution-phase reactions have been more widely studied. The oxidation of carbon monoxide over many metal surfaces in particular seems to have been the subject of deep investigations perhaps because of its technological as well as fundamental significance. Heterogeneous catalysis is widely applied in the chemical industry as a whole—the petrochemical industry in particular—and so there has been both motivation and support for any increase in understanding and control of such reactions. It might also be added that many theoreticians have also taken advantage of the 'grey areas' of uncertainty in the form of appropriate reaction rate laws, using putative surface reactions (and their enzyme equivalents) as an excuse for almost any otherwise unbelievable but convenient polynomial or transcendental form.

As before, we cannot hope to understand the origins of the most complex oscillations and chaos without first paying some attention to bistability and simple periodic solutions. Some of these latter aspects have been thoroughly reviewed, e.g. by Sheintuch and Schmitz (1977), Slin'ko and Slin'ko (1978), Engel and Ertl (1979), Mukesh *et al.* (1983*b*), and Razon and Schmitz (1987). Table 10.1 collects together various studies loosely into 'academic groups'; the subject of many of these papers becomes clear from the titles listed at the end of this chapter. The reactions are typically oxidation processes occurring on metal catalysts in various forms—particularly simple fuels over noble metals such as Pd, Pt, or Rh which may be wires, foils, gauzes, single crystals, or suppoted on 'inert' materials such as α- or γ-alumina or silica. Notice that the earliest reports of oscillations date from 1970, contemporary with the emergence of oscillatory solution-phase reactions.

The earliest, general, modelling studies are perhaps those of Frank-Kamenetskii and Buben (1946) who investigated multistability questions akin to classical thermal explosion theory. More specific to particular chemical processes are the works of Belyaev *et al.* (1974), Dagonnier and Nuyts (1976), Dagonnier *et al.* (1980), Pikios and Luss (1977), Eigenberger (1978*a,b*), Kurtanjek *et al.* (1980), and Ivanov *et al.* (1980). Additional there is an extensive Soviet literature, mainly due to the Institute of Catalysis at Novosibirsk—Bykov *et al.* (1981) and Volokitin *et al.* (1986) summarize some aspects of this work and provide extra references.

TABLE 10.1 Some studies of oscillations and chaos in heterogeneous systems

Barelko *et al.* (1969), Barelko and Volodin (1973, 1976*a,b*), Volodin *et al.* (1982).

Ertl and Rau (1969), Ertl (1980, 1990), Engel and Ertl (1982), Ertl *et al.* (1982). Cox *et al.* (1983, 1985), Behm *et al.* (1983*a,b*), Theil *et al.* (1983), Imbihl *et al.* (1985, 1986, 1988*a,b*) Eiswirth *et al.* (1985, 1988, 1989), Eiswirth and Ertl (1986), Moller *et al.* (1986), Kleinke *et al.* (1987), Schwankner *et al.* (1987), Ladas *et al.* (1988*a,b*).

Hugo (1970), Hugo and Jakubith (1972).

Schmidt and Luss (1971), Flytzani-Stepanopoulos *et al.* (1977, 1980), Flytzani-Stepanopoulos and Schmidt (1979), Takoudis and Schmidt (1983), Klein *et al.* (1985), Lesley and Schmidt (1985*a,b*), Schwartz *et al.* (1986), Schwartz and Schmidt (1987, 1988), Schuth *et al.* (1989), Cordonier *et al.* (1989).

Beusch *et al.* (1972), Wicke (1974), Wicke *et al.* (1980), Keil and Wicke (1980), Bocker and Wicke (1985*a,b*), Onken and Wicke (1986, 1988), Wicke and Onken (1986, 1988), Schuth and Wicke (1988, 1989*a,b*).

Dauchot and van Cakenberghe (1973).

Ecket *et al.* (1973*a,b*), Rathousky *et al.* (1980, 1981), Rathousky and Hlavacek (1982), Kapicka and Marek (1989).

Belyaev *et al.* (1973), Slin'ko and Slin'ko (1978), Slin'ko *et al.* (1988).

McCarthy *et al.* (1975).

Jones *et al.* (1975), Lamb *et al.* (1977), Dabill *et al.* (1978), Galwey *et al.* (1985), Cameron *et al.* (1986), Scott *et al.* (1990).

Dagonnier and Nuyts (1976), Dagonnier *et al.* (1980).

Pikios and Luss (1977), Zunge *et al.* (1978), Rajagopalan *et al.* (1980), Ivanov *et al.* (1980), Kurtanjek *et al.* (1980), Xiao *et al.* (1986), Harold *et al.* (1987*a,b*), Lobban and Luss (1989), Lbbam *et al.* (1989), Dabholkar *et al.* (1989).

Sheintuch and Schmitz (1977), Sheintuch and Pisman (1981), Sheintuch and Schmidt (1986, 1988), Sheintuch (1990)

Varghesse *et al.* (1978), Kaul and Wolf (1984, 1985*a,b*), Regalbuto and Wolf (1986).

Plichta and Schmitz (1979), Schmitz *et al.* (1984), Brown *et al.* (1985), Razon and Schmitz (1986, 1987), Razon *et al.* (1986).

Koval *et al.* (1979), Bykov *et al.* (1981), Blinova *et al.* (1982), Gol'dstein *et al.* (1986), Volokitin *et al.* (1986).

Lagos *et al.* (1979), Turner *et al.* (1981*a,b*), Sales *et al.* (1982), Yates *et al.* (1985), Tsai *et al.* (1986).

Cutlip (1979), Morton and Goodman (1981), Goodman *et al.* (1982), *Mukesh et al.* (1982, 1983*a,b*, 1984), Cutlip *et al.* (1983, 1984), Capsaskis and Kenney (1986). Plath *et al.* (1980, 1990), Jaeber *et al.* (1981*a,b*, 1985*a,b,c*, 1986*a,b*, 1990), Dress *et al.* (1982, 1985, 1987), Kleine *et al.* (1986), Gerhardt *et al.* (1986), Svensson *et al.* (1988), Slin'ko *et al.* (1989), Gerhardt and Schuster (1989).

Norton *et al.* (1981, 1984*a,b*), Jackman *et al.* (1983), Griffiths *et al.* (1984).

Kiss *et al.* (1984*a,b*), McLaughlin McClory and Gonzales (1986), Saymeh and Gonzales (1986), Li and Gonzales (1988).

Lynch and Wanke (1984*a,b*), Lynch *et al.* (1986).

10.1. The catalytic oxidation of carbon monoxide

The most complete story has emerged for the oxidation, either by O_2 or NO, of carbon monoxide. This reaction has been studied, and shows oscillations, from ultra-high-vacuum conditions (10^{-8} Torr) to atmospheric pressure. Two stimuli may have led to the popularity of this system: kinetically it might be hoped to be particularly simple, for there are few obvious intermediate adsorbed states and only one overall reaction; technically, the reaction on supported catalysts at atmospheric pressure is important because of the widespread application of catalytic convertors in car exhausts (the CO + HO reaction removes two of the most significant pollutants), whilst some gas detection systems in mines and elsewhere employ a catalytic oxidation process for both qualitative and quantitative measurements.

The simple stoichiometry for the $CO + O_2$ reaction results from the mechanism

$$CO + S \rightleftharpoons CO_{ad}$$
$$O_2 + nS \rightarrow 2O_{ad}$$
$$CO_{ad} + O_{ad} \rightarrow CO_2(g) + mS$$

where S represents a vacant surface site. The number of sites required for the dissociative adsorption of O_2 and that form during the reaction step have not been specified at this stage, nor has any detail concerning different possible states of the surface. The CO adsorption is the only step treated as reversible.

10.2. The CO + O₂ reaction on single crystals

The most complete study of the catalysed $CO + O_2$ reaction under ultra-high-vacuum (UHV) conditions is that of Ertl and various coworkers (see references from Table 10.1), typically employing single, well-defined platinum crystallographic planes. As well as measuring concentrations of gas-phase reactants and product, of surface species and of the metal workfunction, these investigations also recorded changes in the state of the surface itself using LEED spectroscopy. The workfunction provides an 'overall' measurement of the state of the cayalyst whereas the LEED data can probe small areas locally, of the order of 0.1 mm².

With a Pt(100) crystal plane, oscillations in the reaction have been observed for CO pressures exceeding about 3×10^{-6} Torr (4×10^{-4} N m⁻² or 4×10^{-9} atmospheres). Under favourable conditions, the oscillations in the workfunction and the intensity of the signal corresponding to various LEED spots have a relatively simple periodic waveform (Fig. 10.1), but perhaps more typical are rather complex responses.

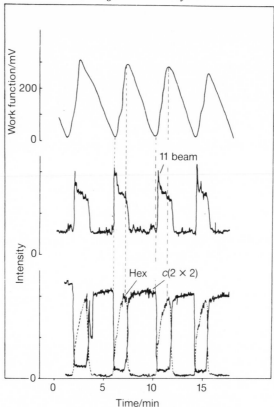

FIG. 10.1. Oscillations in workfunction and LEED spot intensities during the oxidation of CO on Pt(100): $T = 500$ K, $p(O_2) = 4 \times 10^{-4}$ Torr; $p(CO) = 4 \times 10^{-5}$ Torr. (Reproduced with permission from Ertl (1990), © Manchester University Press.)

A 'clean' Pt(100) surface does not have the same arrangement of metal atoms as that found in the bulk of the crystal. The bulk exists as a square (1×1) lattice, but there is an energetically favoured rearrangement (or 'reconstruction') at the surface to a quasihexagonal ('hex') phase (Fig. 10.2). The ('hex') phase has a very low sticking coefficient (probability) for the dissociative adsorption of oxygen, and hence virtually no catalytic activity for the above reaction. The hex and (1×1) phases also have different energies of adsorption for CO, that for the (1×1) being larger (more negative). As the concentration of adsorbed CO increases, the (1×1) phase becomes energetically more favoured and so there is a surface phase transition. This first-order transition occurs sharply at a critical fractional CO coverage of $\theta_{CO} = 0.08$ (on the hex phase). The (1×1) adsorbs more CO and a 'perfect' coverage with every other Pt atom occupied gives a $c(2 \times 2)$ with $\theta_{CO} = 0.5$ which ideally prohibits O_2 adsorption.

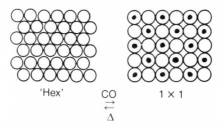

'Hex'   CO   1 × 1
          $\rightleftarrows$
             Δ

Fig. 10.2. The phase transition between the 'hex' and (1 × 1) arrangements of atoms at the Pt(100) surface: the transition to (1 × 1) is favoured by adsorption of CO, which forms a $c(2 \times 2)$ structure (solid circles); the reverse transition is favoured by decreasing adsorption and by increasing temperature. (Reproduced with permission from Ertl (1990), © Manchester University Press.)

On decreasing the fractional CO coverage, e.g. by increasing the temperature of the catalyst, the reverse phase transition from (1 × 1) to 'hex' occurs at a critical value of $\theta_{CO} = 0.3$. Thus there is hysteresis for some range of θ_{CO} with the system existing in either of the two phases for $0.08 \leqslant \theta_{CO} \leqslant 0.3$ depending on the previous history (but note that this hysteresis phenomenon has a slightly different form to that seen elsewhere).

The (1 × 1) phase has a sticking coefficient for dissociative O$_2$ adsorption orders of magnitude higher than that for the 'hex' phase, so the initial adsorption of CO also 'activates' the catalyst, providing a feedback mechanism. Against this activation, however, is the spatial consideration that a high CO coverage cannot provide sufficient vacant sites for the O$_2$ adsorption process. This in turn means that the third step, which produces vacant sites, become autocatalytic.

The basic scenario, then, is starting with a clean surface the CO coverage increases and the hex → (1 × 1) phase transition occurs. More CO is adsorbed and the surface approaches an almost complete $c(2 \times 2)$ layer. (There may also be a tendency for CO$_{ad}$ to migrate to form 'islands' of local high fractional coverage in the (1 × 1) phase—the uncovered sites will not have the driving force for the phase transition and hence remain in the 'hex' form and will not adsorb O$_2$.) Any bare (1 × 1) patches that may arise will have a high sticking coefficient for O$_2$ and any adsorbed O atoms which form can react rapidly with neighbouring CO$_{ad}$, creating extra vacancies. The autocatalysis of this chemical process leads to a sharp reduction in θ_{CO} and a subsequent (1 × 1) → 'hex' transition. No more O$_2$ adsorption occurs on the hex sites and any further CO adsorption leads to removal of O$_{ad}$ by reaction. Eventually the surface concentrations of both species become low and most of the surface is in the 'hex' phase, and the cycle restarts.

In addition to the temporal variations, Ertl and coworkers have observed spatial structures, with waves of phase transition moving across the surface (Fig. 10.3). If there is a single 'pacemaker' site and hence a single wave, the

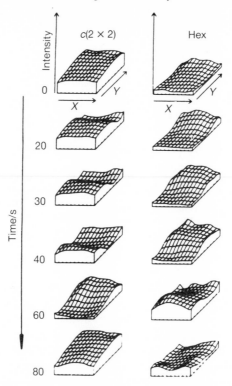

FIG. 10.3. Wave-like propagation of the 'hex' to $(1 \times 1)/c(2 \times 2)$ phase transition during oscillatory oxidation of CO on Pt(100) as detected by scanning LEED for an area of 4 mm $\times$ 7 mm, with $T = 481$ K, $p(O_2) = 4.4 \times 10^{-4}$ Torr and $p(CO) = 3.1 \times 10^{-5}$ Torr. (Reproduced with permission from Ertl (1990), © Manchester University Press.)

temporal oscillations in the workfunction etc. tend to have simple periodicities. More often, however, irregular periodicities occur which seem associated with multiple pacemakers and waves. Schwankner *et al.* (1987) were able to produce simple periodic responses through entrainment by periodically perturbing the partial pressure of O_2.

Similar phase-transition responses arise with the Pt(110) crystal plane and underlie the oscillations and other features observed again by Ertl and his group. The clean surface reconstructs to a (1×2) phase—with missing alternate rows—and converts to the bulk (1×1) phase at high CO coverages. Again the two phases have different sticking coefficients for dissociative O_2 adsorption. Eiswirth and Ertl (1986) found that the range of θ_{CO} for bistability between the phases is much narrower than for Pt(100) and the parameter range for oscillations is also smaller. There appear not to be significant spatial variations (reaction waves) with this surface plane,

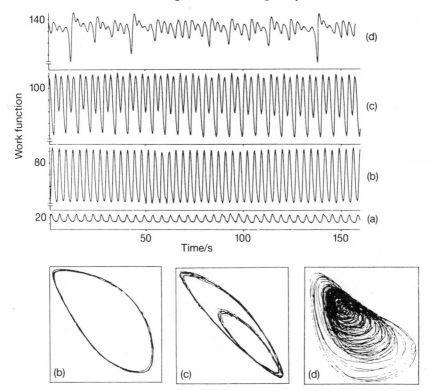

FIG. 10.4. Various oscillatory waveforms, period doubling, and chaos for CO oxidation on Pt(110) on reducing $p(CO)$ from (a) 4.2×10^{-5} Torr to (d) 3.9×10^{-5} Torr, with $T = 530$ K and $o(O_2) = 8 \times 10^{-5}$ Torr. Also shown are the reconstructed attractors for traces (b)–(d). (Reproduced with permission from Ertl (1990), © Manchester University Press.)

perhaps because the individual atoms are synchronized through the gas-phase concentrations of the reactants. However, this system has provided clean examples of a period-doubling cascade to chaos (Fig. 10.4) as the partial pressure of CO is varied. If the system is forced by periodic perturbations of the reactant concentrations various classical forms of entrainment and quasiperiodicity are realized (Eiswirth and Ertl 1986).

The Pt(110) surface, which is the plane with highest density packing and which does not have a reconstructed phase transition, does not appear to support oscillations under UHV conditions.

Strong support for the control exerted by the operation of these mechanisms comes from the successful matching of experiment and computer modelling (Imbohl *et al.* 1985, Möller *et al.* 1986) with kinetic and phase-transition data taken, where possible, from independent studies. In the first of these, a four-variable model was proposed, of the following form:

$$\frac{\partial y_a}{\partial t} = k_1 a p_{CO} - k_2 u_a + k_3 a u_b - \frac{k_4}{a} u_a v_a + k_5 \nabla^2 \left(\frac{u_a}{a} \right) \tag{10.1}$$

$$\frac{\partial u_b}{\partial t} = k_1 b p_{CO} - k_6 u_b - k_3 a u_b \tag{10.2}$$

$$\frac{\partial v_a}{\partial t} = k_7 a p_{O_2} \left\{ \left[1 - 2\left(\frac{u_a}{a} \right) - \frac{5}{3}\left(\frac{v_a}{a} \right) \right]^2 + \alpha \left[1 - \frac{5}{3}\left(\frac{v_a}{a} \right) \right] \right\} - \frac{k_4 u_a v_a}{a} \tag{10.3}$$

and

$$\frac{\partial a}{\partial t} = \begin{cases} \dfrac{a}{u_{a,\,grow}} \dfrac{\partial u_a}{\partial t} & \text{if} \quad u_a > u_{a,\,grow} \quad \text{and} \quad \partial u_a/\partial t > 0 \\[2mm] -k_8 ac & \text{if} \quad c = (u_a/u_{a,\,cr}) + (v_a/v_{a,\,cr}) < 1 \\[2mm] 0 & \text{otherwise.} \end{cases} \tag{10.4}$$

The first three equations describe the rate of change of the surface concentrations of $CO_{ad}(1 \times 1)$, $CO_{ad}(hex)$, and $O_{ad}(1 \times 1)$ whilst in the final equation a is the fraction of the surface in the (1×1) phase. Equation (10.1) has terms for adsorption and desorption, migration from the hex to the (1×1) sites, reaction, and surface diffusion respectively. Equation (10.2) has just adsorption, desorption, and reaction. In eqn (10.3), the form of the adsorption term allows for a coverage-dependent sticking probability, with inhibition for a CO coverage of 0.5 and also by preadsorbed O_2. The final equation describes the phase-transition behaviour and its form depends on the instantaneous surface coverages with CO and O. The various parameters were estimated from literature sources.

To proceed analytically, some further approximations were introduced. Spatial effects were ignored, and the distinction between the two CO_{ad} forms were relaxed and the coverage dependence of O adsorption simplified by allowing $\alpha \to 0$. The resulting three-variable model could then be written as

$$\frac{du}{dt} = k_1'(1 - u - v) - k_4' uv \tag{10.5}$$

$$\frac{dv}{dt} = k_7' a(1 - u - v)^2 - k_4' uv \tag{10.6}$$

$$\frac{da}{dt} = \begin{cases} k_r'(1 - a) & \text{if} \quad c \geq 1 \\ -k_8' a & \text{if} \quad c < 1. \end{cases} \tag{10.7}$$

The form of eqn (10.7) governing the surface structure is such that the fraction in the (1×1) phase changes from an exponentially increasing function of time (approaching a limiting value of unity) for $c \geq 1$ to an exponentially decreasing function of time for $c < 1$. Apart from this rapid switch in form as c passes through unity, a may be a relatively slowly varying function and hence appears as a slowly varying parameter in eqns (10.5) and (10.6). We can thus now look for (pseudo) stationary states of this reduced two-variable system and examine local stabilities of any such states.

Setting $du/dt = dv/dt = 0$, four possible stationary-state solutions exist. For all parameter values there are the states corresponding to complete coverage with CO and with O: $(u = 1, v = 0)$ and $(u = 0, v = 1)$ which are a stable node and a saddle point respectively. The two remaining solutions are given by

$$u^{\pm} = \frac{1}{2}\left\{ 1 - \frac{k_1'}{k_7'a} \pm \left[\left(1 - \frac{k_1'}{k_7'a} \right)^2 - \frac{4k_1'^2}{k_4'k_7'a} \right]^{1/2} \right\} \qquad (10.8)$$

which $v^{\pm} = u^{\mp}$. These two states only exist for certain values of the (time-dependent) parameter a, such that the fraction of the surface in the (1×1) state exceeds a critical value $a > a_{cr}$ given by

$$a_{cr} = \frac{k_1'}{k_1'}\left[1 + \frac{2k_1'}{k_4'} \left(1 + \frac{k_4'}{k_1'} \right)^2 \right]. \qquad (10.9)$$

When these two extra solutions are real, the upper root is a saddle point and the lower either a stable node or focus.

The qualitative behaviour of the system is thus. With the parameter c initially less than unity, only the two (pseudo) stationary states are those of complete coverage with one or other of the reactants. The system will move towards the stable node $(u = 1, v = 0)$. This motion across the u, v phase plane causes c to increase, and c will pass through unity at some point. Then, the form of da/dt changes, with the 'parameter' a becoming an increasing function of time. When a exceeds a_{cr} given above, two new stationary points appear in the phase plane, growing from a saddle–node bifurcation.

If the saddle–node pair is born 'away from' the current position of the system, there will still be an evolution to the stable state of complete CO coverage. However, if the saddle appears 'between' the current point and the complete coverage state, the separatrices will deny access to that state and the system will change course and move towards the new stable node/focus. In the latter case, v may grow at the expense of u (O$_{ad}$ increases and CO$_{ad}$ decreases), as the increasing O$_{ad}$ allows the reaction term to become important. Then c will decrease again, da/dt will switch its form again so a is now decreasing, and when this 'parameter' falls past a_{cr} the two new stationary states vanish. The system begins it trek towards the complete CO coverage state again and the cycle restarts. A computed oscillatory trace evolving under this scenario is shown in Fig. 10.5.

Imbihl *et al.* (1985) also carried out numerical integration of the full equations (10.1)–(10.4), with and without the diffusion term. Fitting their

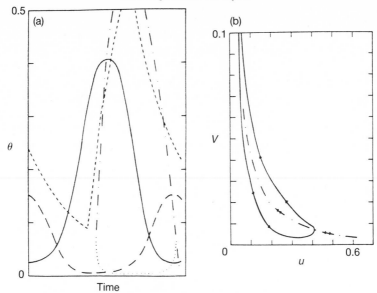

FIG. 10.5. Sustained oscillations for the surface phase-transition model showing (a) time series for CO_{ad} (——), O_{ad} (- - -), and (1 × 1) surface (– – –) and also the coordinates of the non-trivial steady states ($\cdots$ and $-\cdot-\cdot-$); (b) limit cycle in the O_{ad}–CO_{ad} phase plane. (Reproduced with permission from Imbihl *et al.* (1985), © American Institute of Physics.)

computed curve (with no diffusion) to the observed hysteresis loop for CO adsorption as a function of temperature, the remaining unknowns such as the critical coverages for the phase transitions were determined.

Oscillations exist for intermediate p_{CO}/p_{O_2} ratios: with p_{CO} too high a complete CO coverage becomes possible; with p_{CO} too low a stable inhomogeneous state occurs with parts of the surface in the 1 × 1 phase covered by O_{ad} and the remainder of the 'hex' phase and covered with CO_{ad}. When a one-dimensional system with diffusion was studied, propagating wave trains developed, as observed experimentally.

To model a realistic two-dimensional surface, Möller *et al.* (1986) applied a 'cellular automaton' approach to the above mechanistic ideas. A two-dimensional array, typically 17 × 40 or 20 × 78, of 'lattice sites' is constructed and rules are devised which determine the probabilities for the various possible changes in the state of each given site during the next step forward in time. Thus a currently empty 'hex' site may remain empty, adsorb a CO molecule from the gas phase, or may transform to a 1 × 1 site; a CO_{ad}(hex) may undergo desorption, phase transition to a CO_{ad}(1 × 1), or react with a neighbouring O_{ad}. The probabilities are functions of the states of neighbouring sites, so, for instance, a CO_{ad}(hex) will undergo phase transition if its eight nearest neighbours in the square array are occupied.

CO – 1 × 1   Hex

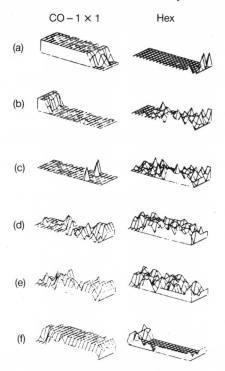

FIG. 10.6. Numerical simulation of three-dimensional phase-transition propagation from (1 × 1) to 'hex' phase during oscillatory oxidation of CO on Pt(100). (Reproduced with permission from Möller *et al.* (1986), © American Institute of Physics.)

Qualitative agreement with the overall (spatially integrated) time evolution of the various states of the surface and of the development of two-dimensional spatial patterns, even from initially homogeneous conditions, appear naturally in this scheme. Typical results are shown in Fig. 10.6.

10.3. The CO + NO reaction at low pressures

UHV and HV studies of the CO + NO reaction, similar to those described above, have been made by Schwartz and Schmidt (1987, 1988) and by Lesley and Schmidt (1985*a,b*) for Pt(100) crystal surfaces and with polycrystalline samples by various workers (Adlhoch *et al.* 1981; Klein *et al.* 1985). Schwartz and Schmidt observed bistability in the stationary-state reaction rate and were able to determine that in the low reaction state the surface was in a reconstructed 'hex' phase whilst the high reaction branch corresponds to a (1 × 1) surface. The surface phase-transition mechanism thus again seems to be important for the non-linear kinetics in this system, suggesting that the two phases have different sticking coefficients for the dissociative adsorption of the oxidant species NO. Schwartz and Schmidt were able to extend their studies up to 1 Torr total pressure and there still found evidence for the reconstructing mechanism. Also, with polycrystalline wires, the Pt(100) plane is important and regions with this orientation of atoms may develop as faceting accompanies reaction (see below). With this in mind, Schwartz and Schmidt asked whether this mechanism might be responsible for oscillations across the complete pressure range from UHV to atmospheric.

10.4. The CO + NO reaction at atmospheric pressure

There are only a few studies of the oscillatory CO + NO reaction at atmospheric pressure, most notably Regalbuto and Wolf (1986) and Schüth and Wicke (1989*a,b*). These authors employed polycrystalline catalysts (Pt or Pd) supported on silica or alumina, or evaporated metal films. The state of the surface was followed by *in situ* infrared (IR) measurements at various wavelengths.

Oscillatory behaviour is observed over a range of catalyst temperatures and reactant composition and may hae simple or complex waveforms (Fig. 10.7). By observing shifts in appropriate IR absorption peaks, Schüth and Wicke showed that the 'hex'/(1×1) phase transition again occurs in appropriate phase with the observed concentration oscillations for it to be the driving mechanism at these atmospheric pressures. It thus seems that the question asked by Schwartz and Schmidt is answered in the affirmative for the CO + NO reaction on Pt(100).

With these polycrystalline catalysts there are other planes that will also contribute to the observed oscillations and there will also be significant temporal (and possibly spatial) variations in the temperature of the catalyst. The temperature excursions may play an important role in synchronizing the discrete catalytic sites in the supported catalyst, although synchronization through the gas-phase concentrations can also play a part (especially for the lower-pressure studies). We return to this point below. The CO coverage of these other planes appears to oscillate out of phase with the reaction rate and $CO_{ad}(1 \times 1)$ (these latter two are in phase).

10.5. The CO + O$_2$ reaction at atmospheric pressure

The CO + O$_2$ reaction over platinum at atmospheric pressure has been studied by a great many workers in a variety of ways. The catalyst has been supported on various oxides, notably silica and alumina, in reactors that vary in their heat transfer characteristics from virtually adiabatic to allegedly isothermal. Additionally, wires and films have been employed or the catalyst has been dispersed in a zeolite matrix.

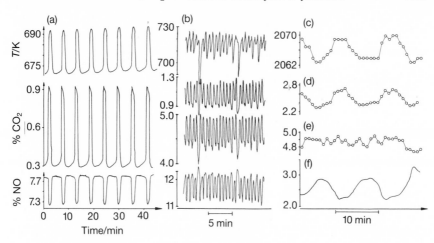

FIG. 10.7. Simple and complex oscillatory waveforms and the changes in IR intensities and wavelengths during CO + NO reaction on supported Pt: (a) 1 per cent Pt/SiO₂, 3.6 per cent CO, 7.4 per cent NO, $T = 658$ K, flow rate $= 11$ ml s^{-1}; (b) 1 per cent Pt/SiO₂, 11 per cent CO, 17 per cent NO, $T = 670$ K, flow rate $= 6.25$ ml s^{-1}, traces are T/K, % N₂O, % CO₂, and % NO; (c) position of L–CO$_{ad}$ band, (d) intensity of absorption band at 2060–2090 cm^{-1}, (e) intensity of absorption band at 2000–2100 cm^{-1}, (f) % CO₂ for 1 per cent Pt/alumina, 12 per cent CO, 10 per cent NO, $T = 609$–656 K. (Reproduced with permission from Schüth and Wicke (1989*a,b*), © VCH Verlagsgesellschaft.).

Böcker and Wicke (1985*a,b*) examined both platinum and palladium catalysts as a foil of individual crystallites in an inert carrier. Figure 10.8 summarizes the typical variations of reaction behaviour on changing the CO partial pressure. For low p_{CO} the reaction rate increases linearly with increasing reactant concentration, but this first-order behaviour eventually gives way to a falling off. A region of oscillations is encountered next. these grow from a supercritical Hopf bifurcation point and end with a hard extinction. At high p_{CO}, the stationary-state reaction rate is much lower and falls with increasing concentration as CO now has an inhibitory effect (McCarthy *et al.* 1975).

Most of the traces reported by Böcker and Wicke have relatively simple waveforms, but more complex responses in the same system have been found (Onken and Wicke 1986; Wicke and Onken 1986). Schüth and Wicke (1989*a*) examined this system with their *in situ* FTIR in a similar way to that described above for the CO + NO reaction. Their most important conclusion is that the phase-transition mechanism that drives the UHV oscillations on Pt(100) does not operate at these higher pressures. The wavelength corresponding to the L–CO absorption band shifts by only two wavenumbers, not

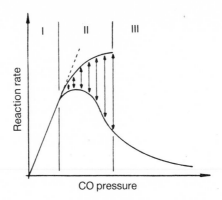

FIG. 10.8. The variation of the steady state and oscillatory reaction rate with CO concentration typical for CO oxidation on supported Pt at atmospheric pressure showing (I) almost first-order behaviour, (II) a fall-off in order and onset of oscillations, (III) inhibition at high CO concentrations. Ignition and extinction may also be found in such responses. (Reproduced with permission from Böcker and Wicke (1984), © VCH Verlagsgesellschaft.)

the 13 observed with CO + NO. There was also no observation of any 'oxidized' surface adsorption peak (expected at $2100 \, cm^{-1}$) in the spectrum, so the interpretation due to Turner *et al.* (1981*a,b*) which has different states of the surface in terms of oxidized and reduced Pt atoms would also appear not to be relevant here. (However, a study under different conditions by Lindstrom and Tsotsis (1984, 1985) gave rise to oscillations of a significantly different character and with an intense band at $2120 \, cm^{-1}$.) Schüth and Wicke suggested that the oscillations in their system may originate from the Pt(110) planes and the reconstructing surface for that plane.

Razón *et al.* (1986) and Kapicka and Marek (1989) report both simple and complex oscillations for fixed bed reactors, although Razón *et al.* operated an 'isothermal' system. Another series of 'isothermal' experiments with supported catalysts is that due to Cutlip, Kenney, and coworkers (Cutlip 1979; Goodman *et al.* 1982; Cutlip *et al.* 1983, 1984; Mukesh *et al.* 1982, 1983*a,b*, 1984). These authors operated a recycle reactor, such that the whole system approximates to a CSTR, with the Pt catalyst supported on alumina. Typical concentrations of 2 per cent CO and 3 per cent O_2 in N_2 give rise to multiple stationary states, but no oscillations are observed in these 'clean' systems. If, however, an inhibitor or catalyst poison is added to the inflow reactants, oscillations occur. A model, known as the Cantabrator, has been developed for this (Morton and Goodman 1981; Capsaskis and Kenney 1986), which builds on the general ideas of Eigenberger (1978*a,b*).

Multiple stationary states arise from a model with competitive chemisorption and reaction, with autocatalysis arising from the requirement for and the creation of vacant sites in the reaction step. The Eigenberger model has been discussed in Section 3.12.4, where it was also shown that an inhibitor that blocks vacant sites through a reversible adsorption/desorption process can induce relaxation-type oscillations.

Other 'isothermal' work has involved the use of an electrically heated catalytic wires (Jones *et al.* 1975; Lamb *et al.* 1977; Cameron *et al.* 1986; Scott *et al.* 1990), following their development for other systems (Davis 1934, 1935; Rader and Weller 1974; Barelko and Volodin 1976*a,b*; Volodin *et al.* 1982; Xiao *et al.* 1986; Sheintuck and Schmidt 1986; Harold *et al.* 1987*a,b*). The Pt catalyst is wound into an 11 turn coil which then forms part of a self-compensating Wheatstone bridge. The wire is heated to a selected resistance (determined by the three other resistances in the bridge circuit) by an electric current: the resistance and hence some measure of the catalyst temperature can be easily monitored in terms of the current passing and the voltage across the catalyst. Any chemical heat evolution that would cause the catalyst to increase in temperature (and hence in resistance) unbalances the bridge and the circuit automatically compensates by reducing the supplied electrical power. The reduction in electrical power gives a direct measure of the rate of chemical heat release and thus, if the stoichiometry is known, of the reaction rate at that temperature.

Figure 10.9 shows the hysteresis phenomena displayed by the reaction rate as a function of catalyst temperature for different $CO:O_2$ mixtures in such a system. As the CO concentration increases, so the hysteresis loop unfolds to yield a monotonic curve. A similar unfolding occurs if the O_2 partial pressure is reduced. Hysteresis and multistability are also displayed as a function of the CO concentration at fixed temperature in Fig. 10.10. Oscillations were not observed in these systems unless an additional species, again an inhibitor, was added. Typical complex waveforms thar arise in the presence of the inhibitor tetraethyl phosphate (TEP) are shown in Fig. 10.11.

These findings appear again to argue for an Eigenberger-like interpretation. There are, however, some additional complicating features. Figure 10.12 shows electron microscope pictures of the surface of a catalyst wire at various stages during operation. The initially smooth surface following annealing undergoes considerable restructuring, with the development of favoured crystallographic planes at the expense of others. In the presence of the inhibitor, this restructuring becomes even more quickly developed, although there is little apparent change in the activity of the catalyst itself.

Luss has also commented on the development of stable and sometimes quite complex spatial structures on electrically heated catalysts (Lobban and Luss 1989; Lobban *et al.* 1989; Dabholkar *et al.* 1989) and similar spatio-temporal structures have been detected during oscillations in other catalytic

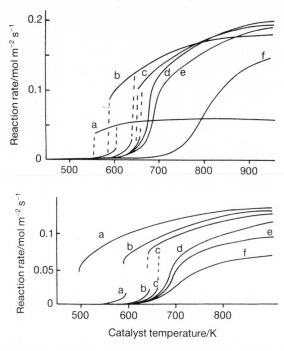

Fig. 10.9. The variation of reaction rate with catalyst temperature for CO oxidation on an 'isothermal' Pt wire showing influence of reactant inflow concentrations. Upper figure, 3.3 per cent O_2 in N_2 with: a, 1.33 per cent CO; b, 5.33 per cent CO; c, 6.0 per cent CO; d, 6.66 per cent CO; e, 8.0 per cent CO; and f, 96.66 per cent CO; the hysteresis loop, with its ignition and extinction points, unfolds when CO is in excess. Lower figure 3.33 per cent CO in N_2 with: a, 19.95 per cent O_2; b, 3.35 per cent O_2; c, 2.47 per cent O_2; d, 2.0 per cent O_2; e, 1.66 per cent O_2; f, 0.83 per cent O_2.

systems (Brown *et al.* 1985). The potential existence of abrupt temperature steps along the catalyst creates problems for quantitative and even, to some extent, qualitative interpretations at present. Jensen and Ray (1980) have proposed a thermal mechanism based partly on the analysis of Uppal *et al.* (see Chapter 3) for the first-order exothermic reaction in a non-isothermal CSTR, and also have a modification in terms of a 'fuzzy wire' model.

Scott and Watts (1981; Cameron *et al.* 1986) have provided a different interpretation. These authors have paid somewhat closer attention to the adsorption processes, arguing that close to the catalyst the gas-phase re-actant concentrations may not be equal to the bulk values. Instead there may be a development of a significant boundary layer. These authors used film theory to relate the concentration at the surface to that in the bulk in terms of parameters that could either be estimated *a priori* or determined from experiment. Additionally they included an Eley–Rideal step whereby gas-phase CO can react directly with O_{ad}. The inclusion of these two effects enabled Cameron *et al.* to match their experimental data for multiple stationary states and hysteresis as a function of p_{CO} and they concluded that in this case the majority of the CO_2 production rate occurs through the Eley–Rideal channel (Fig. 10.13).

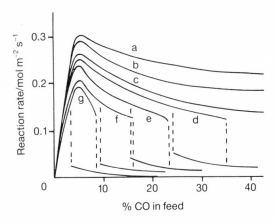

FIG. 10.10. The variation of the steady-state reaction rate with CO inflow concentration for an 'isothermal' Pt wire and the effect of catalyst temperature, $\%$ $O_2 = 3.33$ per cent: a, $T = 775$ K; b, $T = 738$ K; c, $T = 706$ K; d, $T = 684$ K showing onset of hysteresis and bistability; e, $T = 667$ K; f, $T = 647$ K; g, $T = 619$ K.

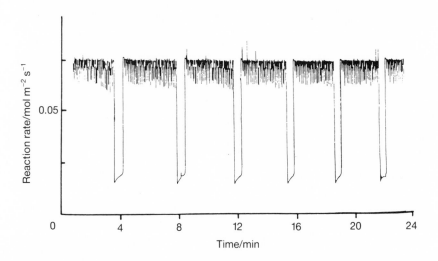

FIG. 10.11. Complex oscillatory oxidation of CO on an 'isothermal' Pt wire in the presence of an inhibitor with $T = 538$ K, 2 per cent CO, and 20 per cent O_2.

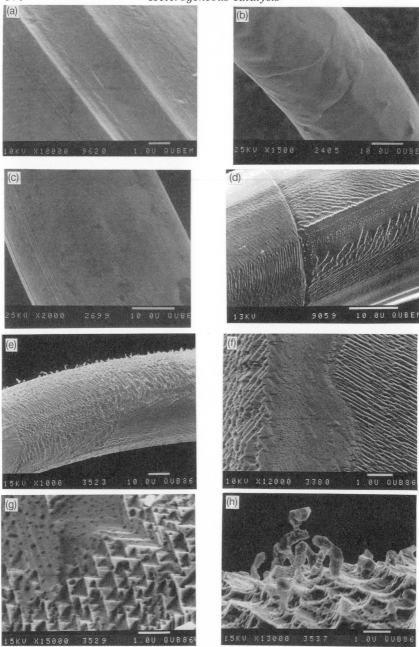

FIG. 10.12. SEM photographs of the surface restructuring that accompanies oxidation on a Pt wire: (a) pristine wire; (b) annealed wire; (c)–(d) restructuring after 15 h without inhibitor; (e)–(h) extensive restructuring from 15 h reaction after treatment with inhibitor.

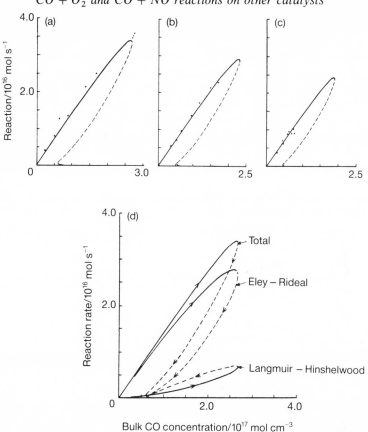

FIG. 10.13. (a)–(c) Comparison of experimental points with numerical predictions (solid and dashed lines) from surface film transport model showing bistability with extinction and ignition; (d) relative importance of Eley–Rideal and Langmuir–Hinshelwood processes for oxidation of CO in this model. (Reproduced with permission from Cameron *et al.* (1986), © Royal Society of Chemistry.)

10.6. CO + O₂ and CO + NO reactions on other catalysts

The CO + O₂ reaction has been studied on a number of other noble metal catalysts (Cant *et al.* 1978; Rathousky *et al.* 1980; Turner *et al.* 1981*a,b*; Cant and Donaldson 1981; Sales *et al.* 1982; Kiss and Gonzales 1984*a,b*; Böcker and Wicke 1984, 1985; Saymeh and Gonzales 1986; McLaughlin McClory and Gonzales 1986; Li and Gonzales 1988; Schüth and Wicke 1989*b*). There is strong evidence again for surface phase transitions induced by CO coverage underlying the Pd Catalysis—although again competing interpretations based on reversible oxidation/reduction of the surface metal atoms have been given (Sales *et al.* 1982).

For the CO + NO reaction on Pd, the situation is less clear and less straightforward. Only recently have oscillations in this system been observed (Schüth and Wicke 1989*b*) and these require much higher reactant concentrations than those typical for the CO + O$_2$ reaction (and higher than those of particular interest to car exhaust conditions). The IR studies have revealed that the surface concentrations of molecular species are rather low during oscillations and that NCO$_{ad}$ (on the support) and N$_{ad}$ apparently play a significant role, as do temperature excursions. The thermokinetic model proposed by Schüth and Wicke includes site blocking by N$_{ad}$ with 'reactivation' due to formation of N$_2$: N$_2$ desorption is slower than the dissociative adsorption of NO at high catalyst temperatures leading to an increase in the number of sited blocked and a decrease in the reaction (and heat release) rate; the relative rates of N$_2$ desorption and N$_{ad}$ formation are reversed at low temperature, leading to a net unblocking of the surface and a consequent increase in the reaction process.

A considerable study of the Pd-catalysed CO + O$_2$ reaction has been undertaken by Jaeger, Plath, and coworkers (Dress *et al.* 1982, 1985; Jaeger *et al.* 1985*a,b,c*, 1986*a,b*, 1990; Gerhardt *et al.* 1986; Svensen *et al.* 1988; Slin'ko *et al.* 1989). These authors have taken advantage of the environment offered by distributing the Pd crystallites within a zeolite matrix. This allows a relative narrow size distribution to be obtained. The reaction can again develop considerable departures for isothermal operation. For other conditions complex and even aperiodic oscillations are regularly observed. The autonomous oscillations are sensitive to forcing of the reactant feeds, and various entrainment and quasiperiodic responses have been recorded—some examples are shown in Fig. 10.14.

10.6. Synchronization and complex oscillation

With supported catalysts, complex oscillations seem to be the order rather than the exception. Similar comments apply for polycrystalline wires and for foils etc. In some special cases, however, relatively simple periodicities can emerge. At the heart of this complexity appears to be the inhomogeneous nature of these systems. There may be many centres of reactions within the whole catalyst and the resulting overall behaviour will depend on how strong or weak is the coupling between these different oscillators. If many oscillators are proceeding virtually independent of each other, the total effect may well seem like 'disorganized chaos' in the non-technical sense of the latter word. As the coupling gets more efficient, so there may be more cooperation and less independence, leading to an overall simpler behaviour.

Some experimental evidence for spatial inhomogeneity has already been mentioned—other studies include those of Schmitz *et al.* (1984), Pawlicky (1987), Kaul and Wolf (1985*a,b*), Sant and Wolf (1988), and Tsai *et al.* (1988).

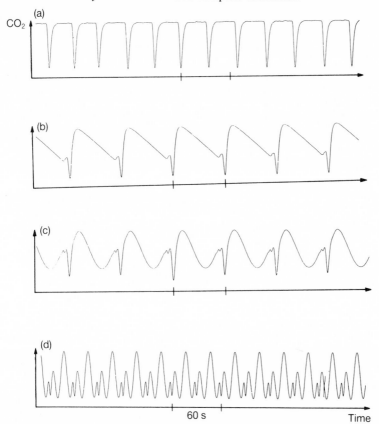

FIG. 10.14. Various responses for periodically forced oxidation of CO on a Pd caalyst supported in a zeolite matrix: (a) autonomous oscillations with $T = 483$ K, 0.5 per cent CO, natural frequency $\omega_0 = 1.746$ min^{-1}; (b) entrainment for triangular forcing waveform with forcing frequency $\omega_f = 1$ min^{-1} and amplitude $= \pm 0.1$ per cent CO (± 20 per cent of inflow concentration); (c) entrainment with sinusoidal forcing; (d) entrainment with sinusoidal forcing with $\omega_f = 4$ min^{-1}. (Reproduced with permission from Jaeger *et al.* (1990), © Manchester University Press.)

Onken and Wicke (1986) showed that individual units in their $CO + O_2$ on Pt system had regular oscillatory time series, but the system as a whole had a complex waveform, and Plath *et al.* (1988) found similar effects when separate plates of catalyst were allowed to have thermal contact. In the various studies described so far, the authors generally comment that the simplest responses arise when temperature excursions are large, postulating that this allows strong thermal coupling to bring about synchronization of otherwise independent oscillators.

Schüth *et al.* (1990) have investigated this idea further using a model derived from the CO + NO on Pd system discussed above, taking a 10×10 array of cells, 10 of which are (randomly) chosen as 'active centres'. This model gives rise to a coupled, non-linear, 200 variable system (two variables in each of the 100 cells). Thermal coupling was allowed through both the catalyst support and the gas phase. Varying the heat transfer coefficient to the gas phase, the periodic oscillations of a single (isolated) cell are modified and chaotic responses can be induced via a period-doubling sequence (Fig. 10.15). For other arrangements, quasiperiodic waveforms were also created.

Jaeger *et al.* (1985; Dress *et al.* 1982, 1985) appear to have initiated the

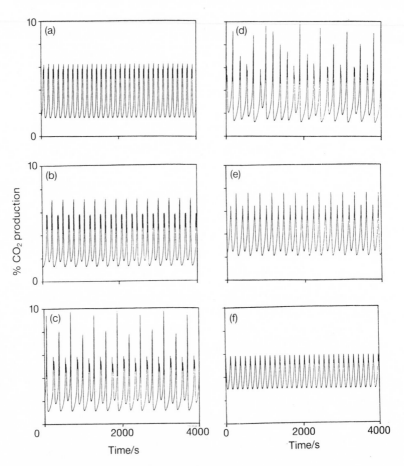

FIG. 10.15. Influence of the coupling between oscillators on the total observed waveform computed numerically for the CO + NO reaction with 10 oscillators on a 10×10 grid: the strength of the coupling through thermal conduction via the gas phase increases from (a) to (f) and induces period doubling to chaos and a subsequent reverse cascade. (Reproduced with permission from Schüth *et al.* (1990), © American Institute of Physics.)

application of cellular automata to heterogeneous catalytic systems. The basic rules for evolution of each cell in the grid have been described briefly above. Gerhardt and Schuster (1989) have extended the application of these approaches to include the spatio-temporal development in solution-phase systems such as the Belousov–Zhabotinskii reaction (see Chapter 8) for which apparently complex spatial patterns can against arise with rather simple oscillatory waveforms at any given location.

10.8. Other systems

Although the CO oxidation reactions are the most widely studied from the present point of interest, the behaviour reported above is not unique to that system. The work of Lobban and Luss (1989) was performed with $H_2 + O_2$ on Ni, whilst Lobban *et al.* (1989) investigated ammonia oxidation on a Pt ribbon. Sheintuck and Schmidt (1988) report a periodic–chaotic bifurcation sequence for ammonia oxidation on a Pt wire and some of their traces are reproduced in Fig. 10.16. Sheintuch (1990) has also recently used this system to study the effects of forcing of a relaxation oscillator.

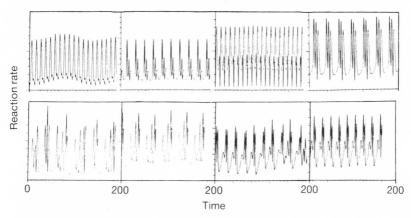

FIG. 10.16. Periodic–chaotic sequences for the oxidation of ammonia on a Pt wire on increasing the ammonia concentration. (Reproduced with permission from Sheintuch and Schmidt (1988). © American Chemical Society.)

Other relatively simple fuels that support oscillations include ethanol (Jaeger *et al.* 1986; Plath *et al.* 1990), methanol (Jaeger *et al.* 1981*b*) and isobutyric aldehyde (Müller and Hofmann 1987*a,b*). The oscillatory processes based on N_2O (Hugo 1970) and on methylamine (Cordonier *et al.* 1990) are of interest in that they involve only a single reactant that decomposes on the catalyst surface. In the N_2O system, this decomposition is exothermic, but for methylamine the reaction is endothermic.

In the case of the latter reaction, the underlying mechanism is believed to be as follows. A Pt ribbon catalyst is heated electrically to a high temperature (typically sufficient to melt the ribbon in the absence of the endothermic reaction). The decomposition proceeds mainly through the stoichiometry

$$CH_3NH_2 \rightarrow HCN + 2H_2.$$

This reduces the local temperature on the catalyst (which is not spatially uniform along its whole length—but that is not apparently a particularly significant feature). There is a minor subsequent reaction that involves the production of strongly adsorbed CN radicals. These block the reaction sites

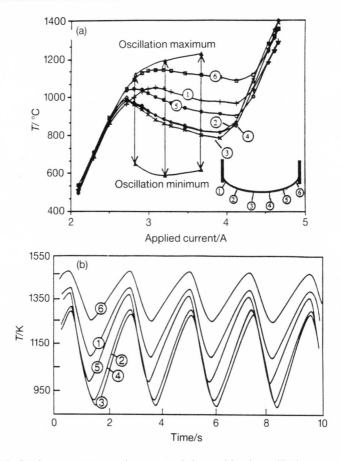

FIG. 10.17. Stationary-state reaction rate and thermokinetic oscillations accompanying the endothermic decompsition of methylamine on a Pt wire: (a) stationary-state wire temperature and amplitude of coexisting oscillation at six locations along the wire; (b) oscillations for $I = 3.21$ A. (Reproduced with permission from Cordonier *et al.* (1990), © American Institute of Physics.)

and hence inhibit the endothermic process. The temperature of the catalyst increases again, and there is a high-temperature desorption of the inhibitor CN and C_2N_2. As the cyanogen departs, vacant sites allow the decomposition reaction above to begin again.

There are some additional operational details, such as the need for a small accompanying flow of O_2 to prevent the build up of a carbonaceous layer, but Cordonier *et al.* were able to model their observations with a thermokinetic model based on the above interpretation. Some of the observed oscillations are shown in Fig. 10.17. Coexisting with this response there is always a stable reaction state. The onset of oscillations does not arise at a Hopf bifurcation, but requires a sufficient perturbation of the system, indicating that the limit cycles lie on an isolated branch—an isola of limit cycles rather than of stationary-state solutions such as discussed in Chapter 3.

References

Abdul-Kareem, H. K., Jain, A. K., Silveston, P. L., and Hudgins, R. R. (1980). Harmonic behavior of the rate of catalytic oxidation of CO under cycling conditions. *Chem. Eng. Sci.*, **35**, 273–82.

Adlhoch, W., Lintz, H. G., and Weisker, T. (1981). Oszillation der Reaktionsgeschwindigkeit bei der Reaktion von NO mit CO an Platin im Knudsengebiet. *Surf. Sci.*, **103**, 576–85.

Barelko, V. V. and Volodin, Yu. E. (1973). Critical phenomena of nonisothermal nature in the ammonia oxidation reaction on platinum. *Dokl. Akad. Nauk. SSR*, **211**, 673–6.

Barelko, V. V. and Volodin, Yy. E. (1976a). The electrothermographic method in heterogenous catalysis. *Kinetic. Catal.*, **17**, 112–18.

Barelko, V. V. and Volodin, Yu. E. (1976b). Nature of the critical phenomena in the oxidation of ammonia on platinum. *Kinet Catal.*, **17**, 593–9.

Barelko, V. V., Abramov, V. G., and Merzhanov, A. G. (1969). Thermographic method for the investigation of the kinetics of gas-phase heterogeneous catalytic reactions. *Russ. J. Phys. Chem.*, **43**, 1589–92.

Barkowski, D., Haul, R., and Kretschmer, U. (1981). Studies on oscillations in the platinum catalysed CO oxidation by means of an omegatron mass probe. *Surf. Sci.*, **107**, L329–33.

Behm, R. J., Thiel, P. A., Norton, P. R., and Ertl, G. (1983a). The interaction of CO and Pt(100). I. Mechanism of absorption and Pt phase transition. *J. Chem. Phys.*, **78**, 7437–47.

Behm, R. J., Penka, V., Cattania, M.-F., Christmann, K., and Ertl, G. (1983b). Evidence for 'subsurface' hydrogen on Pd(110): an intermediate between chemisorbed and dissolved species. *J. Chem. Phys.*, **78**, 7486–90.

Belyaev, V. D., Slin'ko, M. M., Timoshenko, V. I., and Slin'ko, M. G. (1973). Generation of autooscillations in the hydrogen reaction on nickel. *Kinet. Catal.*, **14**, 708–9.

Belyaev, V. D., Slin'ko, M. M., Timoshenko, V. I., and Slin'ko, M. G. (1974).

Autooscillations in the heterogeneous catalytic reaction of hydrogen and oxygen. *Dokl. Akad. Nauk. SSSR*, **214**, 142–4.

Belyaev, V. D., Slin'ko, M. G., and Timoshenko, V. I. (1975). Changes in the contact potential difference in the oscillatory state of the heterogeneous catalytic reaction of hydrogen with oxygen on nickel. *Kinet. Catal.*, **16**, 480.

Berman, A. D. and Krylov, O. V. (1980). Features of the kinetics of heterogeneous reactions with phase transformations on catalyst surfaces. *Int. Chem. Eng.*, **20**, 313–23.

Beusch, H., Fieguth, P., and Wicke, E. (1972). Thermisch und Kinetisch verursachte Instabilitäten in Reaktionsverhalten einzelner Katalysatorkörner. *Chem. Ing. Tech.*, **44**, 445–51.

Blinova, N. A., Yablonskii, G. S., Koval, G. L., Korneichuk, G. P., and Filipov, V. I. (1982). Kinetic models of CO oxidation on Pd-containing catalysts (with and without SO_2). *React. Kinet. Catal. Lett.*, **19**, 207–11.

Böcker, D. and Wicke, E. (1985a). In situ IR study during oscillations of the catalytic CO oxidation. *Ber. Bunsenges, Phys. Chem.*, **89**, 629–33.

Böcker, D. and Wicke, E. (1985b). In-situ IR study during oscillations of the catalytic CO oxidation. In *Temporal order*, (ed. L. Rensing and N. I. Jaeger), pp. 75–85. Springer, Berlin.

Brown, J. R., D'Netto, G. A., and Schmitz, R. A. (1985). Spatial effects and oscillations in heterogeneous catalytic reactions. In *Temporal order*, (ed. L. Rensing and N. I. Jaeger), pp. 86–95. Springer, Berlin.

Burrows, V. A., Dundaresan, S., Chabal, Y. J., and Christman, S. B. (1985). Studies on self-sustained rate oscillations. I. Real time surface infrared measurements during ocillatory oxidation of carbon monoxide on platinum. *Surf. Sci.*, **160**, 122–38.

Bykov, V. I., Yablonskii, G. S., and Elokhin, V. I. (1981). Steady state multiplicity of the kinetic model of CO oxidation reaction. *Surf. Sci.*, **107**, L334–8.

Cameron, P., Scott, R. P., and Watts, P. (1986). The oxidation of carbon monoxide on a platinum catalyst at atmospheric pressure. *J. Chem. Soc. Faraday Trans. 1*, **82**, 1389–403.

Cant, N. W. and Donaldson, R. A. (1981). Infrared spectral studies of reactions of carbon monoxide and oxygen on Pt/SiO_2. *J. Catal.*, **71**, 320.

Cant, N. W., Hicks, P. C., and Lennon, B. S. (1978). Steady state oxidation of carbon monoxide over supported noble metals with particular reference to platinum. *J. Catal.*, **54**, 372.

Capsaskis, S. C. and Kenney, C. N. (1986). Subharmonic response of a heterogeneous catalytic oscillator, the 'Cantabrator', to a periodic input. *J. Phys. Chem.*, **90**, 4631–7.

Cordonier, G. A., Schüth, F., and Schmidt, L. D. (1989). Oscillations in methylamine decomposition on Pt, Rh and Ir: experiments and model. *J. Chem. Phys.*, **91**, 5374–86.

Cox, M. P., Ertl, G., Imbihl, R., and Rüstig, J. (1983). Non-equilibrium surface phase transitions during the catalytic oxidation of CO on Pt(100). *Surf. Sci.*, **134**, L517.

Cox, M. P., Ertl, G., and Imbihl, R. (1985). Spatial self-organization of surface structure during an oscillatory catalytic reaction. *Phys. Rev. Lett.*, **54**, 1725–8.

Cutlip, M. (1979). Concentration forcing of a catalytic surface. *AIChE J.* **25**, 502–8.

Cutlip, M. B., Hawkins, C. J., Mukesh, D., Morton, W., and Kenney, C. N. (1983).

Modelling of forced periodic oscillations of carbon monoxide over a platinum catalyst. *Chem. Eng. Commun.*, **22**, 329–44.

Cutlip, M. B., Kenney, C. N., Morton, W., Mukesh, D., and Capsaskis, S. C. (1984). Transient and oscillatory phenomena in catalytic reactors. *Proc. ISCRE*, **8**, 135–42. Pergamon and Institute of Chemical Engineers, London.

Dabholkar, V. R., Balakotaiah, V., and Luss, D. (1989). Stationary concentration patterns on an isothermal catalytic wire. *Chem. Eng. Sci.*, **44**, 1915–28.

Dabill, D. W., Gentry, S. J., Holland, H. B., and Jones A. (1978). The oxidation of hydrogen and carbon monoxide mixtures over platinum. *J. Catal.*, **53**, 164–7.

Dagonnier, R. and Nuyts, J. (1976). Oscillating CO oxidation on a Pt surface. *J. Chem. Phys.*, **65**, 2061–5.

Dagonnier, R., Dumont, M., and Nuyts, J. (1980). Thermochemical oscillatons in surface reactions. *J. Catal.*, **66**, 130–46.

Dauchot, J. P. and van Cakenberghe, J. (1973). Oscillations during catalytic oxidation of carbon monoxide on platinum. *Nature (Phys. Sci.)*, **246**, 61–3.

Davis, W. (1934). The rate of heating of wires by surface combustion. *Philos. Mag.*, **17**, 233–51.

Davis, W. (1935). Catalytic combustion at high temperatures. *Philos. Mag.*, **19**, 309–25.

Dhalewadikar, S. V., Martinez, E. N., and Varma, A. (1986). Complex dynamic behaviour during ethylene oxidation on a supported silver catalyst. *Chem. Eng. Sci.*, **41**, 1743–6.

Dress, A., Jaeger, N. I., and Plath, P. J. (1982). Zur Dynamik idealer Speicher. Ein einfaches mathematisches Modell. *Theor. Chim. Acta*, **61**, 437–60.

Dress, A. W. M., Gerhardt, M., Jaeger, N. I., Plath, P. J., and Schuster, H. (1985). Some proposals concerning the mathematical modelling of oscillating hetero-geneous catalytic reactions on metal surfaces. In *Temporal order*, (ed. L. Rensing and N. I. Jaeger), pp. 67–74. Springer, Berlin.

Eckert, E., Hlavacek, V., and Marek, M. (1973*a*). Catalytic oxidation of CO on CuO/Al_2O_3. I. Reaction rate model discrimination. *Chem. Eng. Commun.*, **1**, 89–94.

Eckert, E., Hlavacek, V., and Marek, M. (1973*b*). Catalytic oxidation of CO on CuO/Al_2O_3. II. Measurement and description of hysteresis and oscillations in a laboratory catalytic recycle reactor. *Chem. Eng. Commun.*, **1**, 95–102.

Eigenberger, G. (1978*a*). Kinetic instabilities in heterogeneously caralyzed reactions. I. Rate multiplicity with Langmuir-like kinetics. *Chem. Eng. Sci.*, **33**, 1255–62.

Eigenberger, G. (1978*b*). Kinetic instabilities in heterogeneously catalyzed reactions. II. Oscillatory instabilities with Langmuir-like kinetics. *Chem. Eng. Sci.*, **33**, 1263–8.

Eisworth, M. and Ertl, G. (1986). Kinetic oscillations in the catalytic CO oxidation on a Pt(110) surface. *Surf. Sci.*, **177**, 90–100.

Eiswirth, M., Schwankner, R., and Ertl, G. (1985). Conditions for the occurence of kinetic oscillations in the catalytic oxidation of CO on a Pt(100) surface. *Z. Phys. Chem.*, **144**, 59–67.

Eiswirth, M., Krischer, K., and Ertl, G. (1988). Transition to chaos in an oscillating surface reaction. *Surf. Sci.*, **202**, 565–91.

Eisworth, M., Möller, P., and Ertl, G. (1989). Periodic perturbations of the oscillatory CO oxidation on Pt(110). *Surf. Sci.*, **208**, 13.

Engel, T. and Ertl, G. (1979). Elementary steps in the catalytic oxidation of carbon monoxide on platinum metals. *Adv. Catal.*, **28**, 1–78.

Engel, T. and Ertl, G. (1982). The oxidation of carbon monoxide. In *The chemical physics of solid surfaces and heterogeneous catalysis*, (ed. D. A. King and D. P. Woodruff), Vol. 4, p. 73. Elsevier, Amsterdam.

Ertl, G. (1980). Surface science and catalysis. *Pure Appl. Chem.*, **52**, 2051–60.

Ertl, G. (1990). The pscillatory catalytic oxidation of carbon monoxide on platinum surfaces. In *Spatial inhomogeneities and transient behaviour in chemical kinetics*, (ed. P. Gray, G. Nicolis, F. Baras, P. Borckmans, and S. K. Scott), pp. 565–78. Manchester University Press.

Ertl, G. and Rau, P. (1969). Chemisorption und katalytische Reaktion von Sauerstoff und Kohlenmonoxid an einer Palladium(110)-Oberfläche. *Surf. Sci.*, **15**, 443–65.

Ertl, G., Norton, P. R., and Rüstig, J. (1982). Kinetic oscillations in the platinum catalyzed oxidation of CO. *Phys. Rev. Lett.*, **49**, 177–80.

Flytzani-Stepanopoulos, M. and Schmidt, L. D. (1979). Morphology and etching processes on macroscopic metal catalysts. *Prog. Surf. Sci.*, **9**, 83–11.

Flytzani-Stepanopoulos, M., Wong, S., and Schmidt, L. D. (1977). Surface morphology of platinum catalysts. *J. Catal.*, **49**, 51–82.

Flytzani-Stepanopoulos, M., Schmidt, L. D., and Caretta, R. (1980). Steady state and transient oscillations in NH_3 oxidation on Pt. *J. Catal.*, **64**, 346–60.

Frank-Kamenetskii, D. A. (1955). *Diffusion and heat exchange in chemical kinetics.* Princeton University Press.

Frank-Kamenetskii, D. A. and Buben, N. Ya (1946). The absolute velocities of solution. *Zh. Fiz. Khim.*, **20**, 225–38.

Galwey, A. K., Gray, P., Griffiths, J. F., and Hasko, S. M. (1985). Surface retexturing of Pt wires during the catalytic oxidation of CO. *Nature*, **313**, 668–71.

Gerhardt, M. and Schuster, H. (1989). A cellular automaton describing the formation of spatially ordered structures in chemical systems. *Physica*, **36D**, 209–21.

Gerhardt, M., Schuster, H., and Plath, P. J. (1986). Ein diskretes mathematisches Modell für die Dynamik der Methanoloxidation an einem Palladium-Trägerkatalysator. *Ber. Bunsenges. Phys. Chem.*, **90**, 1040–3.

Gol'dstein, V. M., Sobolev, V. A., and Yablonskii, G. S. (1986). Relaxation self-oscillations in chemical kinetics: a model, conditions for realization. *Chem. Eng. Sci.*, **41**, 2761–6.

Goodman, M. G., Kenney, C. N., Morton, W., and Mukesh, D. (1982). Transient studies of carbon monoxide oxidation over platinum catalyst. *Surf. Sci.*, **120**, L453–60.

Griffiths, K., Jackman, T. E., Davies, J. A., and Norton, P. R. (1984). Interaction of O_2 with Pt(100). I. Equilibrium measurements. *Surf. Sci.*, **138**, 113–24.

Harold, M. P., Sheintuch, M., and Luss, D. (1987*a*). Analysis and modelling of multiplicity features. I. Nonisothermal experiments. *Ind. Eng. Chem. Res.*, **26**, 786–94.

Harold, M. P., Sheintuch, M., and Luss, D. (1987*b*). Analysis and modelling of multiplicity features. II. Isothermal experiments. *Ind. Eng. Chem. Res.*, **26**, 794–804.

Hiam, L., Wise, H., and Chaikin, S. (1968). Catalytic oxidation of hydrocarbons on platinum. *J. Catal.*, **9/10**, 272–6.

Hugo, P. (1970). Stabilität und Zeitverhalten von DurchFluß-Kreislauf-Reaktoren. *Ber. Bunsenges. Phys. Chem.*, **74**, 121–7.

Hugo, P. and Jakubith, M. (1972). Dynamisches Verhalten und Kinetic der Kohlenmonoxid-Oxidation am Platin Katalysor. *Chem. Ind. Tech.*, **44**, 383–7.

Imbihl, R., Cox, M. P., Ertl, G., Müller, H., and Brenig, W. (1985). Kinetic oscillations in the catalytic CO oxidation on Pt(100): theory. *J. Chem. Phys.*, **83**, 1578–87.

Imbihl, R., Cox, M. P., and Ertl, G. (1986). Kinetic oscillations in the catalytic CO oxidation on Pt(100): experiments. *J. Chem. Phys.*, **84**, 3519–34.

Imbihl, R., Ladas, S., and Ertl, G. (1988a). The Co induced $1 \times 2 = 1 \times 1$ phase transition of Pt(110) studied by LEED and work function measurements. *Surf. Sci.*, **206**, L903–12.

Imbihl, R., Sander, M., and Ertl, G. (1988b). The formation of new oxygen adsorption states on Pt(110) by facetting induced by catalytic reaction. *Surf. Sci.*, **204**, L701–7.

Ivanov, E. A., Chumalov, G. A., Sin'ko, M. G., Bruns, D. D., and Luss, D. (1980). Isothermal sustained oscillations due to the influence of adsorbed species on the catalytic reaction rate. *Chem. Eng. Sci.*, **35**, 795–803.

Jackman, T. E., Griffiths, K., Davies, J. A., and Norton, P. R. (1983). Absolute coverages and hysteresis phenomena associated with the CO-induced Pt(100) = hex phase transition. *J. Chem. Phys.*, **79**, 3529–33.

Jaeger, N. I., Möller, K., and Plath, P. J. (1981a). Thermische Oszillationen und chaotisches Verhalten bei der Oxidation von Kohlenmonoxid an Pd-Kristalliten in Inneren einer Zeolithmatrix. *Z. Naturforsch.*, **36a**, 1012–15.

Jaeger, N. I., Plath, P. J., and Van Raay, E. (1981b). Chemische Oszillationen bei der Methanoloxidation an einem Pd-Trägerkatalysator. *Z. Naturforsch.*, **36a**, 395–402.

Jaeger, N. I., Möller, K., and Plath, P. J. (1985a). The development of a model for the cooperative behaviour of palladium crystallites during the catalytic oxidation of CO. *Ber. Bunsenges. Phys. Chem.*, **89**, 633–7.

Jaeger, N. I., Möller, K., and Plath, P. J. (1985b). Dynamics of the heterogeneous catalytic oxidation of carbon monoxide on zeolite supported palladium. In *Temporal order*, (ed. L. Rensing and N. I. Jaeger), pp. 96–102. Springer, Berlin.

Jaeger, N. I., Plath, P. J., and Svensson, P. (1985c). Toroidal oscillations during the oxidation of methanol on zeolite supported palladium. In *Temporal order*, (ed. L. Rensing and N. I. Jaeger), pp. 101–2. Springer, Berlin.

Jaeger, N. I., Möller, K., and Plath, P. J. (1986a). Cooperative effects in heterogeneous catalysis. Part 1—Phenomenology of the dynamics of carbon monoxide oxidation on palladium embedded in a zeolite matrix. *J. Chem. Soc. Faraday Trans. 1*, **82**, 3315–30.

Jaeger, N. I., Ottensmeyer, R., and Plath, P. J. (1986b). Oscillations and coupling phenomena between different areas of the catalyst during the heterogeneous catalytic oxidation of ethanol. *Ber. Bunsenges. Phys. Chem.*, **90**, 1075–9.

Jaeger, N. I., Plath, P. J., and Svensson, P. (1990). Concentration forcing of the oscillating heterogeneous catalytic oxidation of carbon monoxode. In *Spatial inhomogeneities and transient behaviour in chemical kinetics*, (ed. P. Gray, G. Nicolis, F. Baras, P. Borckmans, and S. K. Scott), pp. 593–603. Manchester University Press.

Jayaraman, V. K., Ravi Kumar, V., and Kulkarni, B. D. (1981). Isothermal multiplicity on catalytic surfaces: application to CO oxidation. *Chem. Eng. Sci.*, **36**, 1731–4.

Jensen, K. F. and Ray, W. H. (1980). A new view of ignition, extinction and oscillations on supported catalyst surfaces. *Chem. Eng. Sci.*, **35**, 241–8.

Jones, A., Firth, J. G., and Jones, T. A. (1975). Calorimetric bead techniques for the measurement of kinetic data for gas-solid heterogeneous reactions. *J. Phys. E: Sci. Instrum.*, **8**, 37–40.

Kapicka, J. and Marek, M. (1989). Oscillations on individual catalytic pellets in a packed bed: CO oxidation on Pt/Al_2O_3. *J. Catal.*, **119**, 508–11.

Kaul, D. J. and Wolf, E. E. (1984). FTIR studies of surface reaction dynamics. I. Temperature and concentration programming during CO oxidation on Pt/SiO_2. *J. Catal.*, **89**, 348–61.

Kaul, D. J. and Wolf, E. E. (1985a). FTIR studies on surface reaction dynamics. II. Surface coverage and inhomogeneous temperature patterns of self-sustained oscillations during CO oxidation on Pt/SiO_2. *J. Catal.*, **91**, 216–30.

Kaul, D. J. and Wolf, E. E. (1985b). Selected area FTIR studies of surface reaction dynamics. III. Spatial coverage and temperature patterns during self-sustained oscillations of CO oxidation on Pd/SiO_2. *J. Catal.*, **93**, 321–30.

Keil, W. and Wicke, E. (1980). Uber die kinetischen Instabilitäten bei der CO-oxidation an Platin-Katalysatoren. *Ber. Bunsenges. Phys. Chem.*, **84**, 377–83.

Kiss, J. T. and Gonzales, R. D. (1984a). Catalytic oxidation of carbon monoxide over Ru/SiO_2. An in situ infrared and kinetic study. *J. Phys. Chem.*, **88**, 892–7.

Kiss, J. T. and Gonzales, R. D. (1984b). Catalytic oxidation of carbon monoxide over Rh/SiO_2. An in situ infrared and kinetic study. *J. Phys. Chem.*, **88**, 898–904.

Klein, R. L., Schwartz, S., and Schmidt, L. D. (1985). Kinetics of the NO + CO reaction on clean Pt: steady state rates. *J. Phys. Chem.*, **89**, 4908–14.

Kleine, A., Ryder, P. L., Jaeger, N., and Schulz-Ekloff, G. (1986). Electron microscopy of Pt, Pd and Ni particles in a NaX zeolite matrix. *J. Chem. Soc. Faraday Trans. 1*, **82**, 205–12.

Kleinke, G., Scottke, M., Penka, V., Ertl, G., Behm, R. J., and Moritz, W. (1987). Mechanistic and energetic aspects of the H-induced (1×2) reconstructed structures on Ni(110) and Pd(110). *Surf. Sci.*, **189/190**, 177–84.

Koval, G. L., Fenesko, A. V., Filippov, V. I., Korneichuk, G. P., and Yablonskii, G. S. (1979). Kinetic investigation of CO oxidation in the presence of SO_2. *React. Kinet. Catal. Lett.*, **10**, 315–18.

Kurtanjek, Z., Sheintuch, M., and Luss, D. (1980). Reaction rate oscillations during the oxidation of hydrogen on nickel. *Ber. Bunsenges. Phys. Chem.*, **84**, 374–7.

Ladas, S., Imbihl, R., and Ertl, G. (1988a). Microfacetting of a Pt(110) surface during catalytic CO oxidation. *Surf. Sci.*, **197**, 153–82.

Ladas, S., Imbihl, R., and Ertl, G. (1988b). Kinetic oscillations and facetting during the catalytic CO oxidation on Pt(110). *Surf. Sci.* **198**, 42–68.

Lagos, R. E., Sales, B. C., and Suhl, H. (1979). Theory of oscillatory oxidation of carbon monoxide over platinum. *Surf. Sci.*, **82**, 525–39.

Lamb, T., Scott, R. P., Watts, P., Holland, B., Gentry, S., and Jones, A. (1977). Oscillations of the platinum catalyzed oxidation of carbon monoxide induced by inhibitors. *J. Chem. Soc. Chem. Commun.*, 882–3.

Lesley, M. and Schmidt, L. D. (1985a). The NO + CO reaction on Pt(100). *Surf. Sci.*, **155**, 215–40.

Lesley, M. and Schmidt, L. D. (1985*b*). Chemical autocatalysis in the NO + CO reaction on Pt(100). *Chem. Phys. Lett.*, **102**, 459–65.

Li, Y.-E. and Gonzales, R. D. (1988). Catalytic oxidation of CO on Rh/SiO_2: a rapid-response Fourier transform infrared transient study. *J. Phys. Chem.*, **92**, 1589–95.

Lindstrom, T. H. and Tsotsis, T. T. (1984). Experimental observations of isolated surface steady-state branches. *Surf. Sci.*, **146**, L569–75.

Lindstrom, T. H. and Tsotsis, T. T. (1985). Reaction rate oscillations during CO oxidation on Pt/γ-Al_2O_3. *Surf. Sci.*, **150**, 487–502.

Lobban, L. and Luss, D. (1989). Spatial temperature oscillations during hydrogen oxidation on a nickel foil. *J. Phys. Chem.*, **93**, 6530–3.

Lobban, L., Philippou, G., and Luss, D. (1989). Standing temperature waves on electrically heated catalytic ribbons. *J. Phys. Chem.*, **93**, 733–6.

Lynch, D. T. and Wanke, S. E. (1984*a*). Oscillations during CO oxidation over supported metal catalysts. I. Influence of catalyst history on activity. *J. Catal.*, **88**, 333–44.

Lynch, D. T. and Wanke, S. E. (1984*b*). Oscillations during CO oxidation over supported metal catalysts. II. Effect of reactor operating conditions on oscillatory behaviour for a $(Pt–Pd)/Al_2O_3$ catalyst. *J. Catal.*, **88**, 345–54.

Lynch, D. T., Emig, G., and Wanke, S. E. (1986). Oscillations during CO oxidation over supported metal catalysts. III. Mathematical modelling of the observed phenomenon. *J. Catal.*, **97**, 456–68.

McCarthy, E., Zahradnik, J., Kuczynski, G. C., and Carberry, J. J. (1975). Some unique aspects of CO oxidation on supported Pt. *J. Catal.*, **39**, 29–35.

McLaughlin McClory, M., and Gonzales, R. D. (1986). Catalytic oxidation of carbon monoxide over $RuRh/SiO_2$ bimetallic clusters. *J. Phys. Chem.*, **90**, 628–33.

Möller, P., Wetzl, K., Eiswirth, M., and Ertl, G. (1986). Kinetic oscillations in the catalytic CO oxidation on Pt(100): computer simulations. *J. Chem. Phys.*, **85**, 5328–36.

Morton, W. and Goodman, M. G. (1981). Parametric oscillations in simple catalytic reaction systems. *Trans. Inst. Chem. Eng.*, **59**, 253–9.

Mukesh, D., Cutlip, M. B., Goodman, M., Kenney, C. N., and Morton, W. (1982). The stability and oscillations of carbon monoxide over platinum supported catalyst. Effect of butene. *Chem. Eng. Sci.*, **37**, 1807–10.

Mukesh, D., Kenney, C. N., and Morton, W. (1983*a*). Concentration oscillations of carbon monoxide, oxygen and 1-butene over a platinum supported catalyst. *Chem. Eng. Sci.*, **38**, 69–77.

Mukesh, D., Goodman, M., Kenney, C. N., and Morton, W. (1983*b*). Oscillatory phenomena in heterogeneous catalytic oxidation reaction. *Spec. Per. Rep. Chem. Soc.: Catal.*, **6**, 1–26.

Mukesh, D., Morton, W., Kenney, C. N., and Cutlip, M. B. (1984). Island models and the catalytic oxidation of carbon monoxide and carbon monoxide–olefin mixtures. *Surf. Sci.*, **138**, 237–57.

Müller, E. and Hofmann, H. (1987*a*). Dynamic modelling of heterogeneous catalytic reactions—I. Theoretical considerations. *Chem. Eng. Sci.*, **42**, 1695–704.

Müller, E. and Hofmann, H. (1987*b*). Dynamic modelling of heterogeneous catalytic reactions—II. Experimental results—oxydehydrogenation of isobutyric aldehyde to methacrolein. *Chem. Eng. Sci.*, **42**, 1705–15.

Norton, P. R., Davies, J. A., Creber, D. K., Sitter, C. W., and Jackman, T. E. (1981). The Pt(100) (5 × 20) = (1 × 1) phase transition: a study by Rutherford backscattering, nuclear microanalysis, LEED and thermal desorption spectroscopy. *Surf. Sci.*, **108**, 205–24.

Norton, P. R., Bindner, P. E., Griffiths, K., Jackman, T. E., Davies, J. E., and Rüstig, J. (1984*a*). Kinetic oscillations in oxidation of CO over Pt(100): a study by Rutherford backscattering, nuclear microanalysis, LEED and work function techniques. *J. Chem. Phys.*, **89**, 3859–65.

Norton, P. R., Griffiths, K., and Bindner, P. E. (1984*b*). Interaction of O_2 with Pt(100). II. Kinetics and energetics. *Surf. Sci.*, **138**, 125–47.

Onken, H. U. and Wicke, E. (1986). Statistical fluctuations of temperature and conversion at the catalytic CO oxidation in an adiabatic packed bed reactor. *Ber. Bunsenges. Phys. Chem.*, **90**, 976–81.

Onken, H. U. and Wolf, E. E. (1988). Coupled chemical oscillators on a Pt/SiO_2 catalyst disk. *Chem. Eng. Sci.*, **43**, 2251–62.

Pawlocky, P. C. and Schmitz, R. A. (1987). Spatial effects on supported catalysts. *Chem. Eng. Progr.*, **83** (2), 40–5.

Pikios, C. and Luss, D. (1977). Isothermal concentration oscillations on catalyst surfaces. *Chem Eng. Sci.*, **32**, 191–4.

Plath, P. J., Möller, K., and Jaeger, N. I. (1988). Cooperative effects in heterogeneous catalysis. *J. Chem. Soc. Faraday Trans.* 1, **84**, 1751–71.

Plath, P. J., Ottensmeyer, R., and Jaeger, N. I. (1990). Bifurcation analysis and modelling of the heterogeneously catalyzed oxidation of ethanol. In *Spatial inhomogeneities and transient behaviour in chemical kinetics*, (ed. P. Gray, G. Nicolis, F. Baras, P. Borckmans, and S. K. Scott), pp. 605–12. Manchester University Press.

Plichta, R. T. and Schmitz, R. A. (1979). Oscillations in the oxidation of carbon monoxide on a platinum foil. *Chem. Eng. Commun.*, **3**, 387–98.

Rader, C. G. and Weller, S. W. (1974). Ignition on catalytic wires: linetic parameter determination by heated wire technique. *AIChE J.*, **20**, 515–22.

Rajagopalan, K., Sheintuck, M., and Luss, D. (1980). Oscillatory states and slow activity changes during the oxidation of hydrogen by palladium. *Chem. Eng. Commun.*, **7**, 335–43.

Rathousky, J. and Hlavacek, V. (1982). Oscillatory behavior of long and short isothermal beds packed with Pt/Al_2O_3 catalyst. *J. Catal.*, **75**, 122–33.

Rathousky, J., Puszynski, J., and Hlavacek, V. (1980). Experimental observation of chaotic behavior in CO oxidation in lumped and distributed catalytic systems. *Z. Naturforsch.*, **35a**, 1238–44.

Rathousky, J., Kira, E., and Hlavacek, V. (1981). Experimental observations of complex dynamical behavior in the catalytic oxidation of CO on Pt/alumina catalyst. *Chem. Eng. Sci.*, **36**, 776–80.

Ravi Kumar, V., Jayaraman, V. K., and Kulkarni, B. D. (1981). Isothermal multiplicity on caalytic surfaces. *Chem. Eng. Sci.*, **36**, 945–7.

Razon, L. F. and Schmitz, R. A. (1986). Intrinsically unstable behavior during the oxidation of carbon monoxide on platinum. *Cat. Rev. Sci. Eng.*, **28**, 89–164.

Razon, L. F. and Schmitz, R. A. (1987). Multiplicities and instabilities in chemically reacting systems—a review. *Chem. Eng. Sci.*, **42**, 1005–47.

Razón, L. F., Chang, S.-M., and Schmitz, R. A. (1986). Chaos during the oxidation of carbon monoxide on platinum—experiments and analysis. *Chem. Eng. Sci.*, **41**, 1561.

Regalbuto, J. and Wolf, E. E. (1986). FTIR studies of self-sustained oscillations during the carbon monoxide–nitric acid–oxygen reaction on platinum/silica catalysis. *Chem. Eng. Commun.*, **41**, 315–26.

Rieckert, L. (1981). A note on multiplicity of the steady-state and oscillation phenomena in catalytic oxidation. *Ber. Bunsenges. Phys. Chem.*, **85**, 297–9.

Sales, B. C., Turner, J. E., and Maple, M. B. (1982). Oscillatory oxidation of CO over Pt, Pd, Rh and Ir catalysts: theory. *Surf. Sci.*, **114**, 381–94.

Sant, R. and Wolf, E. E. (1988). FTIR studies of catalyst preparation effects on spatial propagation at oscillations during CO oxidation on Pt/SiO_2. *J. Catal.*, **110**, 249–61.

Saymeh, R. A. and Gonzales, R. D. (1986). Catalytic oxidation of carbon monoxide over Ir/SiO_2. An in situ infrared and kinetic study. *J. Phys. Chem.*, **90**, 622–8.

Schmidt, L. D. and Luss, D. (1971). Physical and chemical characterization of platinum–rhodium gauze catalysts. *J. Catal.*, **22**, 269–79.

Schmitz, R. A., D'Netto, G. A., Razon, L. F., and Brown, J. R. (1984). Theoretical and experimental studies of catalytic reactions. In *Chemical Instabilities*, (ed. G. Nicolis and F. Baras), pp. 33–57. Reidel, Dordrecht.

Schüth, F. and Wicke, E. (1988). The formation of isocyanate on noble metal and supported noble metal catalysts. *Ber. Bunsenges. Phys. Chem.*, **92**, 813–9.

Schüth, F. and Wicke, E. (1989a). IR spectroscopic investigations during oscillations of the CO/NO and the CO/O_2 reaction on Pt and Pd catalysts. I: platinum. *Ber. Bunsenges. Phys. Chem.*, **93**, 191–201.

Schüth, F. and Wicke, E. (1989b). IR spectroscopic investigations during oscillations of the CO/NO and the CO/O_2 reaction on Pt and Pd catalysts. II: palladium. *Ber. Bunsenges. Phys. Chem.*, **93**, 491–501.

Schüth, F., Song, X., Schmidt, L. D., and Wicke, E. (1990). Synchrony and the emergence of chaos in oscillations on supported catalysts. *J. Chem. Phys.*, **92**, 745–56.

Schwankner, R. J., Eiswirth, M., Möller, P., Wetzl, K., and Ertl, G. (1987). Kinetic oscillations in the catalytic CO oxidation on Pt(100): periodic perturbations. *J. Chem. Phys.*, **87**, 742–9.

Schwartz, S. B. and Schmidt, L. D. (1987). Is there a single mechanism of catalytic rate oscillations on Pt? *Surf. Sci.*, **183**, L269–78.

Schwartz, S. B. and Schmidt, L. D. (1988). The NO + CO reaction on clean Pt(100): multiple steady-states and oscillations. *Surf. Sci.*, **206**, 169–86.

Schwartz, S. B., Schmidt, L. D., and Fisher, G. B. (1986). Co + O_2 reaction on Rh(III): steady-state rates and adsorbate coverages. *J. Phys. Chem.*, **90**, 6194–200.

Scott, R. P. and Watts, P. (1981). Kinetic considerations of mass transport in heterogeneous, gas–solid catalytic reactions. *J. Phys. E: Sci. Instrum.*, **14**, 100–13.

Scott, S. K., Griffiths, J. F., and Galwey, A. K. (1990). Restructring of catalyst surfaces during oscillatory reactions. In *Spatial inhomogeneities and transient behaviour in chemical kinetics*, (ed. P. Gray, G. Nicolis, F. Baras, P. Borckmans, and S. K. Scott), pp. 579–92. Manchester University Press.

Shanks, B. H. and Bailey, J. E. (1988). Autonomus oscillations in carbon monoxide oxidation over supported rhodium. *J. Catal.*, **110**, 197–202.

Sheintuch, M. (1990). On the forced catalytic oscillator. *J. Chem. Phys.*, **92**, 3340–7.

Sheintuch, M. and Pismen, L. M. (1981). Inhomogeneities and surface structures in oscillatory catalytic kinetics. *Chem. Eng. Sci.*, **36**, 489–97.

Sheintuch, M. and Schmidt, J. (1986). *Chem. Eng. Commun.*, **44**, 33.

Sheintuch, M. and Schmidt, J. (1988). Bifurcations to periodic and aperiodic solutions during ammonia oxidation on a Pt wire. *J. Phys. Chem.*, **92**, 3404–11.

Sheintuch, M. and Schmitz, R. A. (1977). Oscillations in catalytic reaction. *Catal. Rev. Sci. Eng.*, **15**, 107–72.

Slin'ko, M. G. and Slin'ko, M. M. (1978). Self-oscillations of heterogeneous catalytic reaction rates. *Catal. Rev. Sci. Eng.*, **17**, 119–53.

Slin'ko, M. G., Beskov, V. S., and Zharmakhaubetov, F. K. (1988). Oscillations in oxidation of ammonia on a platinum catalyst. *Dokl. Akad. Nauk. SSSR*, **301**, 398–401.

Slin'ko, M. M., Jaeger, N. I., and Svensson, P. (1989). Mechanism of the kinetic oscillations in the oxidation of CO on palladium dispersed within a zeolite matrix. *J. Catal.*, **118**, 349–59.

Suhl, H. (1981). Two-oxidation-state theory of catalyzed carbon dioxide generation. *Surf. Sci.*, **107**, 88–100.

Svensson, P., Jaeber, N. I., and Plath, P. J. (1988). Phase and frequency relations with forces oscillations of the heterogeneous catalytic oxidation of CO on supported Pd. *J. Phys. Chem.*, **92**, 1882–8.

Takoudis, C. G. and Schmidt, L. D. (1983). Reaction between nitric oxide and ammonia on polycrystalline platinum. II. Rate oscillations. *J. Phys. Chem.*, **87**, 964–8.

Theil, P. A., Behm, R. J., Norton, P. R., and Ertl, G. (1983). The interaction of CO and Pt(100). II. Energetic and kinetic parameters. *J. Chem. Phys.*, **78**, 7448–58.

Tsai, P. K. and Maple, M. B. (1986). Oscillatory oxidation of H_2 over a Pt catalyst. *J. Catal.*, **101**, 142–52.

Tsai, P. K., Maple, M. B. and Herz, R. K. (1988). Coupled catalytic oscillators. *J. Catal.*, **113**, 453–65.

Turner, J. E., Sales, B. C., and Maple, M. B. (1981a). Oscillatory oxidation of CO over a Pt catalyst. *Surf. Sci.*, **103**, 54–74.

Turner, J. E., Sales, B. C., and Maple, M. B. (1981b). Oscillatory oxidation of CO over Pd and Ir catalysts. *Surf. Sci.*, **109**, 591–604.

Varghesse, P., Carberry, J. J., and Wolf, E. E. (1978). Spurious limit cycles and related phenomena during CO oxidation on supported platinum. *J. Catal.*, **55**, 76–87.

Vayenas, C. G., Lee, B., and Michaels, J. (1980). Kinetics, limit cycles and mechanism of the ethylene oxidation on platinum. *J. Catal.*, **66**, 36–48.

Volodin, Y. E., Barelko, V. V., and Khalzov, P. I. (1982). Investigation of instability of oxidation of H_2 and NH_3 + H_2 platinum. *Chem. Eng. Commun.*, **18**, 271–85.

Volokitin, E. P., Treskov, S. A., and Yablonskii, G. S. (1986). Dynamics of CO oxidation: a model with two oxygen forms. *Surf. Sci.*, **169**, L321–6.

Wicke, E. (1974). Instabile Reaktionszustände bei der heterogenen Kastalyse. *Chem. Ing. Tech.*, **46**, 365–74.

Wicke, E. and Onken, H. U. (1986). Statistical fluctuations of conversion and temperature in an adiabatic fixed-bed reactor for CO oxidation. *Chem. Eng. Sci.*, **41**, 1681–8.

Wicke, E. and Onken, H. U. (1988). Periodicity and chaos in a catalytic packed bed reactor for CO oxidation. *Chem. Eng. Sci.*, **43**, 2289–94.

Wicke, E., Kummann, P., Keil, W., and Schiefler, J. (1980). Unstable and oscillatory behaviour in heterogeneous catalysis. *Ber. Bunsenges. Phys. Chem.*, **84**, 315–23.

Xiao, R. R., Sheintuch, M., and Luss, D. (1986). Experimental study of bifurcation diagrams and maps of two parallel interacting catalytic reactions. *Chem. Eng. Sci.*, **41**, 2085–92.

Yeates, R. C., Turner, J. E., Gellmann, A. J., and Somorjai, G. A. (1985). The oscillatory behavior of the CO oxidation reaction at atmospheric pressure over platinum single crystals: surface analysis and pressure dependent mechanisms. *Surf. Sci.*, **149**, 175–90.

Yentakakis, I. V. and Vayenas, C. G. (1988). The effect of electrochemical oxygen pumping on the steady-state and oscillatory behavior of CO oxidation on polycrystaline Pt. *J. Catal.*, **111**, 170–88.

Zhou, X., Barshad, Y., and Gulari, E. (1986). CO oxidation on Pd/Al_2O_3. Transient response and rate enhancement through forced concentration cycling. *Chem. Eng. Sci.*, **41**, 1277–84.

Zunge, J. G. and Luss, D. (1978). Kinetic oscillations during the isothermal oxidation of hydrogen on platinum wires. *J. Catal.*. **53**, 312–20.

11

Electrodissolution reactions

Electrodissolution reactions provide perhaps some of the strongest evidence for the widespread applicability of the non-linear principles presented in previous sections. At first sight an experimental situation that can be uncharitably described as 'dangling a rusty nail in a vat of acid' might not be expected to be in anyway regular let alone sufficiently deterministic to hold to the complex but ordered sequences discussed above. Even though the experiments are actually somewhat more sophisticated they might still not be *a priori* an obvious choice for quantitative experimental investigation. But subtle is the Lord, and these do in fact provide some of the best examples of the order within classical bifurcation theory and chaos.

Some 'good features' can be recognized in advance. The sensitivity and precision with which electrical voltages and currents can be controlled and measured is highly developed and fast. As seen previously we do not need to be able to relate any recorded time series to a particular chemical species; reconstructions, etc. can be performed on any quantitative measure of the system. The experiments can be performed in a number of different ways. Most studies employ a potentiostatic system with a fixed operating voltage imposed between the working (dissolving) and reference electrodes. The applied potential then plays the role of an easily changed bifurcation parameter whilst the current is the measured response variable. Simple forcing experiments can also be performed.

Reports of quantitative examinations of bifurcation sequences and occurrence of chaos etc. have appeared only relatively recently, but the existence of oscillations in electrodissolution processes has a long history. Hudson quotes Fechner (1828), Kistiakowsky (1909), Hedges (1929), Bonhoeffer and Gerischer (1948), Bonhoeffer *et al.* (1948), Cooper and Bartlett (1958), and Wojtowicz (1972) as all having observed irregular scillations in such systems. Much of the earlier (and subsequent) work has concentrated on the technologically important corrosion of iron (Bartlett 1945; Franck and FitzHugh 1961; Rius and Lizarbe 1962; Podesta *et al.* 1979; Russell and Newman 1983; Russell 1984; Beni and Hackwood 1984; Russell and Newman 1986; Orazem and Miller 1986; Diem and Hudson 1987) although others have considered different metals: nickel (Osterwaldand Feller 1960; Osterwald 1962; Indira *et al.* 1969; Lev *et al.* 1988); cobalt (Franck and Meunier 1953; Jaeger *et al.* 1985; Hudson *et al.* 1989); and, particularly, copper (Cooper *et al.* 1980;

Kawczynski *et al.* 1984; Lee *et al.* 1985; Glarum and Marshall 1985; Bassett and Hudson 1987, 1988, 1989*a,b*, 1990; Albahadily *et al.* 1989; Schell and Albahadily 1989). Indira *et al.* also refer to oscillatory dissolutions of silver, aluminium, steel, tin, tungsten, zinc, and gold in various environments. In most cases the metals dissolve into an aqueous mineral acid with, perhaps, added electrolytes such as chloride ion.

11.1. Experimental apparatus

In order to lay the 'rusty nail' jibe once and for all, it is worth describing a typical arrangement of the apparatus for such studies. Bassett and Hudson use a rotating disk electrode with an 8.26 mm diameter cylindrical copper rod embedded in a Teflon cylinder, the latter having a 2 cm diameter. As indicated in Fig. 11.1, only the end of the copper rod is thus exposed to the

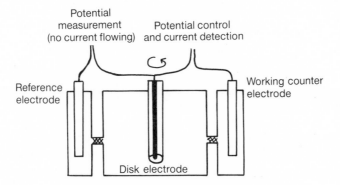

FIG. 11.1. Schematic representation of apparatus for electrodissolution studies.

solution. The electrode is rotated at 200–1000 r.p.m. and operates with two additional electrodes, a platinum sheet counter electrode and a standard calomel reference electrode. (Variations on this basic set-up include the use of an $Hg/HgSO_4$ reference and different diameter or metal rods: in each case, however, the diameter of the Teflon cylinder jacket remains constant.)

The surface of the working electrode needs careful preparation. Bassett and Hudson stress their use of the following procedure: the surface is polished with increasingly fine grades of wet abrasive paper and then cleaned in ethanol. All solutions are rigorously degassed for 2 hours. The electrode may be left in the solution at open circuit before measurement begins to equilibrate.

In the experiment, the working electrode is taken to some initial potential with respect to the reference. This potential may then be varied either continuously or through a series of steps between which the system is monitored under potentiostatic conditions. The current between the working and counter electrode is data captured at typically 120 Hz (about 8 ms between data points). Sweeping the potential across a range provides a rapid indication of any possible bifurcations in the system with respect to this parameter. In theory the fully potentiostatic operation should allow the establishment of transient-free phenomena. This latter ideality is, however, not typically found in practice. There is a continuous evolution of the state of the surface, as we discuss in some detail below, that means that even at constant potential the behaviour of the system is also continuously evolving (though usually rather slowly).

11.2. Potential sweep experiments

Figure 11.2 shows the responses of iron and copper electrodes, in 1 M H_2SO_4 and a solution of 0.1 M Cl^- in 1 N H_2SO_4 respectively, to sweeping the electrode potential. The iron dissolution trace (a), obtained for a sweep rate of 10 mV s^{-1}, shows the current rising with the applied potential until the electrode is about $+0.2$ V with respect to the reference. Two features then emerge. The apparent stationary-state current flattens off considerably to give a characteristic plateau. In the present case, the stationary state also appears to lose stability, undergoing a Hopf bifurcation to give an oscillatory state. At $E = +0.4$ V, the oscillations die out and the current falls discontinuously to zero. We have a hard extinction from a small-amplitude oscillatory state, with many of the features of homoclinic orbit formation seen in other systems in earlier chapters. This extinction is known here as

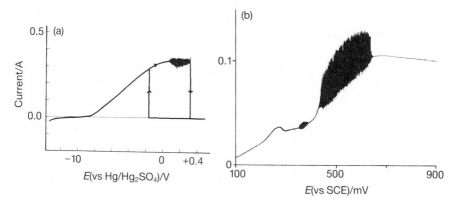

FIG. 11.2. Current–potential bifurcation diagrams from potential sweep experiments indicating oscillatory and stationary-state bifurcation phenomena: (a) Fe showing oscillation and passivation; (b) Cu showing two regions of oscillation. (Reproduced with permission from (a) Diem and Hudson (1987), © American Institute of Chemical Engineers; (b) Bassett and Hudson (1987), © Gordon and Breach.)

'passivation' of the electrode. There is also hysteresis in this response curve. The electrode remains passivated as the potential is decreased until $E \simeq -0.2$ V with respect to the reference electrode. There is then a discontinuous jump characteristic of traversing a saddle–node bifurcation to the previous active state with a current of about 0.3 A.

The copper electrode (Fig. 11.2(b)) does not show passivation or bistability. There are, however, two ranges of oscillatory response, with correspondingly four Hopf bifurcations all of which appear to be supercritical.

Further investigation of the exact location and nature of the various bifurcations and of the forms of any periodic solutions requires steady operating conditions—the potentiostatic mode.

11.3. Potentiostatic responses

In a series of potentiostatic experiments with the copper electrode described above, the behaviour uncovered is slightly more complex than might be inferred from Fig. 11.2(b). For potentials less than 290 mV, a slowly evolving steady-state current is observed. However, oscillations are found for potentials across the whole range $300 \leqslant E/\text{mV} \leqslant 1050$—a much wider extent than revealed in the sweep. Again, the system is continuously evolving. There may be a pseudo-steady, pre-oscillatory period. Oscillations do not last for ever: with $E = 300$ mV, an oscillatory response is observed for 6000 s but at higher potentials this duration tends to decrease; they last only 10 s with $E = 1.05$ V. In the post-oscillatory period, the current decreases to some final steady state. Figure 11.3 characterizes the region in which oscillations are observed and their amplitude, along with the steady final states as a function

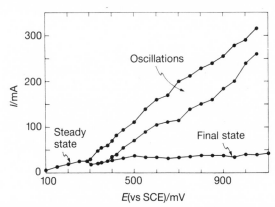

FIG. 11.3. Oscillations and ultimate stationary-state current during Cu dissolution during potentiostatic experiments. (Reproduced with permission from Bassett and Hudson (1987), © Gordon and Breach.)

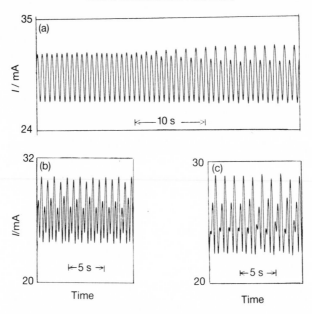

FIG. 11.4. Periodic oscillations during Cu dissolution with $E = 306\,\text{mV}$ (relative to SCE) showing (a) transition from period-1 to period-2, (b) period-4, and (c) transition from period-3 to period-6. Each transition occurs naturally as the experiment proceeds. (Reproduced with permission from Basett and Hudson (1987), © Gordon and Breach.)

of the applied potential. The amplitude is represented by the absolute maxima and minima observed at any given potential. Note that any given system is never oscillating about its final state: the oscillations occur around a slowly evolving unstable state in much the same way as occurs for the simple (but unrelated) model of Chapter 3.

There are also interesting changes in the observed waveform during the oscillatory phase. With $E = 306\,\text{mV}$, Bassett and Hudson (1987) report a development from simple period-1 responses through a sequence of period doublings to a chaotic state, the later emergence of a period-3 waveform that also doubles back into chaos, and later still a period-halving sequence back to period-1 before oscillation ceases. Some of these bifurcations are shown in Fig. 11.4. Although the system is evolving continuously this is a sufficiently slow process so that meaningful time series can be collected and transformed to give reconstructions of pseudo attractors and their power spectra etc., such as that in Fig. 11.5 for a chaotic phase at the above applied potential.

The brief description just given suggests that this is a classic example of a subharmonic cascade to chaos caused by some slowly varying parameter in the system. There are some additional features, however, that reveal a slight complication of that scenario. The system does not appear to go

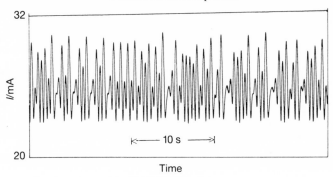

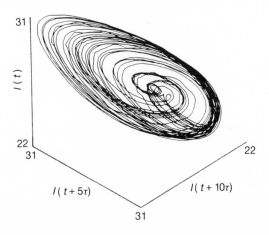

FIG. 11.5. Chaotic evolution during Cu dissolution with $E = 306$ mV showing time series and reconstructed attractor. (Reproduced with permission from Bassett and Hudson (1987), © Gordon and Breach.)

directly from period doubling to chaos, but has an intervening period of 'type III intermittency' (see Section 4.11)—suggesting that later period doublings become subcritical.

Bassett and Hudson also calculated the (correlation) dimension of the strange attractors observed at various applied potentials. This increases as the system passes from intermittency to full chaos (from 2.1 to 2.3 with $E = 306$ mV) and also generally increases with the potential—for a number of results the dimension was found to exceed three (e.g. 3.4 at $E = 335$ mV)—indicating the possibility of the system having more than one positive Lyapunov exponent. Example chaotic time series for some of these higher potentials are shown in Fig. 11.6.

Clean examples of other bifurcations can be found under slightly different operating conditions, With a lower rotation speed of the electrode (200 r.p.m.),

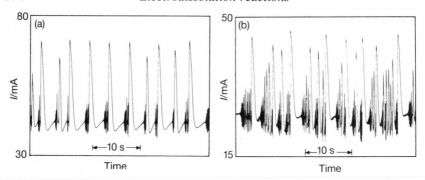

FIG. 11.6. Chaotic evolution for Cu dissolution for (a) $E = 420\,\text{mV}$, (b) $E = 335\,\text{mV}$. (Reproduced with permission from Bassett and Hudson (1987), © Gordon and Breach.)

the oscillations begin at a lower potential (for $250 \leqslant E/\text{mv} \leqslant 650$). A similar bifurcation sequence into chaos as that described previously is observed with $E = 290\,\text{mV}$ (Bassett and Hudson 1989a, 1990), via period doubling and intermittency, but the strange attractor can then evolve into a torus. A typical quasiperiodic time series is shown in Fig. 11.7.

Changes in structure of the strange attractor itself have also been discerned: Fig. 11.8 compares the 'two-band' and 'one-band' chaos. This change also occurs before the transition to a torus. The strange attractors for both small- and large-amplitude chaos (observed at low and high E respectively as discussed above) can also reveal their homoclinic origins: the single loops around the attractors shown in Fig. 11.9 have a definite Shil'nikov 'saddle-focus' form.

The quasiperiodic torus has also been observed to period-double (Bassett and Hudson 1989b), as revealed most clearly by the Poincaré sections and shown in Fig. 11.10.

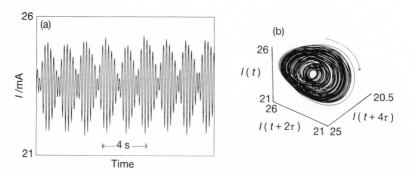

FIG. 11.7. Quasiperiodic time series (a) and reconstructed torus (b) for Cu dissolution. (Reproduced with permission from Bassett and Hudson (1989a), © American Chemical Society.).

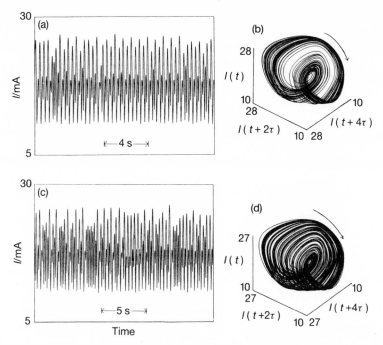

Fig. 11.8. (a)–(b) 'Two-band' chaos during dissolution of Cu; (c)–(d) 'one-band' chaos. (Reproduced with permission from Bassett and Hudson (1989a), © American Chemical Society.)

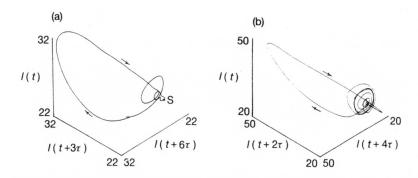

Fig. 11.9. Reconstructed attractors for (a) small-amplitude and (b) large-amplitude chaos during during Cu dissolution showing the Shil'nikov homoclinicity structure from a single loop round the attractor. (Reproduced with permission from Bassett and Hudson (1988), © American Chemical Society.)

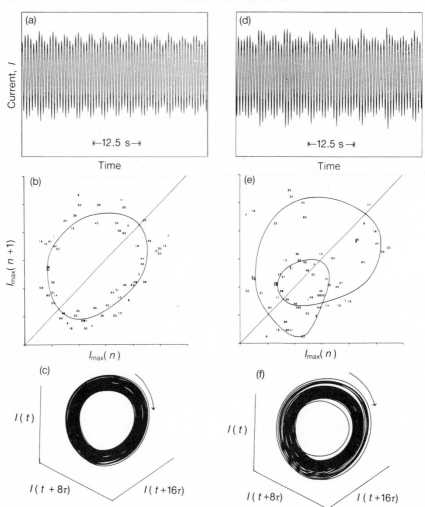

Fɪɢ. 11.10. Period doubling of a torus during the electrodissolution of copper: (a) quasiperiodic trace with (b) next-maximum map and (c) reconstructed three-attractor; (d) period-doubled quasiperiodicity with (e) double loop in next-maximum map and (f) reconstructed attractor. (Reproduced with permission from Bassett and Hudson (1989*b*), © Elsevier.)

11.4. Mixed-mode oscillations and Farey sequences

Albahadily *et al.* (1989; Alabahadily and Schell 1988; Schell and Alabadadily 1989) have observed behaviour very reminiscent of the Belousov–Zhabotinskii reaction, and to the model computations for the non-isothermal autocatalator

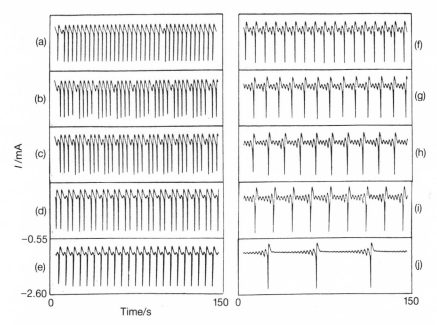

FIG. 11.11. Mixed-mode oscillatory sequence during the electrodissolution of Cu in phosphoric acid: (a) $E = 426.70$ mV, almost 10 behaviour; (b) $E = 427.00$ mV, 4^1; (c) $E = 427.05$ mV, 3^1; (d) $E = 427.25$ mV, 2^1; (e) $E = 428.75$ mV, 1^1; (f) $E = 430.15$ mV, 1^2; (g) $E = 430.30$ mV, 1^3; (h) $E = 430.40$ mV, 1^4; (i) $E = 430.45$ mV, 1^5; (j) $E = 430.49$ mV, 1^{many}, close to 0^1. (Reproduced with permission from Alibahadily *et al.* (1989), © American Institute of Physics.)

scheme of Section 4.1, in their studies of copper dissolution in phosphoric acid. The experimental apparatus is similar to that described above, although they used an 85 per cent acid solution at $-20\,°C$ and a relatively high electrode rotation rate (typically 1600 r.p.m.).

Figure 11.11 shows a sequence of mixed-mode states observed on increasing the potential of the copper electrode. These clearly have the L^s form discussed at length in Chapter 8, bearing a striking resemblance to those time series in Fig. 8.6. A Farey tree constructed from this sequence is given in Fig. 11.12 and indicates the levels of concatenation recorded.

Albahadily *et al.* (1989) constructed a potential–rotation speed parameter plane (Fig. 11.13) which also has regions of chaotic and quasiperiodic responses distinct from the mixed-mode oscillations. Closer investigation of the concatenated sequences (Schell and Albahadily 1989) also revealed the alternating periodicity and chaos of the mixed-mode waveforms reported for the Belousov–Zhabotinskii system.

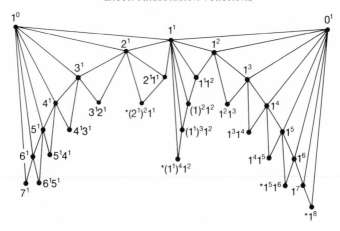

FIG. 11.12. Farey tree associated with the mixed-mode sequence represented in Fig. 11.11. (Reproduced with permission from Alibahadily *et al.* (1989), © American Institute of Physics.)

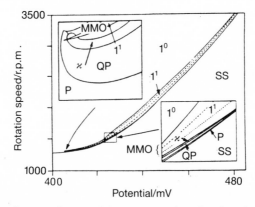

FIG. 11.13. Rotation speed–potential parameter plane for the mixed-mode sequences in Cu dissolution. (Reproduced with permission from Alibahadily *et al.* (1989), © American Institute of Physics.)

11.5. Other systems

Lev *et al.* (1988) have observed many similar sequences to those described above for the electrodissolution of nickel. As well as showing the period doubling to chaos, these authors discuss the evolution of the period-1 limit cycle from its birth at a Hopf bifurcation to extinction at a homoclinic orbit (saddle loop) in the simplest cases, via reconstructed attractors.

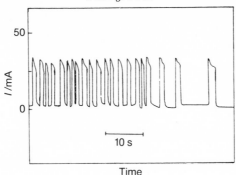

Fig. 11.14. Typical 'relaxation' oscillation waveforms for the electrodissolution of an Fe wire. (Reproduced with permission from Diem and Hudson (1987), © American Institute of Chemical Engineers.)

With iron dissolution, much more complex periodic and, in particular, chaotic evolutions can be observed. Attractors with unusually high fractal dimension—up to 6.0—have been reported (Diem and Hudson 1987). The oscillations observed when an iron wire was used rather than a rotating disk can have the form shown in Fig. 11.14, with significant periods within a given cycle corresponding to an almost completely passivated state (zero current) and rather spiked increases and decreases.

11.6. Film growth

Perhaps the most noticeable physical accompaniment to the dissolution reactions is the growth of a film of insoluble corrosion products on the working metal surface. With copper, this film has a complex composition but appears to be mainly $CuCl$ and Cu_2O resulting from the oxidation reaction

$$Cu + Cl^- \rightarrow CuCl + e^-$$

becoming pure copper(I) chloride at the higher working potentials. As this film grows so it introduces an extra resistance to the dissolution process. The film grows from the outside of the circular surface and can become quite thick.

It is this film growth that provides the slowly varying parameter in the system rather than consumption of the reactants. The limit situation arises when the CuCl formation is matched by additional oxidation processes or complexation

$$CuCl + Cl^- \rightarrow Cu^{2+}(aq) + 2Cl^-(aq) + e^-$$

and

$$CuCl + nCl^- \rightarrow [CuCl_{n+1}(aq)]^{n-}.$$

Film growth also occurs with the other metals investigated, in each case having a passivating influence. With cobalt electrodes, Hudson *et al.* (1989) used a hydrochloric acid/chromic acid electrolyte from which chromium is plated on the electrode surface. The films may remain incomplete allowing some continued oxidation through patches or defects.

11.7. Modelling

There are only a few modelling studies that have addressed the theoretical foundation of the bifurcations described above. Talbot and Oriani (1985) examined a model due to Griffin (1984). This assumes there is an oxidation reaction occurring at the surface of the metal, providing adsorbed ions according to

$$M \rightarrow M^{n+}(ads) + n\,e^- \qquad \text{rate coefficient} = k_{ox}.$$

The dissolution process can occur in either of two ways: the direct dissolution of single adsorbed ions

$$M^{n+}(ads) \rightarrow M^{n+}(aq) \qquad \text{rate coefficient} = k_d'$$

appropriate to low surface coverages, or dissolution from an adsorbed layer appropriate to high coverages and with a different rate coefficient

$$M^{n+}(ads) \rightarrow M^{n+}(aq) \qquad \text{rate coefficient} = k_d''.$$

Both the dissolution steps are assumed to depend on the extent of coverage θ in some exponential form:

$$k_d = k_d^0 \exp(-\beta\theta)$$

for either k_d' or k_d'' and β is an adsorbate interaction parameter. The dissolution current can then be expressed as

$$i_d = k_d^0\theta \exp(-\beta\theta)$$

and this must equal the current due to the oxidation process, given by the Tafel expression

$$i_{ox} = k_{ox}^0(1 - \theta) \exp(nFE/RT)$$

where E is the electrode potential.

Equating the currents leads to an explicit relationship between the stationary-state potential E_{ss} in terms of the surface coverage θ_{ss}:

$$E_{ss} = \frac{RT}{nF}\left[-\beta\theta_{ss} + \ln\left(\frac{k_d^0}{k_{ox}^0}\right) + \ln\left(\frac{\theta_{ss}}{1-\theta_{ss}}\right)\right].$$

This condition allows multiple stationary states provided $\beta > 4$, but it is not cast in a form appropriate for a discussion of local stability of oscillations.

For oscillatory responses, various authors have built on a model suggested by Franck and FitzHugh (1961) which considers the transition of the anodic metal between active and passive states. An important concept here is the Flade potential E_F at which the kinetics of the dissolution process change discontinuously from the active to passive form (Flade 1911). The major carrier of current through the solution in the dissolution cell will be H^+ ions and as dissolution occurs the concentration of H^+ local to the electrode will decrease. The Flade potential is not a constant, but depends on the H^+ concentration, decreasing as the acidity decreases. If E_F falls with $[H^+]$ to a value below the imposed electrode potential E_p, then the electrode becomes passivated. A passive film then forms, decreasing the rate of dissolution. A subsequent increase in $[H^+]$ may then arise if diffusion exceeds consumption sufficiently, and this may cause E_F to rise past E_p again (Wang et al. 1990).

We thus have some feedback mechanism and the idea of a discontinuous change in the electrode kinetics suggests a strong non-linearity. To see the latter, the governing rate equations can be written in terms of two variables: the difference between the electrode potential and the Flade potential $E = E_p - E_F$, and θ the fraction of the surface that is passive. Then the Franck–FitzHugh equations can be expressed in the following dimensionless forms (Talbot and Oriani 1985; Pearlstein et al. 1985; Pearlstein and Johnson 1989):

$$\frac{dV}{dt} = bh - V - b\theta + c^{-1}VG(V, \theta) \tag{11.1}$$

$$\frac{d\theta}{dt} = VG(V, \theta) \tag{11.2}$$

where the dimensionless potential difference V is simply proportional to E; b, c, and h are parameters of which b and c are non-negative; and $G(V, \theta)$ is a function given by

$$G(V, \theta) = \theta + (1 - 2\theta)H(V) \tag{11.3}$$

and $H(E)$ is the Heaviside step function. The form of this step function and its dependence on the potential E is such that $G(V, \theta) = \theta$ for $V < 0$ (and so $E < 0$), i.e. when the electrode potential is less than the Flade potential and electrodissolution is occurring. On the other hand, $G(V, \theta) = 1 - \theta$ for V and $E > 0$, when the potential exceeds E_F and the system is passivating.

In the first of these two cases, $d\theta/dt = V\theta$, and so we have an autocatalytic production of a surface film and also an autocatalytic final term in the potential equation. During passivation, the increasing surface coverage has a deceleratory effect, with $d\theta/dt = V(1 - \theta)$. These aspects will also be reflected in the signs of various terms in the Jacobian matrix which determines the local stability of any given stationary-state solution of eqns (11.1) and (11.2).

The stationary state requires $VG(V, \theta) = 0$ from eqn (11.2). There are three possible situations: case (a) $V_{ss} = 0$ and case (b) $G(V_{ss}, \theta_{ss}) = 0$. The latter of these two divides into two subclasses: (bi) $V_{ss} < 0$ and $G = \theta_{ss} = 0$; (bii) $V_{ss} > 0$ with $\theta_{ss} = 1$ and $G = 1 - \theta_{ss} = 0$. Substituting these into $dV/dt = 0$ from eqn (11.1) we thus have for the full solution:

case (a) $V_{ss} = 0, \theta_{ss} = h$ $0 \leqslant h \leqslant 1$

case (bi) $V_{ss} = bh, \theta_{ss} = 0$ $h \leqslant 0$

case (bii) $V_{ss} = b(h - 1), \theta_{ss} = 1$ $h \geqslant 1$.

If h is taken to be the major bifurcation parameter related to the working electrode potential E_p then there is a unique stationary state for any given h, with the solution moving from case (bi) to (a) to (bii) as h increases—giving a continuous locus but with discontinuities in the gradient as shown in Fig. 11.15. The parameter c does not influence the location of the stationary states but may affect the local stability and dynamics.

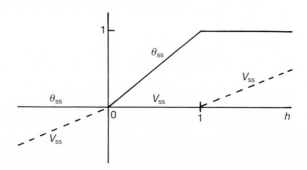

FIG. 11.15. Stationary-state solutions for potential difference and fraction of surface passivated in terms of applied voltage h for Franck–FitzHugh model.

Pearlstein and Johnson (1989) have performed the local stability analysis and full integrations. The stationary state is globally asymptotically stable, so no periodic solutions can exist if any of the conditions $h < 0$, $h > 1$, $c < 0$, or $c > 1$ are satisfied. Any 'interesting' bifurcations must therefore occur inside the parameter square $0 \leqslant h \leqslant 1$, $0 \leqslant c \leqslant 1$ as shown in Fig. 11.16. In fact

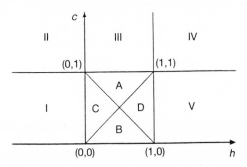

FIG. 11.16. The h–c parameter plane showing regions (I)–(V) and (A)–(D) in which the local stability of the stationary state for the Franck–FitzHugh model may change. (Reproduced with permission from Pearlstein and Johnson (1989), © The Electrochemical Society.)

this square can be further subdivided into four triangles and the stationary state is always stable in A and unstable in B. Pseudo-Hopf bifurcations can (indeed must) occur in regions C and D. These emerge at the discontinuities giving rise to oscillatory responses. The discontinuities also allow for extinction at homoclinic orbits. As the Hopf points may be either subcritical or supercritical, oscillations may coexist with stable stationary states.

Rather than work with the Heaviside function, Wang *et al.* (1990) have smoothed the change in kinetics at the Flade potential by replacing the $G(V, \theta)$ function defined by eqn (11.3) with the form

$$G^*(V, \theta) = \theta + (1 - 2\theta)\left(1 - \frac{1}{\exp(\alpha V) + 1}\right) \tag{11.4}$$

where α is some typically large constant that switches G^* rapidly but continuously from looking like θ for $V < 0$ to $(1 - \theta)$ for $V > 0$.

Equations (11.1), (11.2), and (11.4) admit only one possible stationary-state solution: $V_{ss} = 0$, $\theta_{ss} = h$, so the local H^+ concentration must adjust such that the Flade potential is equal to the fixed electrode potential, and the fraction of the surface covered film is given simply by the parameter h. Note now that only the range $0 \leqslant h \leqslant 1$ is physically acceptable.

The local stability is governed by the Jacobian, which, when evaluated at the stationary state, has the form

$$\mathbf{J} = \begin{pmatrix} -1 + 1/2c & -b \\ \frac{1}{2} & 0 \end{pmatrix}. \tag{11.5}$$

Hopf bifurcation requires $\mathrm{tr}(\mathbf{J}) = (1 - 2c)/2c = 0$ with $\det(\mathbf{J}) = \frac{1}{2}b > 0$. The latter condition is satisfied for positive b, and the trace vanishes with $c = c^* = \frac{1}{2}$. The values of h and b do not influence the Hopf bifurcation

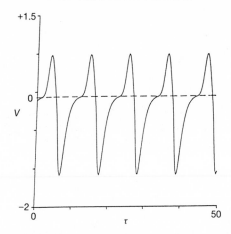

FIG. 11.17. Oscillations in the Wang *et al.* modification of the Franck–FitzHugh model. (Reproduced with permission from Wang *et al.* (1990), © The Electrochemical Society.)

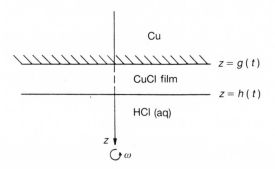

FIG. 11.18. Schematic representation of the model for film growth in Cu electrodissolution showing basis of coordinate system. (Reproduced with permission from Pearlstein *et al.* (1985), © The Electrochemical Society.)

condition, but do affect the waveform of any oscillatory solution emerging for $c < c^*$. Example oscillations are shown in Fig. 11.7. As there are only two variables, no higher-order bifurcations can occur.

The Franck-FitzHugh models may be appropriate to iron dissolution as they predict passivation to play an important role. This is not the observed case experimentally for copper dissolution, as discussed above. Pearlstein *et al.* (1985; Lee *et al.* 1985; Lee and Nobe 1986) considered a different picture for this latter case, as sketched in Fig. 11.18. The model allows one-dimensional diffusion of chloride ion from the bulk across a growing, porous film to the metal surface where reaction occurs. The film is in contact with the bulk at some point $h(t)$ and meets the metal surface at $g(t)$: both these locations may vary with time, i.e. we have a moving boundary-value problem.

The equation governing the diffusion process is simply the appropriate form of Fick's law

$$\frac{\partial c}{\partial t} = D_f \frac{\partial^2 c}{\partial z^2}$$

over $g(t) \leqslant z \leqslant h(t)$, where c is the chloride ion concentration and z the distance. The boundary conditions specify that for a stationary state the flux at the metal surface must match the reaction rate there

$$D_f(\partial c/\partial z) = k_1 c \qquad \text{at} \qquad z = g(t)$$

whilst at the bulk solution–film interface

$$c = k_H c_\infty \qquad \text{at} \qquad z = h(t)$$

where k_H is the solute distribution coefficient and c_∞ is the bulk concentration. Closure requires two more conditions to locate the film boundaries g and h. These are obtained from total Cu balances across the two interaces. Thus at the metal–film interface

$$-\frac{\rho_{Cu}}{M_{Cu}} \frac{dg(t)}{dt} = k_1 c \qquad \text{at} \qquad z = g(t)$$

and at the solution–film surface

$$-\frac{\rho_{CuCl}}{M_{CuCl}} (1 - \varepsilon) \frac{d[h(t) - g(t)]}{dt} = -\frac{\rho_{Cu}}{M_{Cu}} \frac{dg(t)}{dt} - [\alpha(\omega)c_\infty + k_3].$$

In these equations ρ represents density and M the molecular mass, ε is the void fraction in the porous film, α is a mass-transfer-limited reaction rate and a function of the rate of rotation ω of the disk electrode (the assumed form is $\alpha \sim \omega^{1/2}$: k_1 is the rate constant of the anodic dissolution process forming CuCl, whilst k_3 is the rate constant for the further oxidation step forming $Cu^{2+}(aq)$. Again the rate constants are allowed a Tafel dependence on the electrode potential E.

Stationary-state solutions will give time-varying values for the positions $g(t)$ and $h(t)$, but a constant film thickness $\Delta = h(t) - g(t)$. Once the governing equations have been recast in terms of 'film fixed' coordinates (i.e. the origin of the new distance coordinate moves at the same velocity as the film) a steady-state concentration distribution and potential can be determined. Pearlstein et al. derived an explicit solution, finding a unique stationary state

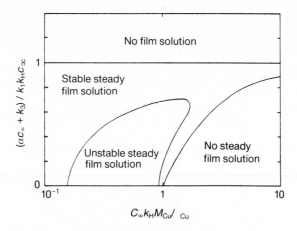

FIG. 11.19. Typical parameter plane for the film growth model showing different regions corresponding to different dynamical responses. (Reproduced with permission from Pearlstein *et al.* (1985), © The Electrochemical Society.)

for all conditions, but one that could lose local stability and give rise to oscillations. Various bifurcation diagrams were constructed; an example is shown in Fig. 11.19, indicating conditions for which oscillations can be expected.

References

Albahadily, F. N. and Schell, M. (1988). An experimental investigation of periodic and chaotic electrochemical oscillations in the anodic dissolution of copper in phosphoric acid. *J. Chem. Phys.*, **88**, 4312–19.

Albahadily, F. N., Ringland, J., and Schell, M. (1989). Mixed-mode oscillations in an electrochemical system. I. A Farey sequence which does not occur on a torus. *J. Chem. Phys.*, **90**, 813–21.

Bartlett, J. H. (1945). Transient anode phenomena. *Trans. Electrochem. Soc.*, **87**, 521.

Bassett, M. R. and Hudson, J. L. (1987). The dynamics of the electrodissolution of copper. *Chem. Eng. Commun.*, **60**, 145–59.

Bassett, M. R. and Hudson, J. L. (1988). Shil'nikov chaos during copper electrodissolution. *J. Phys. Chem.*, **92**, 6963–6.

Bassett, M. R. and Hudson, J. L. (1989a). Quasiperiodicity and chaos during an electrochemical reaction. *J. Phys. Chem.*, **93**, 2731–6.

Bassett, M. R. and Hudson, J. L. (1989b). Experimental evidence of period doubling of tori during an electrochemical reaction. *Physica D*, **35**, 289–98.

Bassett, M. R. and Hudson, J. L. (1990). Homoclinic chaos, quasiperiodicity, noodle maps and other phenomena in an electrochemical reaction. In *Spatial inhomogeneities and transient behaviour in chemical kinetics*, (ed. P. Gray, G. Nicolis, F. Baras, P. Borckmans, and S. K. Scott), pp. 67–75. Manchester University Press.

Beni, G. and Hackwood, S. (1984). Intermittent turbulence and period doubling at the corrosion–passivity transition in iron. *J. Appl. Electrochem.*, **14**, 623–6.

Bonhoeffer, K. F. and Gerischer, H. (1948). Über periodische Chemische Reaktionen V. Das anodische Verhalten von Kupfer in Salzsäure. *Z. Elektrochem.*, **54**, 149–60.

Bonhoeffer, K. F., Brauer, E., and Langhammer, G. (1948). Über periodische Reaktionen. II. *Z. Elektrochem.*, **51**, 29.

Cooper, R. and Bartlett, J. H. (1958). Convection and film instability: copper anodes in hydrochloric acid. *J. Electrochem. Soc.*, **105**, 109–16.

Cooper, J., Muller, R., and Tobias, C. (1980). Periodic phenomena during anodic dissolution of copper at high current densities. *J. Electrochem. Soc.*, **127**, 1733–44.

Diem, C. B. and Hudson, J. L. (1987). Chaos during the electrodissolution of iron. *AIChE J.*, **33**, 218–24.

Fechner, G. Th. (1828). Zur Elektrochemie über Umkehrungen der Polarität in der einfachen Kette. *Schweigg J. Chem. Phys.*, **53**, 29.

Flade, F. (1911). Betrade zur Kenntnis Passivität. *Z. Phys. Chem.*, **76**, 513–59.

Franck, U. F. and FitzHugh, R. (1961). Periodische Elektrodenprozesse und ihre Beschreibung durch ein mathematische Moodell. *Z. Elektrokhem.*, **65**, 156–68.

Franck, U. F. and Meunier, L. (1953). Gekoppelte periodische Elektrodenvorgänge. *Z. Naturforsch.*, **8b**, 396–406.

Glarum, S. H. and Marshall, J. H. (1985). The anodic dissolution of copper into phosphoric acid. *J. Electrochem. Soc.*, **132**, 2872–8.

Griffin, G. L. (1984). A simple phase transition model for metal passivation kinetics. *J. Electrochem. Soc.*, **131**, 18–21.

Hedges, E. S. (1929). An enquiry into the cause of periodic phenomena in electrolysis. *J. Chem. Soc.*, 1028–38.

Hudson, J. L., Bell, J. C., and Jaeger, N. I. (1989). Potentiostatic current oscillations of cobalt electrodes in hydrochloric acid/chromic acid electrolytes. *Ber. Bunsenges. Phys. Chen.*, **92**, 1383–7.

Indira, K. S., Rangarajan, S. K., and Doss, K. G. S. (1969). Further studies on periodic phenomena in passivating systems. *Electroanal. Chem. Interfacial Electrochem.*, **21**, 57–68.

Jaeger, N. I., Plath, P. J., and Quyen, N. Q. (1985). Oscillatory behabiour of cobalt electrodes during transition from the active to the passive state. In *Temporal order*, (ed. L. S. Rensing and N. I. Jaeger), pp. 103–4. Springer, Berlin.

Kawczynski, A. L., Przasnyski, M., and Barampwski, B. (1984). Chaotic and periodic current oscillations at constant voltage conditions in the system $Cu(s)|CuSO_4 + H_2SO_4(aq)|Cu(s)$. *J. Electroanal. Chem.*, **179**, 285.

Kistiakowsky, W. (1909). Ein Wechselstrom Lieferndes Galvanisches Element. *Z. Elektrokhem.*, **15**, 268.

Lee, H. P. and Nobe, K. (1986). Kinetics and mechanisms of Cu electrodissolution in chloride media. *J. Electrochem. Soc.*, **133**, 2035–43.

Lee, H. P., Nobe, K., and Pearlstein, A. J. (1985). Film formation and current oscillations in the electrodissolution of Cu in acidic chloride media. I. Experimental studies. *J. Electrochem. Soc.*, **132**, 1031–7.

Lev, O., Wolfberg, A., Sheintuch, M., and Pismen, L. M. (1988). Bifurcations to periodic and chaotic motions in anodic nickel dissolution. *Chem. Eng. Sci.*, **43**, 1339–53.

Orazem, M. E. and Miller, M. G. (1986). The distribution of current and formation of a salt film on an iron disk below the passivation potential. *J. Electrochem. Soc.*, **134**, 392–9.

Osterwald, J. (1962). Zum stabilitatsverhalten stationarer Elektrodenzustande. *Electrochim. Acta*, **7**, 523–32.

Osterwald, J. and Feller, H. G. (1960). Periodic phenomena at Ni electrode in H_2SO_4. *J. Electrochem. Soc.*, **107**, 473–4.

Pearlstein, A. J. and Johhnson, J. A. (1989). Global and conditional stability of the steady state and periodic solutions of the Franck–FitzHugh model of electrodissolution of Fe in H_2SO_4. *J. Electrochem. Soc.*, **136**, 1290–9.

Pearlstein, A. J., Lee, H. P., and Nobe, K. (1985). Film formation and current oscillations in the electrodissolution of copper in acidic chloride media. II. Mathematical model. *J. Electrochem. Soc.*, **132**, 2159–65.

Podesta, J. J., Piatti, R. C. V., and Arvia, A. J. (1979). The potentiostatic current oscillations at iron/sulfuric acid solution interfaces. *J. Electrochem. Soc.*, **126**, 1363–7.

Rius, A. and Lizarbe, R. (1962). Study of the anodic behaviour of iron at high potentials in solutions containing chloride ions. *Electrochim. Acta*, **7**, 513–22.

Russell, P. P. (1984). Corrosion of iron: the active–passive transition and sustained electrochemical oscillations. *Ph.D. dissertation*. University of California, Berkeley.

Russell, P. P. and Newman, J. (1983). Experimental determination of the passive–active transition for iron in 1 M sulfuric acid, *J. Electrochem. Soc.*, **130**, 547–53.

Russell, P. P. and Newman, J. (1986). Current oscillations observed within the limit current plateau for iron in sulfuric acid. *J. Electrochem. Soc.*, **133**, 2093–7.

Schell, M. and Albahadily, F. N. (1989). Mixed-mode oscillations in an electrochemical ystem. II. A periodic–chaotic sequence. *J. Chem. Phys.*, **90**, 822–8.

Talbot, J. B. and Oriani, R. A. (1985). Application of linear stability analysis to passivation models. *J. Electrochem. Soc.*, **132**, 1545–50.

Wang, Y., Hudson, J. L., and Jaeger, N. I. (1990). On the Franck–FitzHugh model of the dynamics of iron electrodissolution in sulfuric acid. *J. Electrochem. Soc.*, **137**, 485–8.

Wojtowicz, J. (1972). Oscillatory behavior in electrochemical systems. In *Modern aspects of electrochemistry*, (ed. J. O. Bockris and B. Conway), p. 47–120. Plenum, New York.

12

Biochemical systems

No book dealing with chemical non-linearities, oscillations, and chaos would be complete without some mention of biochemical systems. On the other hand, a thorough treatment of this part of the subject would double the length of the book. As some form of compromise, this chapter will make a highly selective (probably somewhat random) foray mainly to indicate some of the scope.

Various recent reviews exist including Olsen and Degn (1985), Rapp (1986), Glass *et al.* (1986), Aihara and Matsumoto (1986), Ross and Schell (1987), and the books by Segel (1980), Babloyantz (1986), Glass and Mackey (1988), and by Murray (1977, 1990). Slightly earlier work is summarized in the collection edited by Chance *et al.* (1973), and the papers by Eigen (1971), by Tyson and Othmer (1978), and by Hess and Boiteux (1980).

At the macroscopic level, oscillatory processes seem relatively common in biological systems. Familiar and immediate examples include rhythmic heartbeat and respiration, each with periods typically of a second. Populations and epidemics can show periodic fluctuations over longer timescales. With the development of more sophisticated monitoring techniques oscillations in nerve functions have also been detected. The discovery of 'laboratory' oscillations in biochemical subsystems dates back to 1964, with the study of glycolysis. Such results indicate the way in which important but complex biochemicals such as enzymes are recycled efficiently rather than produced from new each time they are required—a lesson we should all learn from in a wider context. Before discussing the glycolytic oscillator, it is worth reviewing briefly some basic enzyme kinetics.

12.1. Enzyme rate laws

The simplest interpretation of enzyme catalysis of the conversion of some substrates S to a product P is that due to Michaelis and Menten (1913). A typical dependence of observed (initial) rate V on (initial) substrate concentration has the same form shown in Fig. 12.1 (a). The same data give a linear relationship when $1/V$ is plotted against $1/[S]$ following Lineweaver and Burk (1934; Haldane 1930) (Fig. 12.1(b)).

The kinetic mechanism supposed to operate involves the reversible

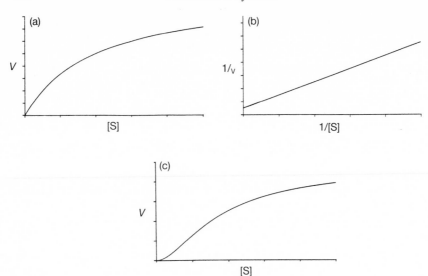

FIG. 12.1. Michaelis–Menten enzyme kinetics: (a) dependence of reaction rate on substrate concentration; (b) Lineweaver–Burk plot; (c) dependence of reaction rate on substrate concentration for an allosteric enzyme process..

formation of an enzyme–substrate complex: competing with the dissociation of the complex is an essentially irreversible reaction step that releases free enzyme and the product.

$$E + S \rightleftharpoons ES \qquad k_1, k_{-1} \qquad (12.1)$$

$$ES \rightarrow E + P \qquad k_2. \qquad (12.2)$$

There are various mathematical justifications of the Michaelis–Menten analysis, based for instance on asymptotic methods and the typical smallness of the initial ratio of enzyme to substrate concentrations (see Segel (1988), Segel and Slemrod (1989) or Murray (1990, Chapter 5.2) for a recent review). For a chemist untroubled by *a priori* questions of rigour the process simply involves making a steady-state approximation to the concentration of the (reactive) intermediate and noting that the total concentration of free and complex enzyme must be some fixed amount, the total enzyme concentration $[E]_0$. This gives two algebraic equations

$$\frac{d[ES]}{dt} = k_1[S][E] - k_{-1}[ES] - k_2[ES] \approx 0 \qquad (12.3)$$

and

$$[E]_0 = [E] + [ES]. \qquad (12.4)$$

Eliminating the free enzyme concentration, we then obtain for $[ES]_{ss}$

$$[ES]_{ss} = \frac{[E]_0[S]}{K_M + [S]} \qquad (12.5)$$

where $K_M = (k_{-1} + k_2)/k_1$ is the 'Michaelis constant'. The rate of product formation V is simply the rate of the second step, given by $k_2[ES]_{ss}$ under steady-state conditions. Thus

$$V = \frac{V_{max}[S]}{K_M + [S]} \qquad (12.6)$$

where the maximum rate $V_{max} = k_2[E]_0$ is that corresponding to the maximum enzyme substrate concentration (all enzyme complexed). Equation (12.6) has the form of Fig. 12.1(a) and inverting we obtain the linear form

$$\frac{1}{V} = \frac{1}{V_{max}} + \frac{K_M}{V_{max}} \frac{1}{[S]} \qquad (12.7)$$

allowing the maximum rate and Michaelis constant to be determined from the intercept and slope of Fig. 12.1(b).

Of more interest in the present context is the behaviour of 'allosteric' enzymes. These are enzymes consisting of more than one catalytically active subunits, each with its own active site where binding and subsequent reaction can occur. The important feature here is that the binding of a substrate to one such site affects the activity of the remaining subunits (if it increases the activity we have positive feedback or activation; if binding reduces the activity of other sites we have negative feedback and inhibition). The allosteric effect can be exemplified by a two-unit enzyme (Monod *et al.* 1965; see also Rubinow 1975). Underlying this interpretation is the idea that such enzymes can exist in different conformations—typically described as tense and relaxed. Only one form, relaxed, can bind substrate, and such binding will help shift the conformational equilibrium further in favour of the relaxed form. All subunits must have the same conformation within a given single enzyme. For a relaxed two-unit enzyme E we can use the following 'mechanism'

$$
\begin{array}{ll}
E + S \rightleftharpoons ES & k_1, k_{-1} \\
ES \rightarrow E + P & k_2 \\
ES + S \rightleftharpoons ES_2 & k_3, k_{-3} \\
ES_2 \rightarrow ES + P & k_4.
\end{array}
$$

Proceeding as above with steady-state approximations for the two complexes and noting that the rate of production of P is now the sum of the rates of of the second and fourth steps, we can obtain an expression of the form

$$V = \frac{V_{max}[S](1 + \beta[S])}{K_{M,1} + [S] + ([S]^2/K_{M,2})} \tag{12.8}$$

where $\beta = (k_4/K_{M,2}k_2)$ and $K_{M,2} = (k_{-3} + k_4)/k_3$. Now the rational expression involves quadratic forms. The resulting dependence of rate on substrate concentration gives a sigmoid curve (Fig. 12.1(c)) typical of systems with feedback.

12.2. Oscillations in glycolysis

The interconversion of the molecules adenosine diphosphate (ADP) and adenosine triphosphate (ATP) plays a principal role in the storage and retrieval of energy within biological systems. A primary source of this energy is glucose, which is metabolized through the glycolytic pathway. The final products depend on the particular organisms and on the availability or otherwise of sufficient oxygen, but in all cases the early stages see the conversion of glucose to pyruvate with a net gain in the energy-rich form ATP.

Glucose molecules are first phosphorylated and then converted to the corresponding fructose-6-phosphate (F6P). In the next step, a second phosphorylation occurs, mediated by the enzyme phosphofructokinase (PFK)

(1) $ATP + F6P \xrightarrow{\text{PFK}} ADP + FDP.$

Later stages in the pathway involve the species nicotinamide-adenine-dinucleotide (NAD) and its hydrogenation to NADH, the formation of phosphoenolpyruvate (PEP), and its conversion to pyruvate under the action of the enzyme pyruvate kinase (PK), which also sees ADP converted to ATP. NADH is a useful indicator of the system as its concentration can be followed by fluorescence measurements.

Process (1) above is not an elementary kinetic step. There is considerable experimental evidence for allosteric inhibition of PFK by ATP, and also of its activation by ADP. These observations are important for model formulation.

The glycolytic pathway provided the first observations of oscillations in biochemical systems (Ghosh and Chance 1964). The original measurements were made *in vivo* for a suspension of yeast cells (see also Betz and Moore 1967; Pye 1969; von Klitzing and Betz 1970; Hess and Boiteux 1971; Boiteux and Hess 1975) and subsequently similar oscillations have been found in yeast cell extracts (Chance *et al.* 1964; Pye and Chance 1966; Hess *et al.* 1966; Hess *et al.* 1969), muscle extracts (Frenkel 1968a,b,c; Tornheim and

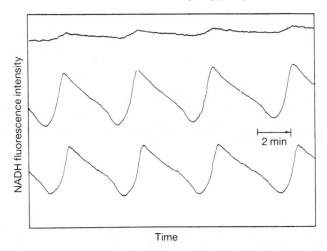

FIG. 12.2. Sustained oscillations in NADH fluorescence intensity accompanying glycolysis. (Reproduced with permission from Hess and Boiteux (1971), © Annual Review Inc.)

Lowenstein 1974, 1975), tumor cells (Ibsen and Schiller 1967), and some progress has been made with reconstituted systems prepared from purified enzymes (Hess and Boiteux 1971; Eschrich *et al.* 1983). An example oscillatory trace is shown in Fig. 12.2.

There are apparently no reports of higher-order complexities or chaos in the autonomous glycolytic system, but much work has been carried out with the yeast cell extracts subject to forcing of the substrate (glucose) inflow fluxes (Boiteux *et al.* 1975; Richter and Ross 1980; Markus and Hess 1984, 1985*a,b*; Markus *et al.* 1984, 1985*a,b*; Hess and Markus 1985*a,b*; Hess *et al.* 1990). Figure 12.3 gives example time series and reconstructed attractors (note that the latter are portrayed in a 'rotating axis' framework; the delays are scaled with $\sin(\omega_f t)$ and $\cos(\omega_f t)$ respectively where ω_f is the forcing frequency). Figure 12.4 compares the experimental excitation diagram with that produced numerically from a model to be described below.

12.2. Mechanism and models for glycolysis

The first model for the glycolytic oscillations also appeared in 1964, due to Higgins. The product FDP in step (1) was allowed to activate that step (autocatalysis) whilst its subsequent removal was given a Michaelis–Menten form. This provides sufficient non-linearity for limit cycle oscillations. Sel'kov (1968) suggested a different form, coupling substrate inhibition with product

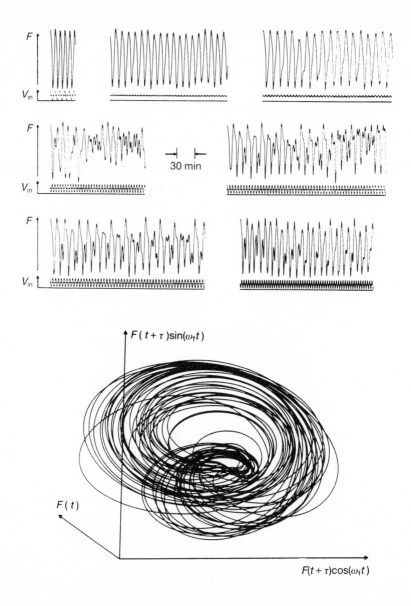

Fɪɢ. 12.3. Various responses of glycolysis to periodic forcing of the substrate inflow ranging from entrainment and quasiperiodicity to chaos; also shown is a reconstructed strange attractor. (Reproduced with permission from Hess *et al.* (1990), © Manchester University Press.)

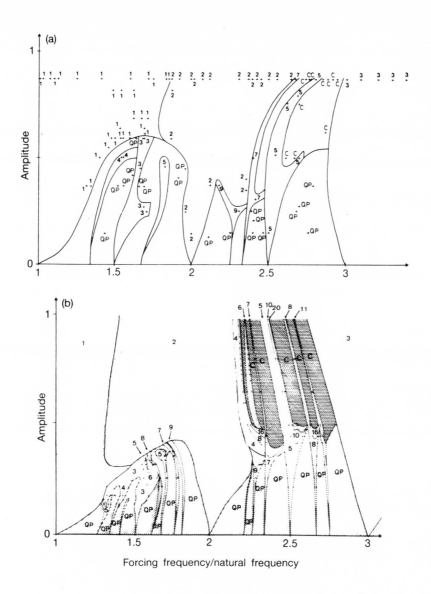

FIG. 12.4. Comparison of (a) experimental and (b) numerically predicted excitation diagrams for forced glycolysis with periodic glucose perturbation. (Reproduced with permission from Hess *et al.* (1990), © Manchester University Press.)

by an external supply or from an earlier reaction in the pathway. The product is then ADP which has an activating effect on the enzyme PFK as well as being removed (or reacting further). The activation process involves the binding of a number γ of ADP molecules with the enzyme. Sel'kov's model can then be written as

$$\rightarrow \text{ATP}$$

$$\text{ATP} + \text{E(ADP)}_\gamma \rightleftharpoons \text{(ATP)E(ADP)}_\gamma$$

$$\text{(ATP)E(ADP)}_\gamma \rightarrow \text{ADP} + \text{E(ADP)}_\gamma$$

$$\text{E} + \gamma\text{ADP} \rightleftharpoons \text{E(ADP)}_\gamma$$

$$\text{ADP} \rightarrow$$

This is, in fact, a basis from which the pool chemical autocatalator scheme of Chapter 3 can be derived with some further approximation. With a steady-state treatment for the complexed enzyme forms, dimensionless equations for the concentrations of the substrate (x) and product (y) can be written in the form

$$\mathrm{d}x/\mathrm{d}t = \mu - f(x, y) \tag{12.9}$$

$$\mathrm{d}y/\mathrm{d}t = f(x, y) - y \tag{12.10}$$

where μ gives the rate of supply of ATP. The function f has the following form:

$$f(x, y) = \frac{xy^\gamma}{1 + y^\gamma(1 + \alpha x)} \tag{12.11}$$

where α is some constant. However, for small μ Sel'kov approximated this, retaining merely the numerator $f(x, y) = xy^\gamma$. For $\gamma = 2$ this gives cubic autocatalysis as seen earlier, with a Hopf bifurcation to limit cycles as μ is decreased through unity.

This model does, however, have considerable physical problems. For $\mu < 0.900\,32$ (with $\gamma = 2$), the system does not settle on to a limit cycle or any other bounded attractor, but instead has $y \rightarrow 0$ and x growing without limit for any initial conditions. The same unbounded growth can in fact occur for any μ, even if the stationary state is stable, from some range of positive initial conditions.

Much analysis has been made of the allosteric model proposed by Goldbeter and Lefever (1972). The model is summarized in Fig. 12.5 where the 'tense' form of the two-compartmental enzyme is represented by a circle and the relaxed form by a square. The two variables are again the substrate ATP and product ADP concentrations, represented by solid and open circles respectively in Fig. 12.5. The key equilibrium is the interconversion of tense

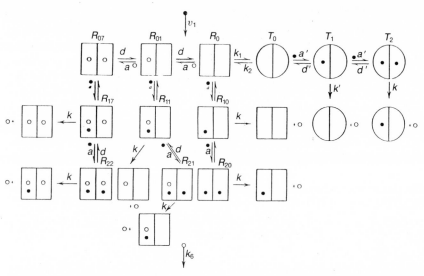

FIG. 12.5. Representation of the Goldbeter–Lefever model for glycolysis involving a two-component enzyme with allosteric effects: the (semi)circles represent low-affinity tense form; squares the relaxed, high-affinity form. Product binds only to the relaxed form, shifting the crucial equilibrium step to the left. (Reproduced with permission from Goldbeter and Levefer (1972), © Biophysical Soc.)

and relaxed forms of the enzyme which occurs only between those with 'empty' sites R_0 and T_0. Subsequent binding of any species to either of these forms shifts the equilibrium further in that direction. The product binds only with the relaxed forms, the substrate with either, but may have different rates of conversion to product in each form.

By performing the usual mass-balance and steady-state manipulations on the enzyme complex species, the dynamics of this system can be reduced to two coupled equations of the form of (12.9) and (12.10). In this case, however, Goldbeter and Lefever have worked with the full form for the function $f(x, y)$:

$$f(x, y) = \frac{\sigma_{\mathrm{M}}\{(1+y)^2(x/1+\varepsilon)[1+(x/1+\varepsilon)]+\theta L(cx/1+\varepsilon')[1+(cx/1+\varepsilon')]\}}{L[1+(cx/1+\varepsilon')]^2+(1+y)^2[1+(x/1+\varepsilon)]^2}$$

(12.12)

where σ_{M}, θ, L, c, ε, and ε' are parameters related to the rate constants in the figure.

Equation (12.12) may appear a bit hideous, but really we still have only two coupled non-linear equations and so expect nothing much worse than limit cycle oscillations (as indeed the experiments show). The two-compartment model has no multiplicity of stationary states, although this may become a feature if there are more subunits (Goldbeter and Nicolis 1976). In their equations, Goldbeter and Lefever have a slight variation in the final term for eqn (12.10)—the product removal. Their choice of dimensionless groups maintains a parameter k_s here, i.e. $-y$ is replaced by $-k_s y$. The condition for Hopf bifurcation can be expressed relatively simply in terms of this parameter if the others are known. The boundaries for oscillation in terms of the other parameters can be determined numerically and these allow various influences to be examined in terms of enhancement or reduction of the oscillatory region. Many qualitative trends in agreement with experiment are observed in both the autonomous system and with the entrainment and chaos patterns found when the model is forced periodically through the parameter μ. Furthermore, Moran and Goldbeter (1984) have found regions in parameter space for which this model shows birhythmicity if the product (ADP) can be recycled into the substrate form (ATP) as might occur elsewhere in the glycolytic pathway.

Markus and Hess (1984) have produced a different model, represented in Fig. 12.6. This involves one feedback process (through ADP) and two enzyme

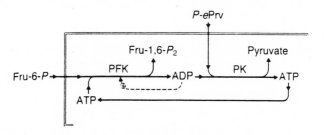

FIG. 12.6. Schematic representation of the Markus–Hess model for glycolysis with ADP feedback to the enzyme PFK. (Reproduced with permission from Markus and Hess (1984), © National Academy of Science, USA.)

species (PFK and PK). Again, the autonomous behaviour can be no more complex than a period-1 limit cycle, but higher dynamic bifurcations arise under periodic forcing.

The equations for this system can be written initially in terms of four species:

$$d[\text{ADP}]/dt = V_{\text{PKF}} - V_{\text{PK}}$$

$$d[\text{ATP}]/dt = V_{\text{PK}} - V_{\text{PKF}}$$

$$d[\text{F6P}]/dt = V_{\text{in}} + A \sin(\omega_f t) - V_{\text{PKF}}$$

$$d[\text{PEP}]/dt = V_{\text{in}} + A \sin(\omega_f t) - V_{\text{PK}}$$

where V_{PKF} and V_{PK} represent the processes involving the two enzymes, V_{in} is the flux of substrates, and the forcing term is included. The system can be reduced to two variables by recognizing two conservations laws: those for phosphate and those for the adenosine group, which give

$$[ATP] + [ADP] = \text{const}_1$$

$$[PEP] + [ATP] - [F6P] = \text{const}_2.$$

The rate functions V_{PK} and V_{PFK} have rather complex forms taken from empirical observations in the literature (e.g. Boiteux *et al.* 1983; Markus *et al.* 1983). Apart from a scaling factor, the rate laws can be written with the forms

$$V_{PK} = \mu[PEP]_f[ADP]_f/(1 + [PEP]_f + [ADP]_f + \mu[PEP]_f[ADP]_f)$$

$$V_{PKF} = ([F6P]_t[ATP]_t R^3 + Lc_{f6p}c_{atp}[F6P]_t[ATP]_t T)/(R^4 + LT^4)$$

with

$$R = 1 + [F6P]_t + [ATP]_t + [F6P]_t[ATP]_t$$

$$T = 1 + c_{f6p}[F6P]_t + c_{atp}[ATP]_t + c_{f6p}c_{atp}[F6P]_t[ATP]_t .$$

$$L = L_0(1 + c_{pep}[PEP])^4(1 + c_{adp}[ADP])^4/(1 + [ADP])^4$$

There are various constants in these expressions, but all can be related to biochemical quantities. The concentrations have here been scaled with respect to their (dimensional) enzyme–complex dissociation constants. The concentrations with subscripts t and f are the total concentration and the concentration of the magnesium-free forms of the given species respectively, μ represents the Mg^{2+} concentration, the c_i are the binding coefficients, and L_0 is the allosteric constant.

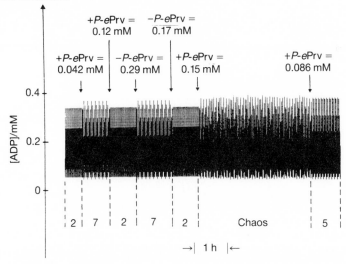

FIG. 12.7. Different possible waveforms and coesixting attractors for Markus–Hess model of glycolysis with periodic forcing: transitions between patterns are induced by transient perturbations to PEP concentration. (Reproduced with permission from Markus and Hess (1984), © National Academy of Science, USA.)

Without forcing, limit cycle solutions are found for some parameter values and, as we expect from Chapter 5, more complex responses emerge when these autonomous oscillations are forced. Markus and Hess observed coexisting limit cycles and strange attractors—Figure 12.7 shows how changes in the waveform occur as PEP is added or removed. The full comparison of the model with experiments has been shown in Fig. 12.4.

12.4. Consecutive allosteric enzyme reactions

Decroly and Goldbeter (1982, 1984, 1985, 1987) have investigated an abstract model related still to metabolic control. This considers two enzyme processes, each activated by its product, with the product of the first process being the substrate for the second:

$$\overset{v}{\rightleftharpoons} S \overset{E_1}{\rightarrow} P_1 \overset{E_2}{\rightarrow} P_2 \overset{k_s}{\rightarrow}$$

Again there is a constant input of the first substrate and a first-order removal of the final product.

If α, β, and γ represent suitable dimensionless forms for the concentrations of S, P_1, and P_2, the rate equations can be written as

$$d\alpha/dt = (v/K_{M,1}) - \sigma_1 \phi(\alpha, \beta, \gamma) \tag{12.13}$$

$$d\beta/dt = q_1 \sigma_1 \phi(\alpha, \beta, \gamma) - \sigma_2 \eta(\alpha, \beta, \gamma) \tag{12.14}$$

$$d\gamma/dt = q_2 \sigma_2 \eta(\alpha, \beta, \gamma) - k_s \gamma \tag{12.15}$$

where the two functions ϕ and η are given by

$$\phi(\alpha, \beta, \gamma) = \frac{\alpha(1 + \alpha)(1 + \beta)^2}{L_1 + (1 + \alpha)^2(1 + \beta)^2} \tag{12.16}$$

and

$$\eta(\alpha, \beta, \gamma) = \frac{\beta(1 + d\beta)(1 + \gamma)^2}{L_2 + (1 + d\beta)^2(1 + \gamma)^2}. \tag{12.17}$$

We thus have a three-variable, highly non-linear system. Again the removal rate coefficient k_s turns out to be a convenient parameter with which to work. There is always a unique stationary state, but this can lose stability. For the particular choice of the other terms $(v/K_{M,1}) = 0.45$, $\sigma_1 = \sigma_2 = 10$, $q_1 = q_2^{-1} = 50$, $L_1 = 5 \times 10^8$, $L_2 = 100$, and $d = 0$, the stationary state is unstable even for $k_s = 0$ (in a sense we have a two-variable system similar to the Sel'kov model). As k_s increases through 0.792, the stationary state regains stability at a subcritical Hopf bifurcation, shedding an unstable limit cycle. Any system at this stationary state can be persuaded over to the original limit cycle with a large enough perturbation, as we have seen previously in other models—this feature is known as 'hard excitation' in biochemistry, unfortunately close to, but not quite the same as, the phenomenon with the same name as used by combusion scientists and engineers.

With $k_s = 1.584$, there is a further Hopf bifurcation at which the stationary state loses its stability again. This is a supercritical event and a second, smaller stable limit cycle LC2 appears in the system. We thus have birhythmicity. When $k_s = 1.82$, the unstable limit cycle from the first Hopf bifurcation collides with and extinguishes the original limit cycle LC1,

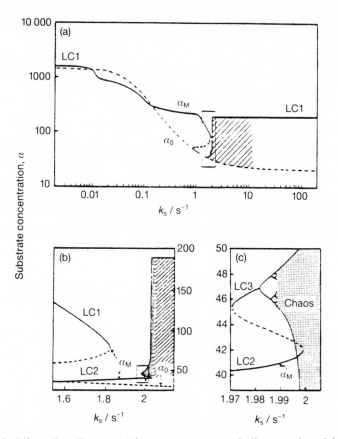

FIG. 12.8. Bifurcation diagram (a) for two-enzyme metabolic control model showing stationary-state and dynamic responses including multiple limit cycles, birhythmicity, and chaos: (b) and (c) show successive detail over range $1.6 \leqslant k_s \text{s}^{-1} \leqslant 2$. (Reproduced with permission from Decroly and Goldbeter (1982), © National Academy of Science, USA.)

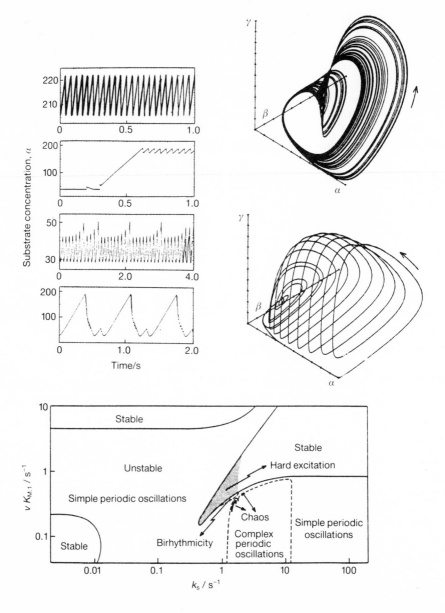

Fig. 12.9. Representative time series, attractors, and a two-parameter plane for the abstract two-enzyme model of metabolic control. (Reproduced with permission from Decroly and Goldbeter (1982), © National Academy of Science, USA.)

leaving a single periodic solution. There is yet more, as indicated in Fig. 12.8. With $k_s = 1.974$, a stable–unstable limit cycle pair is born giving rise to a second region of birhythmicity. The unstable cycle grows to coalesce with the stable limit cycle LC2 that appeared at $k_s = 1.584$. The new stable limit cycle LC3, however, is of more interest as it undergoes a period-doubling sequence into chaos. The chaotic strange attractor appears before the other two cycles have met, and so coexists with these attractors for some range of k_s.

After a reasonably wide range of aperiodicity, periodic windowing, and bursting patterns, there is a final transition to a single, stable period-1 limit cycle that exist for all large k_s $(12.8 < k_s)$. Some example time series, attractors, and the $(v/K_{M,1})$–k_s parameter plane are shown in Fig. 12.9.

There are, as yet, no known experimental or living examples of chaotic biochemical systems that can be categorized as belonging to the above class, but the construction is relatively simple and one suspects consecutive allosteric processes may be quite common occurrences. There are certainly examples of autonomous chaos in biochemistry, and we move on to one of these now.

12.5. The horseradish peroxidase reaction

The enzyme peroxidase catalyses the reduction reaction of molecular oxygen with the consequent oxidation of the species NADH (see Section 12.2 above). The stoichiometry of this process is given by

$$O_2 + 2H^+ + 2NADH \rightarrow 2H_2O + 2NAD^+$$

whilst the supposed mechanism (Degn and Yamazaki 1967) is represented in Fig. 12.10 and summarized in Table 12.1. Species involved include the reduced and oxidized forms ferrousperoxidase per Fe(II) and per Fe(III), three other enzyme forms known as compounds I; II, and III, and H_2O_2, the radicals NAD˙ and O_2^- and O_2.

Again, reaction can be followed relatively easily through fluorescence from NADH and also through the haem group of the peroxidase enzyme and the dissolved oxygen concentration. In an open system the reaction exhibits bistability (Degn 1968), damped and sustained oscillations (Yamazaki *et al.* 1965; Yamazaki and Yokota 1967; Nakamura *et al.* 1969), complex oscillations, and chaos (Olsen and Degn 1977; Olsen 1983). Examples of the observed period-1, chaotic, and mixed-mode traces obtained on varying the enzyme inflow concentration are shown in Fig. 12.11.

Attempts at modelling began with a modified Lotka scheme (Degn and Meyer 1969). Olsen and Degn (1978; Degn *et al.* 1979) suggested a four-variable scheme from the mechanism and were able to reproduce oscillatory responses—but not stability or chaos. Olsen (1983, 1984) obtained good qualitative matching of the sequence from Fig. 12.11 with the following

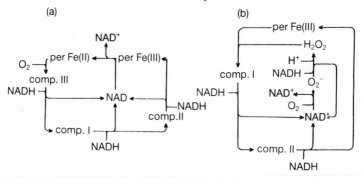

FIG. 12.10. Schematic representation of mechanism for peroxidase catalysed oxidation of NADH involving various species. (Reproduced with permission from Olsen and Degn (1985), © Cambridge University Press.)

TABLE 12.1
Scheme of reactions to model horseradish peroxidase reaction
(reproduced with permission from Aguda and Clarke (1987):
© American Institute of Physics)

$$Per^{3+} + H_2O_2 \rightarrow Co\,I$$

$$Co\,I + NADH \rightarrow Co\,II + NAD^{\cdot}$$

$$Co\,II + NADH \rightarrow Per^{3+} + NAD^{\cdot}$$

$$Co\,II + H_2O_2 \rightarrow Co\,III$$

$$Per^{3+} + NAD^{\cdot} \rightarrow Per^{2+} + NAD^{+}$$

$$Per^{2+} + O_2 \rightarrow Co\,III$$

$$Co\,III + NADH \rightarrow Co\,I + NAD^{\cdot} + H^{+}$$

$$Co\,III + NAD^{\cdot} \rightarrow Co\,I + NAD^{+}$$

$$Co\,I + O_2^{-} \rightarrow Co\,II + O_2$$

$$Per^{3+} + O_2^{-} \rightarrow Co\,III$$

$$NAD^{\cdot} + O_2 \rightarrow NAD^{+} + O_2^{-}$$

$$H^{+} + O_2^{-} + NADH \rightarrow H_2O_2 + NAD^{\cdot}$$

$$2NAD^{\cdot} + H^{+} \rightarrow NADH + NAD^{+}$$

$$2O_2^{-} + 2H^{+} \rightarrow H_2O_2 + O_2$$

$$Per^{2+} + Co\,III \rightarrow Per^{3+} + Co\,I$$

$$\rightarrow O_2$$

$$O_2 \rightarrow$$

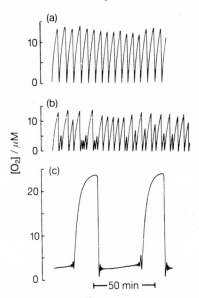

FIG. 12.11. Experimental observations of periodic and aperiodic oscillations in the peroxidase system. (Reproduced with permission from Olsen and Degn (1977), © Macmillan Journals.)

modification of that four-variable model:

(OD1)	$B + X \rightarrow 2X$	k_1
(OD2)	$2X \rightarrow 2Y$	k_2
(OD3)	$A + B + Y \rightarrow 3X$	k_3
(OD4)	$X \rightarrow P$	k_4
(OD5)	$Y \rightarrow Q$	k_5
(OD6)	$X_0 \rightarrow X$	k_6
(OD7)	$A_0 \rightleftharpoons A$	k_7, k_{-7}
(OD8)	$B_0 \rightarrow B$	k_8.

Here A is the dissolved form of O_2, B = NADH, and X and Y are intermediate radical species. The last three steps model the inflow of reactants, with the O_2 adsorption process from the gas phase being reversible. The enzyme itself does not appear as a variable, but will catalyse steps (OD1) and (OD3) and so is implicitly involved through the values of k_1 and k_3.

By varying k_1, the direct branching rate constant, Olsen observed a sequence of periodic and non-periodic states as summarized in Fig. 12.12, with specific examples for comparison with the experimental time series shown in Fig. 12.13.

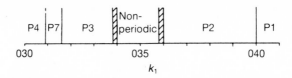

FIG. 12.12. Sequence of oscillatory and chaotic responses observed in Olsen model of the peroxidase system. (Reproduced with permission from Olsen (1983), © Elsevier, North-Holland.)

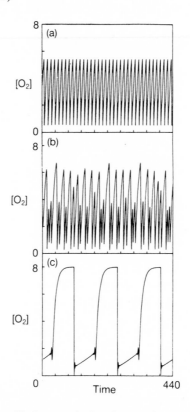

FIG. 12.13. Computed oscillations and chaotic waveforms from Olsen model, for comparison with experimental results in Fig. 12.11. (Reproduced with permission from Olsen (1983), © Elsevier, North-Holland.)

This model and other variations have been scrutinized in much more detail in a sequence of studies (Aguda and Clarke 1987; Larter *et al.* 1987, 1988; Aguda *et al.* 1989). Working with the full set of reactions from Table 12.1, the identified a minimal scheme for bistability and damped oscillations in terms of reactions 1–3, 8, 10–13, 16, and 17 (they assumed that the reaction conditions are such that NAD^+, $NADH$, and H^+ are sufficiently in excess

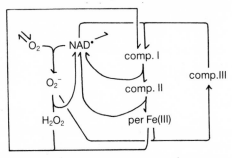

FIG. 12.14. Network of important processes leading to bistability in the Olsen model. (Reproduced with permission from Aguda and Clarke (1987), © American Institute of Physics.)

so that their concentrations can be taken as parameters rather than variables). The network diagram corresponding to this subscheme is sketched in Fig. 12.14, following Aguda and Clarke (1987). An important feature is the inhibition of the enzyme by its substrate O_2. Aarons and Gray (1976) have also considered some general aspects of multiplicity in biochemical systems.

Larter *et al.* (1987) examined in some detail the dynamical behaviour of the Degn–Olsen–Perram model which differs from that of Olsen above only in the first and third steps which are here

(DOP1) $\qquad\qquad$ A + B + X → 2X

(DOP3) $\qquad\qquad$ A + B + Y → 2X.

This biochemical scheme, like the experiments, can exhibit mixed-mode oscillations (see Section 4.8)—an example is shown in Fig. 12.15. The large-amplitude jumps are separated by periods of small oscillations. A full Devil's staircase is observed as k_1 is varied, with an associated Farey tree. Chaotic states also exist between the treads of the staircase in a similar fashion to those observed in the Belousov–Zhabotinskii and other reaction systems (Section 8.3). The waveform for the species B (i.e. [NADH]) is somewhat different from the others and this feature is crucial in the explanation put forward by Larter *et al.* (1988).

The reaction rate equations for the model can be written in the following dimensionless form:

$$da/dt = -abx - \gamma aby + p_2 - p_3 a \qquad (12.18)$$

$$dx/dt = abx - 2x^2 + 2\gamma aby - x + p_1 \qquad (12.19)$$

$$dy/dt = 2x^2 - \gamma aby - \alpha y \qquad (12.20)$$

and

$$db/dt = \varepsilon[-abx - \gamma aby + p_0] \qquad (12.21)$$

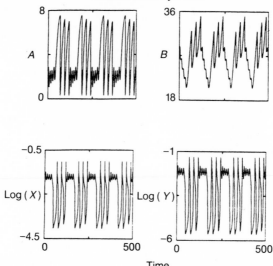

Time

FIG. 12.15. Complex oscillation (3^4) for the four-variable model of the peroxidase system due to Degn *et al.* (DOP) showing the separation into fast and slow variables. (Reproduced with permission from Larter *et al.* (1988), © American Institute of Physics.)

where p_0–p_3, α, γ, and ε are parameters involving the rate constants. Of these, ε is likely to have a particularly small magnitude and hence the timescale on which the concentration b evolves will generally be much slower than that for a, x, and y.

This latter point allows the four-variable equations to be divided into a 'fast' three-variable subset and the slow equation $\mathrm{d}b/\mathrm{d}t$ provides this subset with a slowly evolving 'parameter'. We then become interested in bifurcations of the fast subsystem in terms of the parameter b—which may show saddle–node turning points (a real, zero eigenvalue), Hopf bifurcation (complex pair of eigenvalues with zero real part), and high degenerate bifurcations such as the double zero eigenvalue and the hysteresis–Hopf forms discussed in earlier chapters.

With $p_0 = 1.5625$, $p_1 = 3.125 \times 10^{-3}$, $p_2 = 2.9375$, $p_3 = 5.875 \times 10^{-3}$, $\alpha = 0.0522$, $\gamma = 0.6155$, and $\varepsilon = 9.748 \times 10^{-7}$, the full model shows the mixed-mode forms in Fig. 12.15. The corresponding three-variable scheme typically has a hysteresis loop with at least one Hopf bifurcation point as b is varied. Larter *et al.* followed the turning points and one of the Hopf points as a second unfolding parameter k_1 is changed as shown in Fig. 12.16 (ε is proportional and γ inversely proportional to k_1 and the above values correspond to $k_1 = 0.076\,155\,9$). The crossing of the loci for the Hopf point on one branch and the saddle–node bifurcation of the other appears to be of importance.

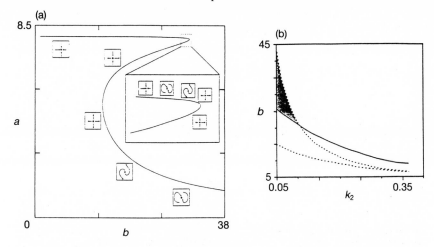

FIG. 12.16. (a) The 'bifurcation diagram for the reduces three-variable DOP model in terms of the 'slow' variable b showing bistability and Hopf bifurcation phenomena along both branches. (b) The loci of saddle–node (- - -) and Hopf bifurcations (——) in the b–k_1 parameter plane for the DOP model. (Reproduced with permission from Larter *et al.* (1988), © American Institute of Physics.)

For small k_1, to the left of this crossing point, lies the region in which complex oscillations occur: these are characterized by having the upper saddle–node occurring for b larger than the Hopf bifurcation on the lower branch (Fig. 12.16(a)), so the lower solution is locally stable as the system 'falls off' the upper branch. If we now decrease b and move along the lower branch underhanging the upper branch, the Hopf bifurcation gives rise to a limit cycle that can become homoclinic with the saddle point at some lower parameter value, at which point the system will try to move back to the upper branch.

For larger k_1, the relative positions of the upper saddle–node point and the Hopf bifurcation on the branch are reversed, the latter occurring at larger b. The lower stationary state is unstable at the parameter value for the upper saddle–node point.

Figure 12.17 shows the complex and simple oscillations for values of k_1 in each of the two ranges just described superimposed on the stationary-state locus for the corresponding three-variable subscheme. In the first case the small-amplitude oscillations can show the typical hourglass effect as they are at first attracted to the stable state of the subsystem on the lower branch. As b varies in the full system, this state loses stability and the small oscillations grow. There is then a move towards the upper branch, perhaps associated with passing the homoclinic orbit of the subsystem. As b varies with this large excursion, so the system passes the upper saddle–node point and falls back to the lower branch. In Fig. 12.17(b), the upper saddle–node

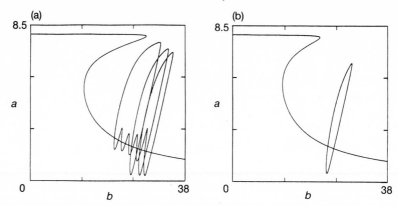

Fig. 12.17. Complex and simple oscillations in the a–b phase plane for the full four-variable DOP model superimposed on the three-variable scheme bifurcation diagram: (a) the 3^4 waveform of Fig. 12.15; (b) with $k_1 = 1.031\,25$, giving a simple 1^0 response. (Reproduced with permission from Larter *et al.* (1988), © American Institute of Physics.)

lies too far away to perturb the simple limit cycle of the three-variable subscheme.

Aguda *et al.* found similar mixed-mode behaviour and chaos in the Olsen scheme given above and were able to resolve the switching that gives the mixed mode in terms of three 'dynamic elements' or networks comprising various selections of the full model.

12.6. Intercell communication

An important characteristic of living systems is the ability of different parts of an organism to communicate with each other. Various mechanisms for this may exist—and the self-organization of inanimate systems also requires a 'communication' process. Here we discuss the role of a messenger species, cyclic adenosine monophosphate (cAMP), and the possible mechanism through which this acts.

The biochemical situation is summarized in Fig. 12.18 which shows a

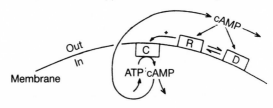

Fig. 12.18. Schematic representation of the Goldbeter–Martiel model for cAMP signalling between cells in *Dictyostelium discoideum*. (Reproduced with permission from Martiel and Goldbeter (1985), © Macmillan Journals.).

membrane separating the cell from its external environment. The cell may 'sense' the presence of the messenger through a series of 'receptor sites' on the membrane to which cAMP binds. The receptor–cAMP complex can activate the enzyme adenylcyclase AC, which is internal to the cell and which catalyses the production of cAMP from its substrate ATP. The new cAMP can then diffuse across the membrane and pass the message on to other cells. We thus see an autocatalysis, with the self-activation via cAMP which provides a biological process similar to the mechanism of inanimate chemical wave propagation. The process may become limited in at least two ways: a second enzyme phosphodiesterase (PDE) catalyses the ring-opening reaction cAMP → AMP that destroys the messenger and also the receptor may become deactivated to form D in Fig. 12.18.

The mechanism for these various component stages has been discussed by Goldbeter and Martiel (1985; Goldbeter and Segel 1977; Martiel and Goldbeter 1984, 1985; Goldbeter *et al.* 1984). In their 'full model' the system is described by seven coupled rate equations, but many of the dynamic features are also shown by a three-variable subscheme (Goldbeter and Martiel 1985) based on the intracellular concentration of ATP, the fraction of the receptor sites that are active (i.e. in the R form), and the extracellular cAMP concentration: the dimensionless measures of these are denoted α, ρ, and γ respectively. The corresponding dimensionless rate equations can be written as

$$d\alpha/dt = v - \sigma\phi(\alpha, \rho, \gamma)$$

$$d\rho/dt = -\rho\mu(\gamma) - (1 - \rho)\eta(\gamma)$$

$$d\gamma/dt = q\sigma\phi(\alpha, \rho, \gamma) - k\gamma$$

with the various functional forms given by

$$\phi(\alpha, \rho, \gamma) = \frac{\alpha[\lambda\theta + \varepsilon\rho\gamma^2/(1 + \gamma^2)]}{(1 + \alpha\theta) + (1 + \alpha)[\varepsilon\rho\gamma^2/(1 + \gamma^2)]}$$

$$\mu(\gamma) = (k_1 + k_2\gamma^2)/(1 + \gamma^2)$$

$$\eta(\gamma) = (k_1 L_1 + k_2 L_2 c^2\gamma^2)/(1 + c^2\gamma)$$

involving a number of identifiable parameters.

This scheme shows almost all of the typical behaviour on non-linear three-variable systems: simple oscillations, complex waveforms and bursting, birhythmicity and chaos (Goldbeter and Martiel 1985; Martiel and Goldbeter 1985). Some examples are shown in Fig. 12.19.

The signalling between cells just described can lead to remarkable phenomena if combined with other influence—that of chemotaxis. Some cells are observed to have the tendency to move up a concentration gradient of some chemical species (Murray 1990). The slime mould *Dictyostelium*

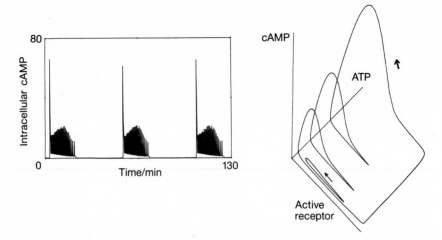

FIG. 12.19. Bursting patterns in the cAMP concentration from the Goldbeter–Martiel model. (Reproduced with permission from Martiel and Goldbeter (1985), © Macmillan Journals.)

discoideum can show both of these features under conditions of starvation (see e.g. Gerish 1982). The individual amoebae move physically, aggregating around a small number of 'pacemakers'. This aggregation is driven by the periodic emission of cAMP from the pacemaker cells, with the message being relayed by each cell as it detects the messenger. The whole process leads to the formation of spatial patterns, generally formed from almost perfect spirals of migrating amoebae—perhaps 10^5 organisms acting cooperatively under this influence.

In addition to the regular periodic pacemakers, some mutant strains show aperiodic pulsing (Durston 1974; Coukell and Chan 1980) and Goldbeter and Martiel have suggested that this may be related to the chaotic behaviour described above.

Another mechanism of intercellular communication has been suggested by Rapp and Berridge (1977; Rapp *et al.* 1985) involving calcium ion (Ca^{2+}). The basis of this is indicated in Fig. 12.20 and again involves the stimulated intracellular production of cAMP by the enzyme AC. A messenger hormone H activates the enzyme and may also increase the permeability of the cell membrane to calcium. The intracellular calcium concentration Ca_i^{2+} may thus increase. Within the cell, calcium is also sequestrated, but the cAMP favours desequestration thus also increasing Ca_i^{2+}. However, calcium is an inhibitor of the enzyme AC.

If the hormone H is simply cAMP this does not differ too strikingly from the model of Goldbeter and Martiel: in the present case the inhibition by Ca_i^{2+} plays the role of the deactivation of the receptor. Othmer *et al.* (1985)

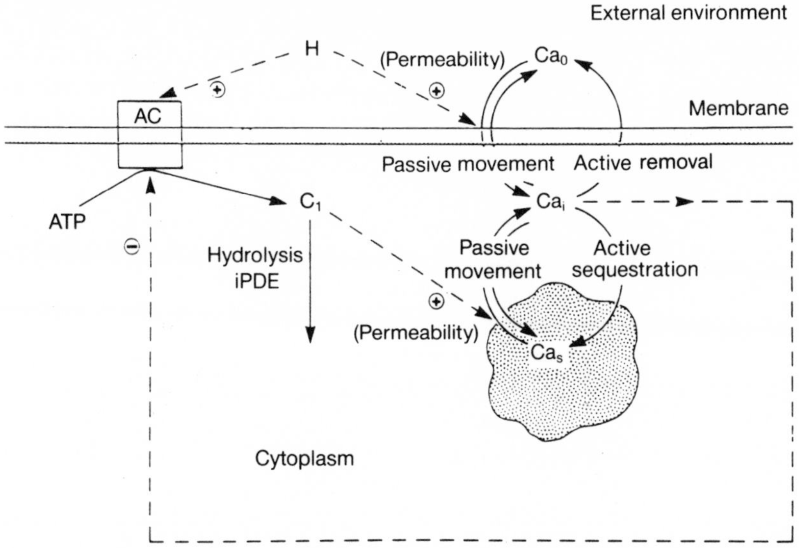

FIG. 12.20. Representation of a model for intercellular communication based on Ca^{2+} and cAMP. (Reproduced with permission from Rapp (1986), © Manchester University Press.)

have used this scheme to interpret the slime mould behaviour. A further twist in the tail arises through the intervention of the protein calmodulin, which also binds Ca^{2+}: this complex formation is a particular important fate for calcium at low concentration (whilst calmodulin is in excess). Furthermore, the calcium–calmodulin complex is an activator of AC rather than an inhibitor. Thus we have effectively calcium activation of the cAMP production at low concentrations and its inhibition at high concentrations.

12.7. Calcium deposition and bone formation

A simple self-oscillating model for some aspects of the complex process through which hard tissue such as bone is formed and renewed has been investigated by Tracqui *et al.* (1985, 1987; Tracqui and Staub 1987; Staub *et al.* 1979; Perault-Staub *et al.* 1984). This is of interest here as it is very similar to the pool chemical autocatalator model examined in Chapter 3. Details of the biological principles involved can be found in Nancollas (1979), Urist (1980), and in Eanes and Termine (1983). The model follows the dynamics for calcium in the extracellular fluid (ECF) and its interface with bone. Global oscillations in the ECF calcium concentration must support the spatial organization that allows a well-ordered deposition of hydroxy-apatite calcium phosphate crystals at the bone–ECF interface.

The two-variable model has a form of 'nucleation and growth' kinetics in which the incorporation of extra calcium ions (X) into the precursor crystallite form Y is 'catalysed' by Y itself, the rate depending on xy^2 where x and y are the local concentrations. The governing equations are written as

$$dx/dt = a - k_1 x + k_2 y - (k + y^2)x$$

$$dy/dt = (k + y^2)x - (1 + k_2)y.$$

The term a represents a constant Ca^{2+} flux into the interface region, while $k_1 x$ constitutes a removal process thar does not produce the bone crystal precursors (e.g. a simple flow away from the interface). The redissolution of the precursor crystallites is represented by $k_2 y$ and there is an additional removal of y—perhaps incorporation into the deep bone structure. The remaining term $(k + y^2)x$ common to both equations gives the uncatalysed and autocatalytic rates of conversion from X to Y.

Tracqui *et al.* (1987) have examined and determined all the bifurcations for this two-variable scheme and located various combinations of multiple stationary states and multiple limit cycles (birhythmicity etc.). They also obtained some non-periodic results if they relaxed their numerical integration techniques: such responses are necessarily artefacts in two-variable schemes but suggest the distinct possibility of higher-order bifurcations if this model is coupled with, say, a slow parameter variation along the lines discussed above for the peroxidase model. Also, the extensions to the model discussed in Chapter 4 and that produced mixed-mode oscillations and chaos there should have similar effects here.

12.8. Nerve signal transmission and heartbeat

Another area in which the theory of chaos has found wide application is that of the signals that can pass along nerve axons. Periodic firing of neurone cells has been observed and the signal is transmitted along the axon—a tubular structure whose surface is a membrane. The signal can be followed by detecting the difference in potential across this membrane. There are a number of species that possess relatively large agons into which micro-electrodes can be inserted—the squid giant axon has been particularly widely studied since the pioneering work of Hodgkin and Huxley (1952). Many of these experiments have involved forcing the axon with periodic electrical stimulation and all of the responses typical of such non-autotonomous systems (see Chapter 5) have been observed. A recent review has been given by Aihara and Matsumoto (1986) and a helpful introduction can be found in Murray (1990, Section 6.5). In typical experiments an intact axon is suspended in a suitable aqueous solution. The potential along the length of the axon is made homogeneous by threading a wire along its length inside the membrane—'space clamping'.

The modelling of nerve axon behaviour is based on either the Hodgkin–Huxley equations or an appropriate model of these due to FitzHugh and Nagumo (FitzHugh 1961; Nagumo *et al.*, 1962; see also Rinzel 1981 and Holden and Muhamad 1986). The potential difference across the membrane is caused by concentration differences, particularly in the cations K^+ which typically has a higher concentration inside the axon and Na^+ which has a higher concentration in the surrounding fluid. These concentration differences are vital to nerve signal transmission and are maintained by ion pumps in the membrane itself.

The current passing through the axon membrane at any moment will have contributions due to ion movement and to capacitance effects. Three ion currents are generally identified: those due to K^+ and Na^+ and a 'leakage current'. Thus, a governing differential equation for the potential difference across the membrane V can be written as

$$C(dV/dt) = I(t) - (I_K + I_{Na} + I_L) \tag{12.22}$$

where C is the membrane capacitance. To eqn ((12.22) we must add specifications for the currents, and these are related to the potential V by the equations

$$I_K = g_K n^4(V - V_K) \qquad I_{Na} = g_{Na} m^3 h(V - V_{Na}) \qquad I_L = g_L(V - V_L)$$

where the g_i and V_i are 'known' constants representing conductances and equilibrium potentials respectively. The other 'parameters' h, m, and n are in fact variables and are determined by the extra differential equations

$$dm/dt = \alpha_m(V)(1 - m) - \beta_m(V)m \tag{12.23}$$

$$dn/dt = \alpha_n(V)(1 - n) - \beta_n(V)n \tag{12.24}$$

$$dh/dt = \alpha_n(V)(1 - h) - \beta_h(V)h. \tag{12.25}$$

The final step is to obtain explicit forms for the six-voltage-dependent functionals α_i and β_i. These are empirical fits to experimental data.

With an applied constant current we have an autonomous four-variable system: if the total current $I(t)$ is forced, we will obtain a non-autonomous system. The rest state is that with no net current passing, $I(t) = 0$. The stationary state of this system is typically stable but excitable (Section 3.13.3)

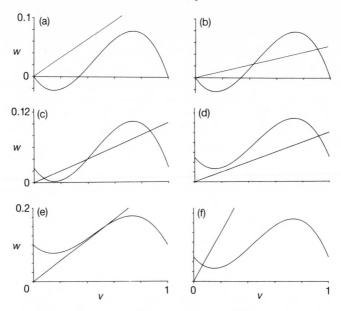

FIG. 12.21. Nullclines for the FitzHugh–Nagumo model of the Hodgkin–Huxley equations: (a) and (b), $I_a = 0$ showing either a single intersection at the origin $(b/\gamma) > \frac{7}{27}$ or three intersections $(b/\gamma) < \frac{7}{27}$; (c)–(f) with non-zero applied current allowing three steady states ($I_a = 0.025$, $b/\gamma = 0.1$) or a unique intersection; (d) $I_a = 0.05$, $b/\gamma = 0.1$, single stable intersecion at high v and w; (e) $I_a = 0.1$, $b/\gamma = 0.28$ showing single unstable intersection between extrema in cubic nullcline; (f) $I_a + 0.05$, $b/\gamma = 0.5$ showing single stable intersections at low v and w.

and this is much of the origin in interest in forced excitable systems (Section 5.4).

The FitzHugh–Nagumo model simplifies the above by noting that the variable m evolves on a much faster timescale and hence can be determined as a function of the other variables by solving the condition $dm/dt = 0$ at any moment in time. If the variable h is also taken as a constant then we have only two variables left: the potential and n. The equations can be 'transformed' into the following form:

$$dv/dt = v(a - v)(v - 1) - w - I_a \qquad (12.26)$$

$$dw/dt = bv - \gamma w \qquad (12.27)$$

where v is the scaled potential and w involves m, n, and h.

These equations are most easily investigated in terms of the nullclines (Fig. 12.21): the two loci in the v–w plane obtained by setting the right-hand sides equal to zero. With zero applied current $I_a = 0$ there may be a single stationary state or multiplicity. Stable states are generally excitable, but no oscillations are found. As the applied current is increased, so a unique unstable stationary state may be obtained, leading to periodic oscillations all in much the same way as occurs for the two-variable Oregonator model for the Belousov-Zhabotinskii reaction as discussed in Section 3.13.3. Murray (1990) has also discussed some aspects of further approximating the model so that the cubic nullcline in Fig. 12.21 is replaced by three straight lines with discontinuous slope—a piecewise linear representation.

In addition to squid axon, studies have been made on other systems, and aperiodic responses in forcing experiments are frequently observed. Irregular oscillations can also be stimulated by the administration of some drugs. Cocaine, strychnine, and benzodiazepines can invoke 'bursting' in snail neurons (Labos and Lang 1978; Klee *et al.* 1978; Hoyer *et al.* 1978): chaos can follow the administration of various convulsants (Holden and Winlow 1982; Holden *et al.* 1982).

There is also considerable (and quite natural) interest in the potential occurrence of chaos in the operation of heart tissue (Glass *et al.* 1986). Heart cells can respond in various typical ways on forcing (Guevara *et al.* 1981; Glass *et al.* 1983, 1984). Chaotic cardiac arrythmias can be detected *in vivo* using ECG traces. These are believed to stem from the development of multiple pacemaker sites on heart. These are not 'a good thing' although not always fatal.

In vitro experiments, typically with embryonic chick heart cells, suffer from the inherently statistical nature of cell extracts and there are problems with imposed noise. Nevertheless, sufficiently many forms of phase locking have been observed for extensive Devil's staircases to be constructed (Glass *et al.* 1983, 1984, 1986). Some phase-locked and chaotic examples are shown in Fig. 12.22. Other aspects such as birhythmicity and quasiperiodicity have also been observed (Guevara and Glass 1981).

The different phenomenon of fibrillation appears to have its origin formly fixed in the existence of wave patterns moving across the surface of the heart tissue. This tissue is another example of an excitable medium—unresponsive to small perturbations but responding to a sufficiently large disturbance by a single large-amplitude transient excursion. For a discussion of this latter behaviour the book and article by Winfree (1980, 1983) and the book by Glass and Mackey (1988) give an introduction.

12.9. Populations and epidemics

The spread of infectious diseases through a susceptible population is, in its most general form, a spatio-temporal process. Early studies of the waves associated with this spread go back to Fisher (1937) and Kolmogorov *et al.* (1937). A wide and rich literature in this area has grown since then and

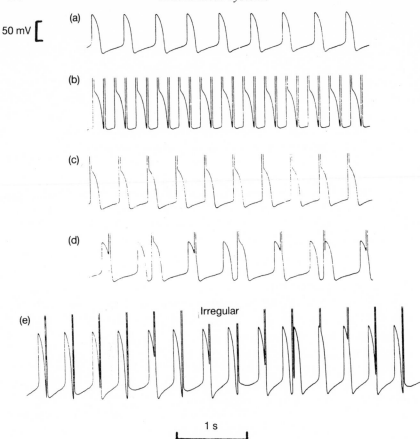

50 mV

1 s

FIG. 12.22. Autonomous oscillations, entrainment, and chaotic responses to forcing of embryonic chick heart preparations. (Reproduced with permission from Glass *et al.* (1986), © Manchester University Press. Adapted from Guevara *et al.* 1981.)

Murray's book (1990) is again an excellent source. If, however, we take a narrower view we might concentrate only on the temporal variations in infected and uninfected populations at a given location.

This local problem becomes, in a sense, now similar to the classic predator–prey population behaviour (although perhaps with rather complex inflow and outflow terms). The simplest example, that of an isolated locality, has been considered since the collection of the classic data on Canadian lynxes and snowshoe hares established from the fur catch records of the Hudson Bay company from 1845 onwards (Murray 1990, p. 68; Williamson 1972). Here oscillations in the implied populations appear reasonably regular, with approximately nine maxima over the 90 year recording period (Fig. 12.23). The classical interpretation of predator–prey data is that of

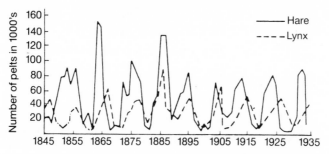

FIG. 12.23. Oscillations in the numbers of lynx and hares trapped by the Hudson Bay Company from 1845 to 1935. (Reproduced from Murray (1990), © Springer.)

Volterra (1926) and Lotka (1910, 1920). Various Lotka–Volterra models have been used with the lynx–hare data (Odum 1953; Elton and Nicholson 1942; Leigh 1968; Gilpin 1973) although Murray remarks on some rather curious interpretational difficulties that arise with closer examination. There have also been suggestions of aperiodicity in the data (Schaffer 1984).

Schaffer and Kot (1985a, 1986) also suggest that chaotic behaviour with consequent underlying strange attractors is exhibited by some data for the

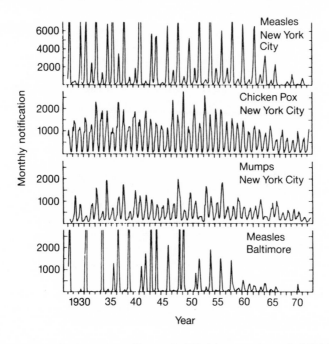

FIG. 12.24. Time series for various diseases in US cities over the period 1927–showing different forms of periodicity. (Reproduced with permission from London and Yorke (1973), © Johns Hopkins University School of Hygiene and Public Health.)

incidence of measles in New York and Baltimore over the period 1928–63 (up to the introduction of efficient vaccination programmes). This is different from the time series for other diseases such as mumps and chickenpox (London and Yorke 1973; Yorke and London 1973) which appear to be a simple oscillatory (yearly) cycle with superimposed noise (Fig. 12.24). The same authors have detected possible chaos in the population of an insect pest *Thrips imaginis* (Schaffer and Kot 1985*b*). Deterministic models for aperiodic behaviour in such systems have been proposed by Gilpin (1979), May (1980), Aron and Schwartz (1984), and by Gardini *et al.* (1989). In many cases these involve 'generalized' Lotka–Volterra predator–prey schemes. These have governing rate equations that can be written for an *n*-variable model in the form

$$\frac{dx_i}{dt} = x_i\left(c_i + \sum_j a_{ij}x_j\right) \qquad i = 1, n \quad j = 1, n$$

so for any given species $\because$ formation and termination steps depend on that species' concentration. With $n = 3$, these are sufficiently non-linear to permit complex oscillations and chaos.

References

Aarons, L. J. and Gray, B. F. (1976). Multistability in open chemical reaction systems. *Chem. Soc. Rev.*, **5**, 359–75.

Aguda, B. D. and Clarke, B. L. (1987). Bistability in chemical reaction networks: theory and application to the peroxidase–oxidase reaction. *J. Chem. Phys.*, **87**, 3461–70.

Aguda, B. D., Larter, R., and Clarke, B. L. (1989). Dynamic elements of mixed-mode oscillations and chaos in a peroxidase–oxidase model network. *J. Chem. Phys.*, **90**, 4168–75.

Aihara, K. and Matsumoto, G. (1986). Chaotic oscillations and bifurcations in squid giant axons. In *Chaos*, (A. V. Holden), pp. 257–70. Manchester University Press.

Aron, J. L. and Schwartz, I. B. (1984). Seasonality and period doubling bifurcations in an epidemic model. *J. Theor. Biol.*, **110**, 665–79.

Babloyantz, A. (1986). *Molecules, dynamics and life*. Wiley, New York.

Betz, A. and Moore, C. (1967). Fluctuating metabolite levels in yeast cells and extracts and the control of phosphofructokinase activity *in vitro*. *Arch. Biochem. Biophys.*, **120**, 268–73.

Boiteux, A. and Hess, B. (1975). Oscillations in glycolysis, cellular respiration and communication. *Faraday Symp. Chem. Sc.*, **9**, 202–14.

Boiteux, A. and Hess, B. (1980). Spatial dissipative structures in yeast extracts. *Ber. Bunsenges. Phys. Chem.*, **84**, 392–8.

Boiteux, A., Golbeter, A., and Hess, B. (1975). Control of oscillating glycolysis of yeast by stochastic, periodic and steady source of substrate: a model and experimental study. *Proc. Natl Acad. Sci. USA* **72**, 3829–33.

Boiteux, A., Markus, M., Plesser, Th., Hess, B., and Malcovati, M. (1983). Analysis of progress curves. Interaction of pyruvate kinase from *Escherichia coli* with fructose 1,6 biphosphate and calcium ion. *Biochem. J.*, **211**, 631–40.

Chance, B., Hess, B., and Betz, A. (1964), DPNH oscillations in a cell-free extract of *S. carlsbergensis*. *Biochem. Biophys. Res. Commun.*, **16**, 182–7.

Chance, B., Pye, E. K., Ghosh, A. K., and Hess, B. (1973). *Biological and biochemical oscillators*. Academic Press, New York.

Coukell, M. B. and Chan, F K. (1980). The precocious appearance and activation of an adenylate cyclase in a rapid developing mutant of *Dictyostelium disciodeum*. *FEBS Lett.*, **110**, 39–42.

Decroly, O. and Goldbeter, A. (1982). Birhythmicity, chaos and other patterns of temporal self-organization in a multiply regulated biochemical system. *Proc. Natl Acad. Sci. USA*, **79**, 6917–21.

Decroly, O. and Goldbeter, A. (1984). Multiple periodic regimes and final state sensitivity in a biochemical system. *Phys. Lett.*, **105A**, 259–62.

Decroly, O. and Goldbeter, A. (1985). Selection between multiple periodic regimes in a biochemical system: complex dynamic behaviour resolved by use of one-dimensional maps. *J. Theor. Biol.*, **113**, 649–71.

Decroly O. and Goldbeter, A. (1987). From a simple to complex oscillatory behaviour: analysis of bursting in a multiply regulated biochemical system. *J. Theor. Biol.*, **124**, 219–50.

Degn, H. (1968). Bistability caused by substrate inhibiion of peroxidase in an open reaction system. *Nature*, **217**, 1047–50.

Degn, H. and Mayer, D. (1969). Theory of oscillations in peroxide catalyzed oxidation reactions in open systems. *Biochem. Biophys. Acta*, **180**, 291–301.

Degn, H., Olsen, L. F., and Perram, J. W. (1979). Bistability, oscillations and chaos in an enzyme reaction. *Ann. NY Acad. Sci.*, **316**, 623–37.

Durston, A. J. (1974). Pacemaker mutants of *Dictyostelium discoideum*. *Dev. Biol.*, **38**, 308–19.

Eanes, E. D. and Termine, J. D. (1983). Calcium in mineralized tissues. In *Calcium in Biology*, (ed. T. G. Spiro), pp. 203–33. Wiley, New York.

Eigen, M. (1971). Selorganization of matter and the evolution of biological macromolecules. *Naturwissenschaft*, **33a**, 465–523.

Elton, C. S. and Nicholson, M. (1942). The ten-year cycle in numbers of lynx in Canada. *J. Anim. Ecol.*, **191**, 215–44.

Eschrich, K., Schellenberger, W., and Hofmann, E. (1983). Sustained oscillations in a reconstituted enzyme system containing phosphofructokinase and fructose 1,6-biphosphate. *Arch. Biochem. Biophys.*, **222**, 657–60.

Fisher, R. A. (1937). The wave of advance of advantageous genes. *Ann. Eugen.* **7**, 353–69.

FitzHugh, R. (1961). Impulses and physiological states in theoretical models of nerve membrane. *Biophys. J.*, **1**, 445–66.

Frenkel, R. (1968a). Control of reduced disphosphopyridine nucleotide oscillations in beef heart extracts. I. Effects of modifiers of phosphofructokinase activity. *Arch. Biochem. Biophys.*, **125**, 151–6.

Frenkel, R. (1968b). Control of reduced diphosphopyridine nucleotide oscillations in beef heart extracts. II. Oscillations of glycolytic intermediates and adenine nucleotides. *Arch. Biochem. Biophys.*, **125**, 157–65.

Frenkel, R. (1968c). Control of reduced diphosphopyridine nucleotide oscillations in beef heart extracts. III. Purification and kinetics of Beef Heart phosphofructokinase. *Arch. Biochem. Biophys.*, **125**, 166.

Gardini, L. Lupini, R., and Massia, M. G. (1989). Hopf bifurcation and transition to chaos in Lotka–Volterra equations. $N = 3$. *J. Math. Biol.*, **17**, 259–72.

Gerisch, G. (1982). Chemotaxis in *Dictyostelium*. *Ann. Rev. Physiol.*, **44**, 535–52.

Ghosh, A. and Chance, B. (1964). Oscillations of glycolytic intermediates in yeast cells. *Biochem. Biophys. Res. Commun.*, **16**, 174–81.

Gilpin, M. E. (1973). Do hares eat lynx? *Am. Nat.*, **107**, 727–30.

Goldbeter, A. and Segel, L. A. (1977). Unified mechanism for relay and oscillation of cyclic AMP in *Dictyostelium discoideum*. *Proc. Natl Acad. Sci. USA*, **74**, 1543–7.

Goldbeter, A., Martiel, J.-L., and Decroly, O. (1984). From excitability and oscillations to birhythmicity and chaos in biochemical systems. In *Dynamics of biochemical systems*, (ed. J. Ricard and A. Cornish-Bowden), pp. 173–212. Plenum, New York.

Guevara, M. R. and Glass, L. (1982). Phase-locking, period doubling bifurcations and chaos in a mathematical model of periodically driven oscillators. *J. Math. Biol.*, **14**, 1–23.

Guevara, M. R., Glass, L., and Shrier, A. (1981). Phase-locking, period-doubling bifurcations and irregular dynamics in periodically stimulated cardiac cells. *Science*, **214**, 1350–3.

Guevara, M. R., Ward, G., Shrier, A., and Glass, L. (1984). Electric alternans and period-doubling bifurcations. *Comp. Cardiol.*, 167–70.

Haldane, J. B. S. (1930). *Enzymes*. Longman, London.

Hess, B. and Boiteux, A. (1968). In *Regulatory functions of biological membranes*, (ed. J. Järnfelt), p. 148, Elsevier, Amsterdam.

Hess, B. and Boiteux, A. (1971). Oscillatory phenomena in biochemistry. *Ann. Rev. Biochem.*, **40**, 237–58.

Hess, B. and Boiteux, A. (1980). Oscillations in biochemical systems. *Ber. Bunsenges. Phys. Chem.*, **84**, 346–51.

Hess, B. and Markus, M. (1985a). The diversity of biochemical time patterns. *Ber. Bunsenges. Phys. Chem.*, **89**, 642–51.

Hess, B. and Markus, M. (1985b). Dynamic coupling and time patterns in biochemical processes. In *Temporal order*, (ed. L. Rensing and N. I. Jaeger), pp. 179–90. Springer, Berlin.

Hess, B., Brand, K., and Pye, E. K. (1966). Continuous oscillation in a cell-free extract of *S. carlsbergensis*. *Biochem. Biophys. Rev. Commun.*, **23**, 102–8.

Hess, B., Boiteux, A., and Krüger, J. (1969). Cooperation of glycolytic enzymes. *Adv. Enzyme Regul.*, **7**, 149–67.

Hess, B., Markus, M., Müller, S.. C., and Plesser, T. (1990). From homogeneity towards the anatomy of a chemical spiral. In *Spatial inhomogeneities and transient behaviour in chemical kinetics*, (ed. P. Gray, G. Nicolis, F. Baras, P. Borckmans, and S. K. Scott), pp. 353–69. Manchester University Press.

Higgins, J. (1963). A chemical mechanism for oscillation of glycolytic intermediates in yeast cells. *Proc. Natl Acad. Sci. USA*, **51**, 989–94.

Hodgkin, A. L. and Huxley, A. F. (1952). A quantiative description of membrane current and its application to conduction and excitation in nerve. *J. Physiol.*, **117**, 500–44.

Holden, A. V. and Muhamad, M. A. (1986). A graphical zoo of strange and peculiar attractors. In *Chaos*, (ed. A. V. Holden), pp. 15–36. Manchester University Press.

Holden, A. V. and Winlow, W. (1982). Bifurcation of periodic activity from periodic activity in a molluscan neurone. *Biol. Cybern.*, **42**, 189–94.

Holden, A. V., Winlow, W., and Haydon, P. G. (1982). The induction of periodic and chaotic activity in a molluscan neurone. *Biol. Cybern.*, **43**, 169–73.

Hoyer, J. Park, M. R., and Klee, M. R. (1978). Changes in ionic currents associated with flurazepam-induced abnormal discharges in Aplysia neurone. In *Abnormal neuronal discharges*, (ed. N. Chalazonitis and M. Boisson), pp. 310–10. Raven, New York.

Ibsen, K. H. and Schiller, K. W. (1967). Oscillations of nucleotides and glycolytic intermediates in aerobic suspensions of Ehrlich ascites tumor cells. *Biochim. Biophys. Acta*, **131**, 405–7.

Klee, M. R., Faber, D. S., and Hoyer, J. (1978). Doublet discharges and bistable states induced by strychnine in a neuronal soma membrane. In *Abnormal neuronal discharges*, (ed. N. Chalazonitis and M. Boisson), pp. 287–300. Raven, New York.

Kolmogorov, A., Petrovsky, I., and Piscounov, N. (1937). Étude de l'équation de la diffusion avec croissance de la quantité de matière et son application à un problème biologique. *Moscow Univ. Bull. Math.*, **1**, 1–25.

Labos, E. and Lang, E. (1978). On the behavior of snail neurons in the presence of cocaine. In *Abnormal neuronal discharges*, (ed. N. Chalazonitis and M. Boisson), pp. 177–88. Raven, New York.

Larter, R., Bush, C. L., Lonis, T. R., and Aguda, B. D. (1987). Multiple steady states, complex oscillations and the devil's staircase in the peroxidase–oxidase reaction. *J. Chem. Phys.*, **87**, 5765–71.

Larter, R., Steinmetz, C. G., and Aguda, B. D. (1988). Fast–slow variable analysis of the transition to mixed-mode oscillations and chaos in the peroxidase reaction. *J. Chem. Phys.*, **89**, 6506–14.

Leigh, E. (1968). The ecological role of Volterra's equations. In *Some mathematical problems in biology*, (ed. M. Gerstenhaber), pp. 1–64. American Mathematical Society, Providence, RI.

Lineweaver, H. and Burk, D. (1934). The determination of enzyme dissociation constants. *J. Am. Chem. Soc.*, **56**, 658–64.

London, W. P. and Yorke, J. A. (1973). Recurrent outbreaks of measles, chickenpox and mumps. I and II. *Am. J. Epidemiol.*, **98**, 453–82.

Lotka, A. J. (1910). Contribution to the theory of periodic reactions. *J. Phys. Chem.*, **14**, 271–4.

Lotka, A. J. (1920). Undamped oscillations derived from the law of mass action. *J. Am. Chem. Soc.*, **42**, 1595–9.

Markus, M. and Hess, B. (1984). Transition between oscillatory modes in a glycolytic model system. *Proc. Natl Acad. Sci. USA* **81**, 4394–8.

Markus, M. and Hess, B. (1985a). Input-response relationships in the dynamics of glycolysis. *Arch. Biol. Med. Exp.*, **18**, 261–71.

Markus, M. and Hess, B. (1985b). Dimension and Liapounov exponents of a strange attractor from biochemical data. In *Temporal order*, (ed. L. Rensing and N. I. Jaeger), pp. 191–3. Springer, Berlin.

Markus, M., Plesser, Th., Boiteux, A., Hess, B., and Malcovati, M. (1983). Analysis of progress curves. Rate law of pyruvate kinase type I from *Escherichia coli*. *Biochem. J.*, **189**, 421–33.

Markus, M., Kuschmitz, D., and Hess, B. (1984). Chaotic dynamics in yeast glycolysis under periodic substrate input flux. *FEBS Lett.*, **172**, 235–8.

Markus, M., Kuschmitz, D., and Hess, B. (1985a). Properties of strange attractors in yeast glycolysis. *Biophys. Chem.*, **22**, 95–105.

Markus, M., Müller, S. C., and Hess, B. (1985b). Observations of entrainment, quasiperiodicity and chaos in glycolyzing yeast extracts under periodic glucose input. *Ber. Bunsenges. Phys. Chem.*, **89**, 651–4.

Martiel, J.-L. and Goldbeter, A. (1984). Oscillations et relais des signaux d'AMP cyclique chez *Dictyostelium discoideum*: analyse d'un modèle fondé sur la modification du receptor pour l'AMP cyclique. *CR Acad. Sci. Ser. III*, **298**, 549–52.

Martiel, J.-L. and Goldbeter, A. (1985). Autonomous chaotic behaviour of the slime mould *Dictyostelium discoideum* predicted by a model for cyclic AMP signalling. *Nature*, **313**, 590–2.

May, R. M. (1980). Non-linear phenomena in ecology and epidemiology. *Ann. NY Acad. Sci.*, **357**, 267–81.

Michaelis, L. and Menten, M. I. (1913). Die Kinetic der Invertinwirkung. *Biochem. Z.*, **49**, 333–69.

Monod, J., Wyman, J., and Changeux, J. P. (1965). On the nature of allosteric transitions: a plausible model. *J. Mol. Biol.*, **12**, 88–118.

Moran, F. and Goldbeter, A. (1984). Onset of birhythmicity in a regulated biochemical system. *Biophys. Chem.*, **20**, 149–56.

Murray, J. D. (1977). *Nonlinear differential equations models in biology*. Clarendon, Oxford.

Murray, J. D. (1990). *Mathematical biology*. Springer, New York.

Nagumo, J. S., Arimoto, S., and Yoshizawa, S. (1962). An active pulse transmission line simulating nerve axon. *Proc. IRE*, **50**, 2061–71.

Nakamura, S., Yokota, K., and Yamazaki, I. (1969). Sustained oscillations in a lactoperoxidase NADPH and O_2 system. *Nature*, **222**, 794.

Nancollas, G. H. (1979). The growth of crystals in solution. *Adv. Colloids Interface Sci.*, **10**, 215–52.

Odum, E. P. (1953). *Fundamentals of ecology*. Saunders, Philadelphia, PA.

Olsen, L. F. (1978). The oscillating peroxidase–oxidase reaction in an open system: analysis of the reaction mechanism. *Biochim. Biophys. Acta*, **527**, 212–20.

Olsen, L. F. (1979). Studies of the chaotic behaviour in the peroxidase–oxidase reaction. *Z. Naturforsch.*, **34a**, 1544–6.

Olsen, L. F. (1983). An enzyme reaction with a strange attractor. *Phys. Lett.*, **94A**, 454–57.

Olsen, L. F. (1984). The enzyme and the strange attractor–comparisons of experimental and numerical data for an enzyme reaction with chaotic motion. In *Stochastic phenomena and chaotic behaviour in complex systems*, (ed. P. Schuster), pp. 116–23. Springer, Berlin.

Olsen, L. F. and Degn, H. (1977). Chaos in an enzyme reaction. *Nature*, **267**, 177–8.

Olsen, L. F. and Degn, H. (1978). Oscillatory kinetics of the peroxidase–oxidase reaction in an open system. Experimental and theoretical studies. *Biochim. Biophys. Acta*, **523**, 321–34.

Olsen, L. F. and Degn, H. (1985). Chaos in biological systems. *Q. Rev. Biophys.*, **18**, 165–225.

Othmer, H. G., Monk, P. B., and Rapp, P. E. (1985). A model for signal relay and adaptation in *Dictyostelium discoideum*. II. Analytical and numerical results. *Math. Biosci.*, **77**, 79–139.

Perault-Staub, A. M., Brezillon, P., Tracqui, P., and Staub, J. F. (1984). A self-organized temporal model for rat calcium metabolism: mineral nonlinear reaction, circadian oscillations. In *Physiological systems: dynamics and controls*, pp. 55–61. Institute of Measurement and Control, Oxford.

Pye, E. K. (1969). Biochemical mechanisms underlying the metabolic oscillations in yeast. *Can. J. Bot.*, **47**, 271–85.

Pye, E. K. and Chance, B. (1966). Sustained sinusoidal oscillations of reduced pyridine nucleotide in a cell-free extract of *Saccharomyces carlsbergensis*. *Proc. Natl. Acad. Sci.*, **55**, 888–94.

Rapp, P. E. (1986). Oscillations and chaos in cellular metabolism and physiological systems. In *Chaos*, (ed. A. V. Holden), pp. 179–208. Manchester University Press.

Rapp, P. E. and Berridge, M. J. (1977). Oscillations in calcium–cyclic AMP control loops form the basis of pacemaker activity and other high frequency biological rhythms. *J. Theor. Biol.*, **66**, 497–525.

Rapp, P. E., Othmer, H. G., ad Monk, P. B. (1985). A model for signal-relay and adaptation in *Dictyostelium discoideum*. I. Biological processes and the model network. *Math. Biosci.*, **77**, 35–78.

Richter, P. H. and Ross, J. (1980). Oscillations and efficiency in glycolysis. *Biophys. Chem.*, **12**, 285–97.

Rinzel, J. (1981). Models in neurobiology. In *Nonlinear phenomena in physics and biology*, (ed. R. H. Enns, B. L. Jones, R. M. Miura, and S. S. Rangnekar), pp. 345–67. Plenum, New York.

Ross, J. and Schell, M. (1987). Thermodynamic efficiency in nonlinear biochemical reaction. *Ann. Rev. Biophys. Biophys. Chem.*, **16**, 401–22.

Rubinow, S. I. (1975). *Introduction to mathematical biology*. Wiley, New York.

Schaffer, W. M. (1984). Stretching and folding in lynx fur returns: evidence for a strange attractor in nature? *Am. Nat.*, **124**, 798–820.

Schaffer, W. M. and Kot, M. (1985a). Nearly one dimensional dynamics in an epidemic. *J. Theor. Biol.*, **112**, 03–27.

Schaffer, W. M. and Kot, M. (1985b). Do strange attractors govern ecological systems? *Bioscience*, **35**, 342–50.

Schaffer, W. M. and Kot, M. (1986). Differential systems in ecology and epidemiology. In *Chaos*, (ed. A. V. Holden), pp. 158–78. Manchester University Press.

Segel, L. A.)1980). *Mathematical models in molecular and cellular biology*. Cambridge University Press.

Segel, L. A. (1988). On the validity of the steady state asumption of enzyme kinetics. *Bull. Math. Biol.*, **50**, 579–93.

Segel, L. A. and Slemrod, M. (1989). The quasi-steady state assumption: a case study in perturbation. *SIAM Rev.*, **31**, 446.

Sel'kov, E. E. (1968). Self-oscillations in glycolysis. 1. A simple kinetic model. *Euro. J. Biochem.*, **4**, 79–86.

Staub, J. F., Perault-Staub, A. M., and Milhaud, G. (1979). Endogenous nature of circadian rhythms in calcium metabolism. *Am. J. Physiol.*, **237**, R311–7.

Tornheim, K. and Lowenstein, J. M. (1974). The purine nucleotide cycle. IV. *J. Biol. Chem.*, **249**, 3241–7.

Tornheim, K. and Lowenstein, J. M. (1975). The purine nucleotide cycle. Control of phosphofructokinase and glycolytic oscillations in muscle extracts. *J. Biol. Chem.*, **250**, 6304–14.

Tracqui, P. and Staub, J. F. (1987). Solutions stables périodiques multiples d'un champ de vecteurs sur R^2 caractérisé par un monome de degré 3. *C.R. Acad. Sci. Paris*, **305**, 331–6.

Tracqui, P., Brezillon, P., Staub, J. F., Perault-Staub, A. M., and Milhaud, G. (1985). Diversité et complexité des comportements dynamiques temporels d un modèle autocatalytique simple. *CR Acad. Sci. Paris*, **300**, 253–8.

Tracqui, P., Perault-Staub, A. M., Milhaud, G., and Staub, J. F. (1987). Theoretical study of a two-dimensional autocatalytic model for calcium dynamics at the extracellular fluid–bone interface. *Bull. Math. Biol.*, **49**, 597–613.

Tyson, J. J. and Othmer, H. G. (1978). The dynamics of feedback control circuits in biochemical pathways. *Prog. Theor. Biol.*, **5**, 1–62.

Urist, M. R. (1980). *Fundamental and clinical bone physiology.* Lippincott, Philadelphia, PA.

Volterra, V. (1926). Variatzione fluttuazioni del numero d'individui in species animali conviventi. *Mem. Acad. Lincei.*, **2**, 31–113.

von Klitzing L. and Betz A. (1970). Metabolic control in a flow system. *Arch. Mikrobiol.*, **71**, 22–5.

Williamson, M. (1972). *The analysis of biological populations.* Arnold, London.

Winfree, A. T. (1980). *The geometry of biological time.* Springer, New York.

Winfree, A. T. (1983). Sudden cardiac death: a problem in topology. *Sci. Am.*, **248**, 144–61.

Yamazaki, I. and Yokota, K. (1967). Analysis of the conditions causing the oscillatory oxidation of reduced nicotinamide-adenine dinucleotide by horseradish peroxidase. *Biochim. Biophys. Acta*, **132**, 310–20.

Yamazaki, I., Yokota, K., and Nakajima, R. (1965). Oscillatory oxidations of reduced pyridine nucleotide. *Biochem. Biophys. Res. Commun.*, **21**, 582–6.

Yokota, K. and Yamazaki, I. (1965). Reaction of peroxidase with reduced nicotinamide–adenine dinucleotide phosphate. *Biochim. Biophys. Acta*, **105**, 301–12.

Yorke, J. A. and London, W. P. (1973). Recurrent outbreaks of measles, chickenpox and mumps: systematic differences in contact rates and stochastic effects. *Am. J. Epidemiol.*, **98**, 469–82.

Epilogue

Chaos seems to be one of those subjects that demands a grand philosophical summary setting out all of its implications, importance, etc. In their excellent book Berge *et al.* provide the script of a discussion amongst the authors to this end whilst other authors have attempted their own solutions to this. There are many unanswered, and perhaps unanswerable, questions of this sort. To paraphrase Degn and Olsen, what use does an innocuous plant better known as an accompaniment to roast beef and Yorkshire pudding gain from having an enzyme that can support chaotic behaviour in its roots? Even if biological and biochemical systems have found it useful to evolve aperiodic processes (and there are some arguments along these lines), it does not seem likely that the inorganic systems such as the BZ reaction have also 'evolved' in this way. Are there any practical applications that will allow industry, say, to exploit aperiodicity (a disproportionately important consideration in British universities at this time)? There is much evidence that economic gains are to be made by any manufacturer sufficiently brave to operate chemical reactors periodically under some circumstances—but there seems a general reluctance in industry for anything other than steady states. With chaos, such advantages are less clear. The incredible sensitivity of such systems to the initial conditions is such that even if we have a good day today, there is no chance of repeating it tomorrow when we start up again from what must be at least marginally different conditions.

Perhaps we need simply to learn to recognize that the existence of chaos implies a new view of science. Some chemical processes (and many from other disciplines) are inherently unpredictable. This is not a consequence of 'randomness' or imprecision over modelling. Even if we can obtain the exact reaction kinetic rate laws from a fully determined mechanism and can evaluate all the required rate constants to unheard-of precision, predictability will still be beyond us. This problem will not be solved by throwing more money at it: bigger and more expensive computers or more extensive (and again phenomenally expensive) multi-laser kinetic studies will not get us round the real unpredictability. Only if we could determine all the initial (or some other instantaneous) concentrations to a ridiculous (strictly infinite) number of significant figures simultaneously will we be able to run our integrations forward with full confidence. In this way, God does not need to play dice to confound us. Even were we able to fathom all the rules of our universe, we will still be unable to predict the future weather, course of history, or winner of next year's cup final, as we do not have infinitely precise knowledge of the initial state (which presumably an omniscient He—or She—would have).

All is not lost; we can make limited predictions of the future, at least in mildly chaotic regimes—and under the most favourable of such circumstances we can even gain some measure, based on the magnitude of the Lyapounov exponents, of how long we should 'trust' our predictions.

If aperiodic behaviour does obtain classification as 'a bad thing' then we need to learn to recognize the warning signs of its impending onset. Chaos rarely arrives 'out of the blue', rather it normally develops from steady state and then simple oscillatory states proceeding along a limited number of more or less well-documented scenarios as some parameter or experiment condition changes. An observation of a couple of period doublings is a very strong hint that chaos is just around the corner (although like all rules of thumb there are exceptions). Similarly, mixed-mode oscillations can herald the existence of delicate concatenations and non-periodic evolution.

In a similar way, if we wish to confirm that a particular system is behaving chaotically, we should not attempt to do so from an isolated experimental record. Although various tests exist, such as evaluating the Fourier spectrum, reconstructing the attractor, and then calculating the Lyapounov exponent and dimension, these are rarely conclusive on their own from 'real' data (as opposed to computed output from small models). The system should always be studied, where possible, over as wide a parameter range as can be achieved and the recognizable scenarios described above sought.

But is there really 'chemical chaos'? Chaos is seen in other scientific fields—fluid mechanics and physics (almost) certainly. Oscillations and chaos, like multistability, are not complicated phenomena, or at least they do not require particularly complicated building blocks. Some of the simplest mathematical equations are proven aperiodics if certain (not particularly demanding) minimum requirements on the non-linearity, feedback, and number of variables are met. Such behaviour seems almost the rule rather than the exception in chemistry (really it is truly interdisciplinary—Nature does not recognize the artificial boundaries erected by Man in his attempt to 'understand' the world). In the foregoing chapters we have seen experimental examples of complex oscillations and classic routes to chaos in almost the complete range of chemical systems: homogeneous gas-phase, solution-phase, and solid-phase reactions, heterogeneous processes involving gas–solid or solid–liquid interfaces, and biochemical examples. Very similar qualitative aspects are found in each. In the earlier, theoretical chapters a whole host of simple and yet chemically recognizable schemes were shown to possess all the necessary requirements for chaos. If two first-order reactions, or even a single autocatalytic process, in a non-isothermal CSTR can support chaos, why then should not the much greater mechanisms of real-life combustion systems which have all these features and more? In fact one would almost be tempted to conclude that there would need to be some additional and strong 'thermodynamic' principle operating if all chemical systems were able *de facto* to escape the chaotic net. In earlier times, the second law of thermodynamics was (incorrectly) invoked to 'forbid' oscillations, but fortunately chemical reactions did not listen to such ideas and have continued to oscillate until scientists began to think a bit more carefully.

And yet there are still sufficient hiding places for the disbeliever. For those of us advocates there can still be uncomfortable questions of homogeneity and external effects. No experiment is 'ideal': there are problems with incomplete mixing for both solutions and gases, and inevitable slight perturbations on the inflow and outflow. Many of the solution-phase reactions either evolve a gaseous product (such as CO_2 or Br_2) or a precipitate (such as colloidal sulphur or MnO_2). For 'homogeneous' gas-phase reactions, extra contributions from surface reactions are almost always present and generally little understood. These often provide unduly sensitive steps in the modelling studies. Heterogeneous processes are almost notorious for unexplainable side features, and yet the UHV single crystal plane work is some of the technically most advanced. The sheer wealth of data from various sources—including the almost enigmatic reproducibility of the unrepeatableness of chaos—has led me to come down on the side of the believers, that chaos would survive in even the most perfect chemical experiment of the classes described in the main text, but the issue is not cut and dried by any means. Fortunately there is no need for cut-and-dried answers at this stage, and there is much more exciting research to be followed in non-linear chemical kinetics—especially, perhaps, in the spatio-temporal aspects that have been much less well developed to date. I suspect that any second edition of this book will not just require correction of mistakes, but many new sections, even chapters, on many new developments in the subject.

Index